MTP International Review of Science

Volume 10

Thermochemistry and Thermodynamics

Edited by **H. A. Skinner**
University of Manchester

Butterworths · London
University Park Press · Baltimore

THE BUTTERWORTH GROUP

ENGLAND
Butterworth & Co (Publishers) Ltd
London: 88 Kingsway, WC2B 6AB

AUSTRALIA
Butterworths Pty Ltd
Sydney: 586 Pacific Highway 2067
Melbourne: 343 Little Collins Street, 3000
Brisbane: 240 Queen Street, 4000

NEW ZEALAND
Butterworths of New Zealand Ltd
Wellington: 26–28 Waring Taylor Street, 1

SOUTH AFRICA
Butterworth & Co (South Africa) (Pty) Ltd
Durban: 152–154 Gale Street

ISBN 0 408 70271 0

UNIVERSITY PARK PRESS

U.S.A. and CANADA
University Park Press Inc
Chamber of Commerce Building
Baltimore, Maryland, 21202

Library of Congress Cataloging in Publication Data
Skinner, Henry Alistair, 1916–
Thermochemistry and thermodynamics
(Physical chemistry, series one, v. 10) (MTP international review of science)
1. Thermochemistry. 2. Thermodynamics.
I. Title. [DNLM: 1. Chemistry, Analytical–Period. 2. Chemistry, Physical–Period. 3. Crystallography–Period. W1 PH683K]
QD453.2.P58 vol. 10 [QD511] 541′.3′08s [541′.36]
ISBN 0–8391–1024–3 72–5214

MTP MEDICAL AND TECHNICAL PUBLISHING CO. LTD.
Seacourt Tower
West Way
Oxford, OX2 OJW
and
BUTTERWORTH & CO. (PUBLISHERS) LTD.

Filmset by Photoprint Plates Ltd., Rayleigh, Essex
Printed in England by Redwood Press Ltd., Trowbridge, Wilts
and bound by R. J. Acford Ltd., Chichester, Sussex

MTP International Review of Science

Thermochemistry and Thermodynamics

MTP International Review of Science

Publisher's Note

The MTP International Review of Science is an important new venture in scientific publishing, which we present in association with MTP Medical and Technical Publishing Co. Ltd. and University Park Press, Baltimore. The basic concept of the Review is to provide regular authoritative reviews of entire disciplines. We are starting with chemistry because the problems of literature survey are probably more acute in this subject than in any other. As a matter of policy, the authorship of the MTP Review of Chemistry is international and distinguished; the subject coverage is extensive, systematic and critical; and most important of all, new issues of the Review will be published every two years.

In the MTP Review of Chemistry (Series One), Inorganic, Physical and Organic Chemistry are comprehensively reviewed in 33 text volumes and 3 index volumes, details of which are shown opposite. In general, the reviews cover the period 1967 to 1971. In 1974, it is planned to issue the MTP Review of Chemistry (Series Two), consisting of a similar set of volumes covering the period 1971 to 1973. Series Three is planned for 1976, and so on.

The MTP Review of Chemistry has been conceived within a carefully organised editorial framework. The over-all plan was drawn up, and the volume editors were appointed, by three consultant editors. In turn, each volume editor planned the coverage of his field and appointed authors to write on subjects which were within the area of their own research experience. No geographical restriction was imposed. Hence, the 300 or so contributions to the MTP Review of Chemistry come from many countries of the world and provide an authoritative account of progress in chemistry.

To facilitate rapid production, individual volumes do not have an index. Instead, each chapter has been prefaced with a detailed list of contents, and an index to the 13 volumes of the MTP Review of Physical Chemistry (Series One) will appear, as a separate volume, after publication of the final volume. Similar arrangements will apply to the MTP Review of Organic Chemistry (Series One) and to subsequent series.

Butterworth & Co. (Publishers) Ltd.

Physical Chemistry Series One

Consultant Editor
A. D. Buckingham
Department of Chemistry University of Cambridge

Volume titles and Editors

1 **THEORETICAL CHEMISTRY**
Professor W. Byers Brown, *University of Manchester*

2 **MOLECULAR STRUCTURE AND PROPERTIES**
Professor G. Allen, *University of Manchester*

3 **SPECTROSCOPY**
Dr. D. A. Ramsay, F.R.S.C., *National Research Council of Canada*

4 **MAGNETIC RESONANCE**
Professor C. A. McDowell, *University of British Columbia*

5 **MASS SPECTROMETRY**
Professor A. Maccoll, *University College, University of London*

6 **ELECTROCHEMISTRY**
Professor J. O'M Bockris, *University of Pennsylvania*

7 **SURFACE CHEMISTRY AND COLLOIDS**
Professor M. Kerker, *Clarkson College of Technology, New York*

8 **MACROMOLECULAR SCIENCE**
Professor C. E. H. Bawn, F.R.S., *University of Liverpool*

9 **CHEMICAL KINETICS**
Professor J. C. Polanyi, F.R.S., *University of Toronto*

10 **THERMOCHEMISTRY AND THERMODYNAMICS**
Dr. H. A. Skinner, *University of Manchester*

11 **CHEMICAL CRYSTALLOGRAPHY**
Professor J. Monteath Robertson, F.R.S., *University of Glasgow*

12 **ANALYTICAL CHEMISTRY – PART 1**
Professor T. S. West, *Imperial College, University of London*

13 **ANALYTICAL CHEMISTRY – PART 2**
Professor T. S. West, *Imperial College, University of London*

INDEX VOLUME

Inorganic Chemistry
Series One
Consultant Editor
H. J. Emeléus, F.R.S.
Department of Chemistry
University of Cambridge

Volume titles and Editors

1 **MAIN GROUP ELEMENTS—HYDROGEN AND GROUPS I–IV**
Professor M. F. Lappert, *University of Sussex*

2 **MAIN GROUP ELEMENTS—GROUPS V AND VI**
Professor C. C. Addison, F.R.S. and Dr. D. B. Sowerby, *University of Nottingham*

3 **MAIN GROUP ELEMENTS—GROUP VII AND NOBLE GASES**
Professor Viktor Gutmann, *Technical University of Vienna*

4 **ORGANOMETALLIC DERIVATIVES OF THE MAIN GROUP ELEMENTS**
Dr. B. J. Aylett, *Westfield College, University of London*

5 **TRANSITION METALS—PART 1**
Professor D. W. A. Sharp, *University of Glasgow*

6 **TRANSITION METALS—PART 2**
Dr. M. J. Mays, *University of Cambridge*

7 **LANTHANIDES AND ACTINIDES**
Professor K. W. Bagnall, *University of Manchester*

8 **RADIOCHEMISTRY**
Dr. A. G. Maddock, *University of Cambridge*

9 **REACTION MECHANISMS IN INORGANIC CHEMISTRY**
Professor M. L. Tobe, *University College, University of London*

10 **SOLID STATE CHEMISTRY**
Dr. L. E. J. Roberts, *Atomic Energy Research Establishment, Harwell*

INDEX VOLUME

Organic Chemistry
Series One
Consultant Editor
D. H. Hey, F.R.S.
Department of Chemistry
King's College, University of London

Volume titles and Editors

1 **STRUCTURE DETERMINATION IN ORGANIC CHEMISTRY**
Professor W. D. Ollis, F.R.S., *University of Sheffield*

2 **ALIPHATIC COMPOUNDS**
Professor N. B. Chapman, *Hull University*

3 **AROMATIC COMPOUNDS**
Professor H. Zollinger, *Swiss Federal Institute of Technology*

4 **HETEROCYCLIC COMPOUNDS**
Dr. K. Schofield, *University of Exeter*

5 **ALICYCLIC COMPOUNDS**
Professor W. Parker, *University of Stirling*

6 **AMINO ACIDS, PEPTIDES AND RELATED COMPOUNDS**
Professor D. H. Hey, F.R.S. and Dr. D. I. John, *King's College, University of London*

7 **CARBOHYDRATES**
Professor G. O. Aspinall, *University of Trent, Ontario*

8 **STEROIDS**
Dr. W. D. Johns, *G. D. Searle & Co., Chicago*

9 **ALKALOIDS**
Professor K. F. Wiesner, F.R.S., *University of New Brunswick*

10 **FREE RADICAL REACTIONS**
Professor W. A. Waters, F.R.S., *University of Oxford*

INDEX VOLUME

Physical Chemistry
Series One

Consultant Editor
A. D. Buckingham

Consultant Editor's Note

The MTP International Review of Science is designed to provide a comprehensive, critical and continuing survey of progress in research. The difficult problem of keeping up with advances on a reasonably broad front makes the idea of the Review especially appealing, and I was grateful to be given the opportunity of helping to plan it.

This particular 13-volume section is concerned with Physical Chemistry, Chemical Crystallography and Analytical Chemistry. The subdivision of Physical Chemistry adopted is not completely conventional, but it has been designed to reflect current research trends and it is hoped that it will appeal to the reader. Each volume has been edited by a distinguished chemist and has been written by a team of authoritative scientists. Each author has assessed and interpreted research progress in a specialised topic in terms of his own experience. I believe that their efforts have produced very useful and timely accounts of progress in these branches of chemistry, and that the volumes will make a valuable contribution towards the solution of our problem of keeping abreast of progress in research.

It is my pleasure to thank all those who have collaborated in making this venture possible – the volume editors, the chapter authors and the publishers.

Cambridge A. D. Buckingham

Preface

The emphasis in this volume is directed towards recent developments and new results in the essentially experimental science of chemical thermodynamics—thus excluding pure theory at one extreme, and applied thermodynamics at the other. Classical areas of chemical thermodynamics—represented here by chapters on thermochemistry, crystal properties, liquid mixtures, and electrolyte solutions—are seen to be quick to adapt to improved techniques of measurement, and are accumulating results of higher precision at an ever increasing rate on a wider range of compounds and systems. The concomitant 'data explosion' has raised problems not only for the producers of critical tables of thermodynamic data, but also for the users. Zwolinski and Chao attempt to guide us, as users, through the present confusion by giving full details of the tabulation programs now available, and those in preparation.

Two areas relatively new to the enticement of calorimetric and equilibrium studies are reviewed by Wadsö (biochemical thermochemistry) and Jameson (coordination complexes). The impact of conduction and flow microcalorimetry on biochemical investigations has been steadily mounting in recent years, and can be expected to continue; the increasing availability of commercial calorimeters and scanning calorimeters of high performance has greatly assisted in this. The two remaining chapters in this volume are concerned with high temperature studies. De Maria deals with the application of the highly successful Knudsen cell–mass spectroscopic techniques to investigations of high temperature (1500 K) equilibria between atoms and small inorganic molecules. Beckett and Cezairliyan report on the development of pulse calorimeters for specific heat measurements at temperatures reaching 3000 K and beyond.

The coverage of experimental chemical thermodynamics here achieved, although representative, is not total; significant gaps which remain will be singled out for attention in later volumes.

Manchester H. A. Skinner

Contents

Biochemical thermochemistry 1
I. Wadsö, *Lund University, Sweden*

Thermodynamics of metal-complex formation 45
R. F. Jameson, *The University, Dundee*

Thermochemistry of chemical compounds 57
G. Pilcher, *University of Manchester*

Critically evaluated tables of thermodynamic data 93
B. J. Zwolinski and Jing Chao, *Texas A & M University*

Thermodynamics of organic mixtures 121
H. V. Kehiaian, *Université de Provence, Marseilles*

Pulse calorimetry of solids at high temperatures 159
A. Cezairliyan and C. W. Beckett, *National Bureau of Standards, Washington*

Thermodynamics of electrolyte solutions 177
K. P. Mishchenko, *Leningrad Institute of Pulp and Paper Technology*

Equilibrium studies at high temperatures 209
Giovanni De Maria and Giovanni Balducci, *Università di Roma, Rome*

Thermodynamics of crystals 231
E. F. Westrum, Jr., *University of Michigan*

1
Biochemical Thermochemistry

I. WADSÖ
Thermochemistry Laboratory, Chemical Centre, Lund University, Sweden

---

1.1 INTRODUCTION 3

1.2 MODEL SYSTEMS AND SIMPLE BIOCHEMICAL COMPOUNDS 4
1.2.1 *Solution and transfer processes* 4
1.2.1.1 *Hydrocarbons* 5
1.2.1.2 *Simple alcohols, carboxylic acids, amines and amides* 5
1.2.1.3 *Amino acids* 6
1.2.1.4 *Various simple organic compounds* 6
1.2.1.5 *Ionic compounds* 7
1.2.2 *Some association reactions* 7
1.2.3 *Ionisation reactions* 8
1.2.4 *Hydrolysis reactions* 9
1.2.5 *Formation of metal–ion complexes* 9
1.2.6 *Heat capacity measurements* 10
1.2.7 *Heat of combustion studies* 11
1.2.8 *Miscellaneous studies* 11

1.3 BINDING OF LOW MOLECULAR WEIGHT COMPOUNDS TO PROTEINS 11
1.3.1 *General comments* 11
1.3.1.1 *Lysozyme, saccharide inhibitors* 12
1.3.1.2 *Chymotrypsin, inhibitors* 12
1.3.1.3 *Glutamine synthetase, tryptophan and* AMP 13
1.3.1.4 *Aldolase,* D-*hexitol* 1,6-*diphosphate* 13
1.3.1.5 *Biotin, avidin* 15
1.3.1.6 *Ribonuclease, nucleotide inhibitors* 15
1.3.1.7 *Glyceryl aldehyde 3-phosphate dehydrogenase,* NAD 16
1.3.1.8 *Phosphorylase b,* AMP 17
1.3.1.9 BSA, *lysolecitin* 17
1.3.1.10 *Albumin, warfarin* 17
1.3.1.11 *β-lactoglobulin, dodecyl sulphate* 17
1.3.1.12 *Hemerythrin, anion ligands* 17
1.3.1.13 *Apocarbonic anhydrase, transition metal ions* 18
1.3.1.14 *Oxygenation and oxidation of hemerythrin* 18

1.3.1.15 *Oxygenation of myoglobin* 18
1.3.1.16 *Antibody, hapten binding* 19

1.4 PROTEIN–PROTEIN (POLYPEPTIDE) BINDING 20
1.4.1 *Binding between different molecules* 20
1.4.1.1 *Haemoglobin, haptoglobin* 20
1.4.1.2 Fc *fragment of* IgC, *protein A* 20
1.4.1.3 *Association of α and β chains of haemoglobin* 21
1.4.1.4 S-*protein,* S-*peptide* 21
1.4.2 *Protein self-association* 22
1.4.2.1 *α-Chymotrypsin* 22
1.4.2.2 *Hemerythrin* 23
1.4.2.3 *Virus protein* 23

1.5 MISCELLANEOUS PROTEIN REACTIONS 23
1.5.1 *Oxidation of cytochrome c* 23
1.5.2 *Activation of chymotrypsinogen A* 24
1.5.3 *Reconstitution of virus from protein and nucleic acid components* 24

1.6 UNFOLDING PROCESSES IN BIOPOLYMERS 24
1.6.1 *Proteins, thermally induced unfolding processes* 25
1.6.1.1 *The chymotrypsinogen family of proteins* 25
1.6.1.2 *The* RNaseA *family* 27
1.6.1.3 *Lysozyme* 27
1.6.1.4 *Chymotrypsinogen,* RNaseA, *lysozyme in more concentrated solutions* 28
1.6.1.5 *Myoglobin* 29
1.6.1.6 *Tropocollagen and collagen* 29
1.6.2 *Proteins and polypeptides, chemically induced unfolding processes* 30
1.6.2.1 *The chymotrypsinogen family of proteins* 30
1.6.2.2 *Lysozyme* 31
1.6.2.3 *Poly-*L*-lysine* 32
1.6.2.4 *Polypeptides in organic solvents* 32
1.6.3 *Conformational transitions in polynucleotides and nucleic acids* 33
1.6.3.1 Poly U, *monomeric adenine derivatives* 33
1.6.3.2 Poly C, poly I 34
1.6.3.3 Poly A, poly U 34
1.6.3.4 Poly I, poly C 35
1.6.3.5 d(AT) 35
1.6.3.6 Poly A 35
1.6.3.7 DNA 36

1.7 CALORIMETRIC STUDIES OF THE INTERACTION BETWEEN WATER AND BIOPOLYMERS, HEAT CAPACITY MEASUREMENTS 37
1.7.1 *Degree of hydration of proteins and nucleic acids* 37
1.7.2 C_p *for solid insulin and chymotrypsinogen* 39
1.7.3 *Vaporisation of water from some proteins* 39
1.7.4 *Water vapour sorption on biopolymers* 39

1.1 INTRODUCTION

Biochemical thermodynamics covers a vast field and includes a large number of experimental techniques as well as theoretical work. It is not feasible to cover all these activities in a brief review of the present type. The aim has been rather to concentrate on recent thermochemical work performed on well defined systems.

Biochemical thermochemistry is conducted in areas extending from general chemistry into the field of biology (Figure 1.1).

Thermochemical, or calorimetric, results from measurements on complex biochemical compounds cannot usually be discussed in such detail as is possible with those compounds most frequently dealt with in general thermochemistry. An obvious approach towards deeper understanding of bio-thermochemical data is to study biochemical model systems. Within a series of model compounds the structure can be varied systematically and it is possible to make correlations between thermochemical data and structural

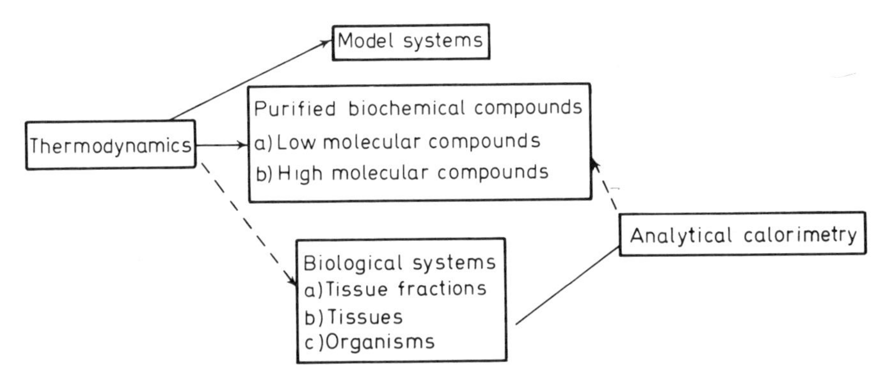

Figure 1.1 The field of bio-thermochemistry

details. One very important part of current model studies involves interactions between water and other solvents and solutes.

For simple biochemical compounds the accuracy of thermochemical measurements can often approach that obtained in model work. For more complex biochemical compounds, however, it is usually not possible to define the state of the sample to better than a few per cent. Nevertheless, it is often desirable to perform the thermochemical measurements with a high precision. Studies can be performed to compare the interactions of a series of well defined compounds with a given biopolymer, or experimental conditions such as pH, the buffer system, and temperature may be systematically varied. The poorly defined state of a given macromolecule is maintained throughout the series of experiments, and thus serves as an internal reference. While the highest precision is desirable in this kind of work, the thermodynamic data so determined are not expected to be accurate on an absolute scale.

Biological systems with more or less intact life functions are very poorly defined from a physico-chemical point of view. Calorimetric investigations on such systems will therefore not usually provide chemical-thermodynamic results which can be discussed seriously on a molecular level as may be possible for data derived from high purity systems.

Calorimetric studies on biological systems undoubtedly have their main importance for general analytical purposes. This is a very promising experimental development which can be expected to become the most important single application in bio-calorimetry. However, this kind of work lies outside the field of physical chemistry and will not be treated in this review.

There are many different types of calorimetric instrument. Of particular interest in the biochemical field are the various kinds of 'microcalorimeters'. During recent years these have been developed into very sensitive and precise instruments which, furthermore, are often easy to operate.

Commercial manufacturers are now marketing calorimeters suitable for bio-thermochemical studies, evidenced by the fact that more than 50% of the papers discussed in this chapter were based on experimental results obtained using commercially available calorimeters.

Calorimetric principles and calorimetric instrumentation with special reference to biochemical work has recently been treated in another review article[1] by the present author. In this chapter, therefore, the subject of calorimetric techniques will not be emphasised.

This review concentrates on work published from 1969 to about August 1971. Some earlier work has also been discussed and in a few cases more recent work has been included.

During the preparation of the manuscript for this chapter it was very helpful to have available a compilation of titles in biochemical thermodynamics which has been prepared by Professor G. Kresheck for a part of the *Bulletin of Thermodynamics and Thermochemistry* (1971)[2]. It contains references to calorimetric as well as to non-calorimetric thermodynamic work. Papers on analytical calorimetry in biochemistry and biology are also listed.

1.2 MODEL SYSTEMS AND SIMPLE BIOCHEMICAL COMPOUNDS

1.2.1 Solution and transfer processes

Thermodynamic data for the transfer of simple compounds from water to other media are needed in order to interpret thermochemical results obtained with complex biochemical compounds.

From the difference between the measured enthalpies of solution in different solvents the transfer value between the solvents is directly obtained. Enthalpy of solution measurements are often very simple to perform calorimetrically with high accuracy and a fair amount of work has already been done, but by no means enough. In particular it seems that more attention should be given to the determination of ΔC_p values for solution and transfer processes.

$$\frac{\mathrm{d}\Delta H}{\mathrm{d}t} = \Delta C_p \qquad (1.1)$$

A very important thermodynamic function in this connection is the (apparent) partial molar heat capacity for the compound in solution, $\Delta\overline{C}_{p2}$. This is easily arrived at by combining the ΔC_p value for the solution process with the heat capacity value for the pure compound, C_p^0

$$\Delta\overline{C}_{p2} = C_p^0 + \Delta C_p \tag{1.2}$$

$\overline{C}_{p2}$ may also be determined by direct heat capacity measurement, but for a dilute solution the experimental route according to equation (1.2) will normally give more accurate results. Unfortunately we must note that even for very simple compounds only a few accurate C_p^0 data have been determined.

1.2.1.1 Hydrocarbons

Thermodynamic data for solutions of hydrocarbons in water have been of basic importance for most theoretical work concerning water–solute interactions. Reid *et al.*[3] have determined the enthalpies of solution of benzene, n-hexane and cyclohexane by an intricate microcalorimetric technique. Krishnan and Friedman[4] determined enthalpies of solution of benzene and toluene in water. Their value for benzene is in reasonable agreement with the more precise microcalorimetric results. The calorimetric studies indicate that existing literature ΔH values for solution of hydrocarbons in water based on the van't Hoff isochore are very unreliable.

1.2.1.2 Simple alcohols, carboxylic acids, amines and amides

Arnett *et al.*[5] have reported ΔH_{soln}, ΔC_p and $\overline{C}_{p2}$ for a number of alcohols, and Arnett and McKelvey[6] measured ΔH_{soln} in H_2O and D_2O for several amides and *N*-alkylamides. Krishnan and Friedman measured ΔH_{soln} for selected alcohols in water, propylene carbonate and in dimethyl sulphoxide[4].

In a recent study Konicek and Wadsö[7] determined enthalpies of solution in water at different temperatures and concentrations for a number of carboxylic acids, primary amines and *N*-alkylamides. They also determined heat capacities for the pure compounds and calculated $\overline{C}_{p2}^0$ values according to equation (1.2).

ΔH_{soln} values are generally very sensitive to the type of substituent in the carbon chain. This is in contrast to ΔC_p values for the solution process and in particular to the $\overline{C}_{p2}^0$ values. The CH_2 increment for the $\overline{C}_{p2}^0$ values of straight chain alcohols, carboxylic acids, amines and *N*-substituted amides is essentially constant, *c.* 20 cal deg^{-1} mol^{-1}, see Figure 1.2. Thus the value is not affected by the nature or the position of the hydrophilic group, which suggests[7] (cf. reference 8) that the water structure around the hydrophobic groups is only slightly affected by the biochemically important functional groups —OH, —COOH, —NH_2 and —CO—NH—.

Kresheck and Klotz[9] have investigated the thermodynamics of transfer of *N*-methylacetamide and *N*,*N*-dimethylacetamide from CCl_4 to water. The water–methylacetamide transfer process was chosen as a model for the transfer of a peptide group from a hydrophobic environment to the aqueous solution, such as may take place in protein unfolding reactions. By combination of calorimetric heats of solution and dilution, they determined the enthalpy change for the transfer of monomeric methylacetamide from CCl_4 to water to be -7.8 kcal mol^{-1}. The calorimetric results com-

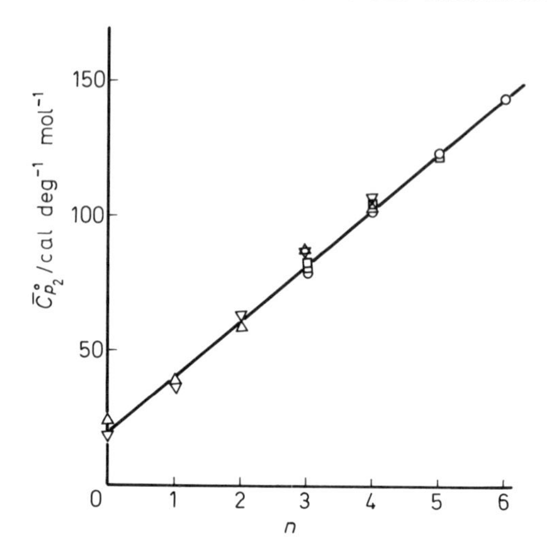

Figure 1.2 Partial molar heat capacities for some straight chain alcohols (∇), carboxylic acids (△), amines (O) and *N*-substituted amides (□) *v.* total number of carbon atoms in the alkyl groups, *n*
(From Konicek, J. and Wadsö, I.[7], by courtesy of *Acta Chem. Scand.*)

bined with data from previous equilibrium studies indicate that there is a negative free energy change when a peptide hydrogen bond in an apolar environment is exposed to aqueous solvent.

1.2.1.3 Amino acids

Spink and Auker[10] determined the enthalpy of solution of glycine, DL-alanine, DL-α-aminobutyric acid and L-valine in water and in aqueous ethanol solutions. Combination of the calorimetric enthalpy values with known free energy data leads to the suggestion that the polar zwitterion portion of the molecules cause a structural breakdown in the mixed solvent. Results further support the idea that non-polar side-chains, in proportion to their size, will promote the structure of mixed aqueous solvents, but less markedly than in pure water.

1.2.1.4 Various simple organic compounds

A number of recent calorimetric aqueous solution measurements involving a number of simple organic compounds are judged to be of biochemical interest. Corkhill *et al.*[11] measured ΔH and ΔC_p for the solution of n-alkyl methyl sulphoxides, di-n-alkyl sulphoxides and α,ω diols. Franks and co-workers have been measuring heats of aqueous solution and ΔC_p values for dialkylamines[12] and for cyclic ether derivatives[13], and Sarma and Ahluwalia[14] measured enthalpies of solution in water for Bu_4NBr, Pr_4NI and $NaBPh_4$ between 3 and 35 °C.

Subramanian *et al.*[15] report results from calorimetric measurements of enthalpies of solution in water of urea, thiourea, guanidine hydrochloride, 1,3-dimethylurea and tetramethylurea. Values for integral heats of solution as well as ΔC_p are given. For urea ΔC_p^0 is close to zero. This observation, together with the observed small n.m.r. chemical shifts suggested to the authors that this denaturant does not in fact alter the water structure appreciably. Guanidine hydrochloride should by the same criteria decrease the structure of water (negative ΔC_p^0). It is concluded that no simple correlation can be made between the effects of denaturants on water structure (as expressed in terms of ΔC_p values and n.m.r. shifts) and their denaturation ability.

As was remarked earlier, much work remains to be done on solution and transfer studies for simple model compounds. In particular this is the case for compounds containing phenyl and indol groups, which are often thought of as important groups at hydrophobic bonding sites.

1.2.1.5 Ionic compounds

Krishnan and Friedman have reported results from extensive calorimetric measurements of heats of solution for a large number of ionic compounds in different solvents. They have determined enthalpies of solution of R_4X^+ salts[16] ($X = N$, P, As, Sb; $R =$ alkyl, phenyl) as well as salts of R_3NH^+, $R_2NH_2^+$, RNH_3^+ and NH_4^+ [17] in H_2O, D_2O, propylene carbonate and dimethyl sulphoxide. They have also studied enthalpies of solution of various other ionic compounds in H_2O and D_2O [18]. Enthalpies of single-ion transfer between the solvents have been calculated and correlated with structural features of the ionic groups and of the solvents.

1.2.2 Some association reactions

Nucleotides tend to associate both in organic solvents and in water. Studies of this kind of association reaction are of interest as models for those molecular structures which contribute to the stability of nucleic acid.

No very recent calorimetric work seems to have been done. Gill and co-workers have investigated the association of several purine and pyrimidine derivatives in water solutions. For pyrimidine riboside compounds[19], the enthalpy of association was found to be about $-2.5\ \text{kcal mol}^{-1}$ which is nearly the same as found for purine ribosides and for deoxyadenosine[20]. More negative values were obtained for purine[17] ($-4.2\ \text{kcal mol}^{-1}$) and 6-methylpurine[21] ($-6.0\ \text{kcal mol}^{-1}$). Comparison with non-calorimetric equilibrium data leads to negative entropy values.

Binford and Holloway[22] studied base pair formation between 1-cyclohexyluracil and 9-ethyladenine in chloroform. From calorimetric data they calculated both the equilibrium constant ($K = 98 \pm 10 \, l \, mol^{-1}$) and the ΔH value (-6.2 ± 0.3 kcal mol^{-1}). The values for the base pair formation are considerably larger than corresponding self-association data for the two nucleotides.

Shiao and Sturtevant[23], in connection with their work on chymotrypsin inhibitor binding (see Section 1.3.1.2), investigated the self-association of proflavin by calorimetric dilution experiments. Assuming stepwise aggregation with identical equilibrium constants and enthalpy values for each step they calculated the values $\Delta G' = -4.0$ and $\Delta H = -7.13$ kcal mol^{-1} for the polymerisation in water at pH 7.8.

1.2.3 Ionisation reactions

Thermodynamic data for biochemically important ionisation processes are not only of great interest *per se* but also for their use in making appropriate corrections in connection with reaction calorimetric work. During the last 2 years a large number of ionisation reactions have been studied calorimetrically and for many important ionisation processes reliable thermodynamic quantities (ΔG, ΔH, ΔS, ΔC_p) are now available. Depending on the experimental technique employed and on the value for the ionisation constant, a calorimetric or a non-calorimetric value for the K value (ΔG) may be preferred. However, unless the equilibrium measurement is done with very high accuracy, calorimetrically determined ΔH and ΔC_p values are to be preferred. In particular this holds true for the ΔC_p values for which many data in older compilations are judged to be very uncertain.

Thermodynamic quantities for ionisation reactions have been compiled by Izatt and Christensen[24], by Jencks and Regenstein[25] (K values only) and by Larson and Hepler[26].

Among more recent work we note in particular the large number of studies made by Christensen and Izatt and their co-workers who make use of a calorimetric titration technique. Since their compilation[24] was completed they have reported ΔG and ΔH values for a large number of protonated amines[27]. In this report they also compiled earlier ionisation data for this group of compounds (work by Öjelund and Wadsö[28] on ΔH_i values for protonated alkylamines and hydroxyalkylamines was not included). In other recent work the Christensen–Izatt group reports data (ΔG_i, ΔH_i, ΔS_i, ΔC_{pi}) from extensive studies on aliphatic carboxylic acids[29,30], purines and their nucleosides[31], pyrimidines and their nucleosides[32] and on monosaccharides[33].

In a very recent review article[34] Izatt *et al.* have summarised data for sites and thermodynamic properties associated with proton (and metal ion) binding to ribonucleic acids, deoxyribonucleic acid and their constituent bases, nucleosides and nucleotides.

Woolley *et al.*[35] have determined enthalpies of ionisation for imidazolium ion, 6-uracil carboxylic acid and hypoxanthine·H^+. They also report potentiometrically determined pK_a values for these compounds.

Watt and Sturtevant[36] have reported ΔH_i values at 25 °C for phosphate (second ionisation), $\Delta H_i = 1.13$ kcal mol^{-1}, borate, $\Delta H_i = 2.80$ kcal mol^{-1} and carbonate (second ionisation), $\Delta H_i = 4.60$ kcal mol^{-1}.

Beres and Sturtevant[37] have measured ΔH_i at 25 °C (0.1 mol l^{-1} NaCl) for the biochemical buffer substances PIPES, $\Delta H_i = 2.74$ kcal mol^{-1} and HEPES, $\Delta H_i = 5.01$ kcal mol^{-1} [piperazine-*N*,*N*′-bis(2-ethanesulphonic acid) and *N*-2-hydroxyethylpiperazine-*N*′-2-ethanesulphonic acid, respectively].

Leung and Grunwald[38] have reported measurements of ΔH_i and ΔC_{pi} for water, acetic acid and benzoic acid. Calorimetric measurements were made over the temperature range 0–55 °C.

1.2.4 Hydrolysis reactions

In the field of thermochemistry of simple biochemical compounds and models there has been a marked interest for some time in hydrolysis reactions, in particular those involving phosphate compounds and peptides. A much smaller proportion of current work belongs to this category.

Greengard *et al.*[39] and Rudolph *et al.*[40] measured enthalpies of hydrolysis of several cyclic nucleotides. They showed that in all cases the hydrolysis for 3′,5′-nucleotides is more exothermic than for the corresponding 2′,3′ compounds (AMP, GMP, UMP). The nature of the base showed only a small effect on the enthalpy values for the 2′,3′-nucleotides whereas the value for corresponding 3′,5′-compounds varied markedly with the nature of the base. Among the compounds investigated cyclic 3′,5′-AMP has the highest exothermic heat of hydrolysis, $\Delta H = -14.1$ kcal mol^{-1}, as well as having a very negative ΔG value.

Strack and Müller[41] measured the enthalpy of hydrolysis of *O*-acetyl-carnitine, $(CH_3)_3N^+CH_2CH(OAc)CH_2COOH$ and some of its carboxyl group derivatives. They found that substitution of the carboxyl group by an amide or a nitrile group made the *O*-acetyl hydrolysis process rather strongly endothermic, $\Delta H = 4.42$ and 5.27 kcal mol^{-1} for the amide and for the nitrile, respectively, which should be compared with that of the carboxyl compound $\Delta H = -1.38$ kcal mol^{-1}. In contrast, esterification of the carboxyl group with a methyl group led to an intermediate value for the *O*-acetyl group, $\Delta H = 0.27$ kcal mol^{-1}.

Rajender *et al.*[42] have determined the heat of hydrolysis at 25 °C of ATEE (*N*-acetyl-L-tryptophan ethyl ester) to form the un-ionised acid and ethanol, $\Delta H = 0.8 \pm 0.5$ kcal mol^{-1}.

1.2.5 Formation of metal–ion complexes

Letter and Bauman[43] using calorimetric titration, have studied complexes between copper(II) and a series of tyrosine isomers, while Meyer and Bauman[44] did work on the formation of copper(II) complexes with alanine, histidine and related amino acids.

Lim and Nancollas[142] have made calorimetric measurements on the formation of nickel complexes of diglycine, triglycine and glycyl-γ-aminobutyric acid and of subsequent proton dissociation from the triglycinate species at higher pH. The calorimetric data were combined with results of potentiometric measurements to provide ΔG, ΔH and ΔS values for each of the equilibria.

Beleich and Sari[45] determined by a microcalorimetric technique both K and ΔH for complex formation between Mg ion and AMP, ADP and ATP. Their results are summarised in Table 1.1.

The results are not in a very good agreement with earlier van't Hoff estimates. In this connection the detailed analysis by Alberty[46,47] and by Phillips *et al.*[48] concerning the thermodynamic quantities of hydrolysis reactions and Mg^{2+} binding for adenine nucleotides, should be pointed out.

Table 1.1 Thermodynamic data for Mg^{+} complex formation with adenine nucleotides at 30 °C and at ionic strength of 0.2[45]
(From Belaich, J. P. and Sari, J. C.[45] by courtesy of National Academy of Science)

Reaction	ΔG/kcal mol^{-1}	ΔH/kcal mol^{-1}
$ATP^{4-} + Mg^{2+} \rightleftarrows ATP\text{–}Mg^{2-}$	−6.51	4.47
$ADP^{3-} + Mg^{2+} \rightleftarrows ATP\text{–}Mg^{-}$	−5.12	3.15
$AMP^{2-} + Mg^{2+} \rightleftarrows ATP\text{–}Mg$	−2.50	1.78

The thermochemistry of cation binding by 'carrier antibiotics' and similar compounds has recently attracted considerable interest.

Möschler *et al.*[49] and Früh *et al.*[50], both using microcalorimetric techniques, studied binding of K^+ to valinomycin in ethanol and methanol, respectively. Derived enthalpy values were quite different, −8.9 and −4.5 kcal mol^{-1}, respectively. Früh *et al.* also studied complex formation with nonactin (K^+, Na^+), monactin (Na^+) and monensin (K^+). Calorimetrically derived K values reported by both these groups of workers are in good agreement with values obtained through other techniques.

Izatt *et al.*[51,52], using their calorimetric titration technique, have studied cation binding to the two isomers A and B of the cyclic polyether dicyclohexyl-18-crown-6. (See also review article by Christensen *et al.*[141].)

Experiments were conducted at 10–40 °C in aqueous solution. From the calorimetric results K, ΔH, ΔS and ΔC_p values were calculated for the binding of the following cations: K^+, Rb, Cs^+, NH_4^+, Ag^+, Sr^+ and Ba^{2+}. The alkali metal ion stability sequence is identical with the permeability sequences for these metal ions with the structurally related antibiotics valinomycin and monactin.

1.2.6 Heat capacity measurements

Kresheck[53] has determined values for the relative apparent molal heat contents at different concentrations for *N,N*-dimethylacetamide (in water and in CCl_4) and for *N*-methylacetamide, acetamide and urea (in water). Results are discussed in terms of possible dipole–dipole interactions as a cause for association of this group of molecules in solution.

In a subsequent communication Kresheck reported $\overline{C}^0_{p2}$ values for some amino acids in H_2O, D_2O, 10% ethanol, 6 mol l^{-1} urea, 2 mol l^{-1} KI and 2 mol l^{-1} NaCl. Direct measurements were made of the heat capacities of the solutions. His results support the idea that non-polar groups stabilise the structure in H_2O to a larger extent than in D_2O. It further appears that useful correlations can be made between salting in and salting out effects and the $\overline{C}^0_{p_2}$ values in aqueous solution.

As discussed above, C_p and $\overline{C}^0_{p_2}$ values have recently been reported[7] for several alcohols, carboxylic acids, amines and amides.

1.2.7 Heat of combustion studies

Very few combustion calorimetric measurements of biochemical significance are being made at present. A recent study of interest is the one by Longo *et al.*[55] on the heat of formation of porphin. Using bond enthalpy data they arrived at a resonance stabilisation value as high as 419 kcal mol^{-1}, and attributed 75% of this to resonance energies of the four 2-vinylpyrrole groups which form the porphin system.

1.2.8 Miscellaneous studies

Rodewald and Baumeister made a calorimetric study of the formation of complexes between caffein and sodium salts of aromatic acids. K_c and ΔH values for the formation of 1:1 complexes were reported.

Papenmeier and Campagnoli[57] used a microcalorimetric technique to study the thermodynamics of interaction between two oppositely charged detergents: sodium-dodecyl sulphate and cetyl-pyridinium chloride. The process actually measured was quite complex due to dilution and micellisation processes. Using different models, different sets of thermodynamic data (ΔG, ΔH, ΔS) could be evaluated. Results obtained were in quite close agreement with earlier data from studies of the binding of various anions to serum albumin.

Pilcher *et al.*[61] have studied the enthalpy of micellisation of sodium dodecylsulphate (SDS) as a function of concentration. Experiments were performed as dilution experiments in a microcalorimeter. ΔH for the micellisation process was found to change in the negative direction with increasing concentration of SDS, with increasing temperature and on addition of salt.

Under all experimental conditions ΔH values were found to be small, $\Delta H = 0.087 \pm 0.015$ and $\Delta H = -0.153 \pm 0.003$ kcal mol^{-1} in water and in 0.02 mol l^{-1} NaCl, respectively (25 °C). The experimental results were discussed in terms of transfer of the hydrocarbon chain of SDS from water to the micelle and of the binding of ions to the micelle surface.

1.3 BINDING OF LOW MOLECULAR WEIGHT COMPOUNDS TO PROTEINS

1.3.1 General comments

Studies of specific binding of low molecular weight compounds to proteins form one of the most active fields of biochemical thermochemistry today. The first calorimetric study of this kind was probably the one by Canady and Laidler[58] (1958) on the inhibition of chymotrypsin by hydrocinnamate ion. Further calorimetric work of this type does not seem to have been undertaken until quite recently when suitable microcalorimeters became readily available. It is believed, however, that the interest will continue to grow. It seems clear that results from this kind of work will contribute decisively to our understanding of reactions taking place at the active sites of proteins.

Less specific binding, such as occurs between proteins and buffer substances, neutral salts etc., should also be considered. So far little calorimetric work has been done in this important area.

1.3.1.1 Lysozyme, saccharide inhibitors

The x-ray crystallographic work by Phillips and co-workers on the structure of lysozyme and its complex with saccharide inhibitors [*N*-acetyl-D-glucosamine, GlcNAc, and its trimer $(GlcNAc)_3$] triggered a great interest in the thermodynamic properties of these systems. Equilibrium studies were carried out by several groups, and from two laboratories calorimetric results (ΔG, ΔH, ΔS) have been reported. Vichutinskij *et al.*[59] reported briefly on the binding of GlcNAc to lysozyme while Bjurulf *et al.*[60,61] have studied the binding of GlcNAc, $(GlcNAc)_2$ and $(GlcNAc)_3$. Experimental results by Bjurulf *et al.* were in agreement with the formation of 1:1 complexes and the thermodynamic data calculated were in good agreement with data from equilibrium studies.

The experiments by Vichutinskij *et al.*, determined under slightly different conditions, indicated the presence of two different binding sites. However, recent studies by Bjurulf and Wadsö (unpublished) do not confirm this. Bjurulf *et al.*[61] also reported ΔC_p values for binding of GlcNAc and $(GlcNAc)_3$ and found them to be small (-50 and $+30$ cal mol^{-1} deg^{-1}, respectively). This is in accordance with results of the x-ray work which indicates that the binding reaction does not cause any pronounced conformational change of the enzyme molecule. These low ΔC_p values are in drastic contrast to the strongly negative ΔC_p values for other protein-inhibitor binding reactions (see below).

1.3.1.2 Chymotrypsin, inhibitors

Shiao and Sturtevant[23] and Shiao[62] report results from extensive flow calorimetric measurements on the binding of various inhibitors to α-chymotrypsin, Table 1.2. In the table are also given equilibrium data which are combined with calorimetric values to give the corresponding entropy changes.

Results from measurements at pH 7.8 are complex because the enzyme at this pH forms aggregates. Calorimetric dilution experiments (cf. Section 1.4.2.1) were consistent with the formation of a monomer–dimer equilibrium mixture. Results of the binding studies suggested that only the monomer was capable of binding the inhibitors. Proflavin was shown to exist partly as polymeric species (see Section 1.2.2). Under the conditions of the binding experiments the calorimetric results surprisingly enough did not indicate any difference in binding of the enzyme to monomers and polymers.

From the values in Table 1.2 it is evident that ΔH and ΔS compensate each other to a very large extent at pH 7.8 (For a general discussion of enthalpy–entropy phenomena of α-chymotrypsin, see the recent article by Lumry and Rajender[63].) In the lower pH range the compensation effect is much less pronounced. On the basis of the results it was suggested that the large ΔH and ΔS values at pH 7.8 are due largely to conformation changes of the protein whereas this effect is much less marked at pH 3.6.

Table 1.2 **$\Delta G'$, ΔH and ΔS for the binding of various inhibitors to α-chymotrypsin at pH 7.8 and 25°**[23, 62]

(From Shiao, D. D. F. and Sturtevant, J. M.[23] and Shiao, D. D. F.[62] copyright 1969 and 1970 by the American Chemical Society. Reprinted by permission of the copyright owners)

Compound	$\Delta G'$/kcal mol^{-1}	ΔH/kcal mol^{-1}	$\Delta S'$/cal deg^{-1} mol
Benzoate	−2.64	−18.2	−52
Hydrocinnamate	−3.71	−15.7	−40
β-Naphthoate	−5.30	−17.5	−40
N-Acetyl-L-tryptophan	−2.96	−21.2	−61
N-Acetyl-D-tryptophan	−3.29	−19.0	−52
Proflavin	−6.04	−11.3	−18
Phenol	−3.05	−13.5	−35
Indole	−4.31	−15.2	−37

Later Beres and Sturtevant[37] presented results from binding of hydrocinnamate to chymotrypsinogen A and to π-, δ- and α-chymotrypsins. Enthalpies of binding to the chymotrypsins were similar to the value determined for α-chymotrypsin under slightly different conditions (see Table 1.2). The mixing of solutions of chymotrypsinogen A and hydrocinnamate was not accompanied by any significant heat evolution.

1.3.1.3 Glutamine synthetase, tryptophan and AMP

Ross and Ginsburg[64] studied the binding of feedback inhibitors, L-tryptophan and adenosine 5′-monophosphate (AMP), to glutamine synthetase, by microcalorimetry.

Under saturation conditions they obtained $\Delta H = -7.35$ and -2.00 kcal mol^{-1} subunit for tryptophan and AMP, respectively. (The enzyme forms a dodecamer, molecular weight = 600 000.) Under saturation conditions, for a mixture of the two inhibitors, they obtained $\Delta H = -9.6$ kcal mol^{-1} which is nearly equal to the sum of the two individual ΔH values. The calorimetric results thus suggest that the enzyme has separate and non-interacting sites for the two feedback inhibitors investigated.

1.3.1.4 Aldolase, D*-hexitol 1,6-diphosphate*

Hinz *et al.*[65] have studied the thermochemistry of binding of the substrate analogous inhibitor D-hexitol 1,6-diphosphate to rabbit muscle aldolase at pH 7.5. Measurements were conducted by flow microcalorimetric titrations in the temperature range 5–45 °C. It was shown that the protein undergoes a change in its extent of protonation after binding the inhibitor. By use of different buffers with different heats of ionisation it was shown that 1.4 ± 0.2 moles of H^+ per mole of enzyme is adsorbed in the binding reaction. This number did not vary with the temperature. Enthalpies of binding were calculated from the titration curves whereas corresponding equilibrium constants were evaluated using the assumption that there are 2.7 independent binding sites per molecule, cf. reference 66. A summary of the thermodynamic data they derived is given in Figure 1.3. It is very striking that ΔG values are

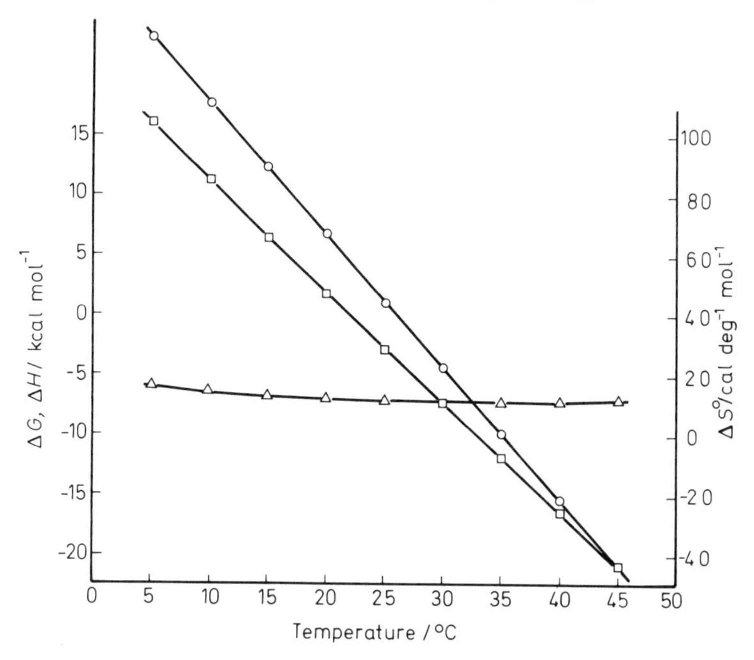

Figure 1.3 ΔG (△), ΔH (O) and ΔS (□) *v.* temperature for the binding of hexitol 1,6-diphosphate to rabbit muscle aldolase at pH 7.5
(Reprinted from Hinz, H. J., Shiao, D. D. F. and Sturtevant, J. M.[65], *Biochemistry*, **10**, 1347. Copyright 1970, by the American Chemical Society. Reprinted by permission of the copyright owner)

essentially constant over the investigated temperature range whereas enthalpies and entropies show a very marked temperature dependence. We further note the very strong enthalpy–entropy compensation. Figure 1.2 dramatically illustrates the fact that ΔG values alone are not well suited for elucidation of energy–structure relationships in biochemical reactions. Furthermore, arguments used in discussions based on ΔH and ΔS values obtained at one temperature only, may be contradicted by values obtained at other temperatures.

The essentially linear relationship of ΔH with temperature leads to a constant, strongly negative ΔC_p value of -1100 cal deg^{-1} mol^{-1} of enzyme (-410 cal deg^{-1} mol^{-1} inhibitor).

The thermodynamic data and their relationship with the nature of the enzyme–inhibitor binding cannot be analysed in any detail at present. However, Hinz *et al.* suggest that the large negative ΔC_p value must be explained in terms of a conformational change of the enzyme leading to a decrease in the exposure of hydrophobic groups.

Lehrer and Barker[67] have reported that various binding and kinetic parameters for this enzyme show a strong temperature dependence. They suggested that the enzyme showed a temperature dependent transition between two different forms having markedly different binding and catalytic properties. Hinz *et al.* were able to show that data by Lehrer and Barker for the binding of the inhibitor D-arabinitol 1,5-diphosphate could be explained by assuming a ΔC_p value of -1200 cal deg^{-1} mol^{-1} of enzyme. It is thus not necessary to postulate two different forms of the enzyme.

1.3.1.5 *Biotin, avidin*

Suurkuusk and Wadsö, as briefly reported in reference 61, have confirmed Green's earlier microcalorimetric determination of ΔH for the biotin–avidin binding reaction and report $\Delta H(25°) = -21.3 \pm 0.1$ kcal mol^{-1} biotin (4 molecules biotin are bound per avidin molecule.) They further determined ΔC_p to be -232 ± 12 cal deg^{-1} mol^{-1} biotin.

Making reasonable assumptions, based on model compound data, most of this fairly large negative value can be accounted for if it is assumed (as has been suggested) that the biotin molecule is transferred from the aqueous solution to the hydrophobic environment within the protein.

1.3.1.6 *Ribonuclease, nucleotide inhibitors*

Bolen *et al.*[68] have very recently reported conclusive evidence that for the system ribonuclease A–3′-cytidine monophosphate (3′CMP) the number of binding sites is 1 per protein molecule. Their calorimetric procedure is based on two series of calorimetric titrations. In one series of experiments the concentration of inhibitor is varied (enzyme concentration constant) whereas in the other the concentration of enzyme is varied (inhibitor concentration constant).

From the microcalorimetric measurements (25 °C, pH 5.5, ionic strength 0.5) the binding constant was calculated to be $K = 6.1 \cdot 10^3$ mol^{-1} and the enthalpy value $\Delta H = 9.8$ kcal mol^{-1}.

Results of the measurements further suggested that Cl^- and OAc^- do not bind at the inhibitor binding site, whereas a phosphate ion presumably interacts with the two histidine residues at the inhibitor site.

In a brief preliminary report Maurer *et al.*[69] have communicated results from a very extensive calorimetric investigation on the inhibition of ribonuclease A and ribonuclease T by a variety of mononucleotides. Experiments were carried out using a batch microcalorimeter in which very small volumes of inhibitor solution were successively added to the enzyme–inhibitor reaction mixture. In most cases measurements were conducted at more than one pH value. From the calorimetric measurements ΔH as well as ΔG and ΔS values were derived.

In most cases the calorimetric results indicated the formation of 1:1 complexes. However, in the cases of 2′CMP-*N*-oxide and 2′CMP binding to RNaseA, and in the case of 2′GMP binding to RNaseT_1 anomalous behaviour was observed.

The following nucleotides were reacted with RNaseA: 2′CMP-*N*-oxide, 2′CMP, 3′CMP, 2′AMP, 2′UMP, 3′UMP, 3′AMP and 3′CMP-*N*-oxide, which are listed in order of decreasing binding constant at pH 5.50.

The inhibitors reacted with RNaseT were 2′GMP, 5′GMP, 3′GMP, 3′XMP and 3′IMP, likewise listed in order of decreasing binding constant.

1.3.1.7 *Glyceryl aldehyde 3-phosphate dehydrogenase,* NAD

Velick *et al.*[70] studied the thermodynamics of binding of NAD to glyceryl aldehyde 3-phosphate dehydrogenase from yeast and from rabbit skeletal muscle by fluorimetric equilibrium titrations and by flow microcalorimetric titrations. The investigations were made over wide temperature ranges, 5–40 °C and 5–25 °C for the two enzymes, respectively. From the data, ΔG, ΔH, ΔS and ΔC_p values were evaluated. Fluorimetric and calorimetric measurements were found to be completely consistent with each other.

Results obtained by Velick *et al.* indicated that for the yeast enzyme the four binding sites are identical and independent. Thermodynamic data for the yeast enzyme are summarised in Table 1.3.

Table 1.3 Thermodynamic quantities for the binding of NAD to yeast glyceraldehyde phosphate dehydrogenase at pH 7.3

(From Velick, S. F., Eaggort, J. P. and Sturtevant, J. M.[70]. Copyright 1971 by the American Chemical Society. Reprinted by permission of the copyright owner)

Temperature/K	ΔG°/kcal mol^{-1} of NAD*	ΔH/kcal mol^{-1} of NAD	ΔS°/cal deg^{-1} mol^{-1} of NAD
278.2	−7.40	−1.9	+19.9
298.2	−7.33	−12.4	−16.8
313.2	−7.19	−20.1	−41.6

* Results from equilibrium measurements. Corresponding calorimetric values were −7.43 (298.2 K) and −6.99 (313.2 K).

It is seen that the ΔG values are nearly constant whereas ΔH and ΔS values are strongly temperature dependent. From the temperature dependency of ΔH a strongly negative and essentially constant ΔC_p value was calculated, $\Delta C_p = -520$ cal deg^{-1} mol^{-1} NAD. There is no close relationship between this enzyme–coenzyme system and the aldolase–inhibitor system discussed earlier, but the very similar trend in change of ΔH and ΔS values with temperature as well as the strongly negative ΔC_p values are noteworthy.

Results for the muscle enzyme make it evident that there is a strong interaction among sites. Apparent thermodynamic properties (ΔG, ΔH, ΔS) were calculated for each site. The sums of the enthalpy changes (which can be given with much higher accuracy than the individual values) were $\Delta H =$

-51.8 ± 1.0 kcal mol^{-1} and -69.8 ± 1.0 kcal mol^{-1} at 5 and 25 °C, respectively. The corresponding ΔC_p value will thus be -220 cal deg^{-1} mol^{-1} of NAD.

As for most of the studies reviewed in this article it is not yet possible to give any straightforward interpretation of the thermodynamic data. The large negative ΔC_p values do suggest 'hydrophobic binding' or conformational changes which result in a decreased exposure of hydrophobic groups. However, for a process of this nature one would expect only minor changes in ΔH and large positive ΔS values.

1.3.1.8 Phosphorylase b, AMP

Wang *et al.*[71] used a microcalorimeter to study the binding of AMP to glycogen phosphorylase b. On the basis of results from the calorimeter, the ultracentrifuge and kinetics they suggested the existence of two sets of AMP binding sites. However, the calorimetric data do not seem to be precise enough to give strong support to results obtained by the other techniques. It was stated that further thermochemical work on this system is in progress.

1.3.1.9 BSA, *lysolecitin**

Klopfenstein[72, 73] measured the binding of lysolecitin to bovine serum albumin. Calorimetric data, which indicated formation of a 1:1 complex, led to values for both ΔG (-6.2 kcal mol^{-1}) and ΔH (-18 kcal mol^{-1}).

1.3.1.10 Albumin, warfarin

O'Reilly *et al.*[74] studied the interaction between the common anticoagulant drug warfarin and human plasma albumin. Heat effects were determined calorimetrically and the percentage of warfarin found was determined by equilibrium dialysis and ultracentrifuge techniques.

1.3.1.11 β-lactoglobulin, dodecyl sulphate

Lovrien and Anderson[75], using a microcalorimetric method, studied the binding of dodecyl sulphate to β-lactoglobulin and showed that there are two strong binding sites at acidic pH. ΔG, ΔH and ΔS values were reported. The authors made the relevant comment that, as shown by calorimetric studies, proteins are likely to bind at least a few anions of the types commonly used in buffers and supporting electrolytes (cf. Bolen *et al.*[68] Section 1.3.1.6).

1.3.1.12 Hemerythrin, anion ligands

Langerman and Sturtevant[77] in their microcalorimetric study on hemerythrin (see Section 1.3.1.4, 1.4.2.2) measured the enthalpy of complexation of

---

*Note added in proof: Calorimetric data for the binding of iodide and chloride ions and of dodicyl sulphate were recently reported by Lovrein and Sturtevant[145].

this protein with azide and thiocyanate. In both cases enthalpy changes were about -15 kcal mol^{-1}.

1.3.1.13 *Apocarbonic anhydrase, transition metal ions*

Carbonic anhydrase is a metallo-enzyme containing one atom of very tightly bound zinc(II) per molecule. Henkens *et al.*[78] converted bovine carbonic anhydrase A and B to the zinc-free apoenzymes and measured enthalpies of binding of Zn^{2+}, Co^{2+}, Cu^{2+}, Cd^{2+} and Ni^{2+} by batch and flow microcalorimetry. Comparison with earlier equilibrium data enabled them to calculate ΔG, ΔH and ΔS values. In all cases the binding was found to be endothermic although very strongly exergonic. Comparison with thermodynamic data for the formation of zinc complexes with amino acids and other compounds led to the suggestion that carboxylic acid groups rather than amino or imidazole nitrogen atoms are involved in the Zn–apoenzyme binding reaction.

1.3.1.14 *Oxygenation and oxidation of hemerythrin*

Langerman and Sturtevant[77] have measured the enthalpy value for the oxygenation of hemerythrin

$$HrFe_2^{II} + O_2 \rightarrow HrFe_2^{II}O_2 \qquad (1.3)$$

A solution of the deoxygenated protein was mixed in a flow calorimeter with buffer containing dissolved oxygen. During the measurement the calorimetric system was protected from atmospheric oxygen by being enclosed in a N_2-box. The enthalpy determined for reaction (1.3) was $\Delta H = -9.2 \pm 0.4$ kcal mol^{-1} monomer, pH 7.0, 25 °C (cf. Section 1.4.2.2).

The enthalpy of oxidation of the Fe^{II} protein was determined in a batch microcalorimeter by mixing protein solution with a solution of $K_3Fe(CN)_6$. Comparison with available auxiliary data leads to $\Delta H = 24.5 \pm 2.3$ kcal mol^{-1} of monomer for the reaction

$$2\,e^- + [HrFe_2^{III}(H_2O)_m] \rightarrow [HrFe_2^{II}] + m\,H_2O \qquad (1.4)$$

The enthalpy value based on iron, -12.2 ± 1.7 kcal mol^{-1}, is similar to the corresponding value determined for cytochrome *c* by Watt and Sturtevant[36], $\Delta H = 14.1$ kcal mol^{-1} (see Section 1.5.1).

1.3.1.15 *Oxygenation of myoglobin*

Keyes *et al.*[79] have measured the enthalpy of oxygen binding to sperm whale myoglobin. A solution of O_2-free myoglobin was mixed in a flow microcalorimeter with buffer containing oxygen. The enthalpy value obtained, $\Delta H = -19.0 \pm 1.0$ kcal mol^{-1} (25 °C, pH 8.5, 10^{-3} mol l^{-1} tris, ascorbate present) agreed with the van't Hoff value obtained from their spectrophotometric measurements, $\Delta H = 18.1 \pm 0.4$ kcal mol^{-1}.

1.3.1.16 Antibody, hapten binding

Barisas *et al.*[80] have recently reported extensive microcalorimetric and fluorimetric titration studies of rabbit anti-2,4-dinitrophenyl antibodies with dinitrophenyl (DNP) haptens. The calorimetric titrations, which were made by flow microcalorimetry, showed that about 30% of the antibody sites will bind ε-N-DNP-L-lysine with $\Delta H = -22.8 \pm 0.4$ kcal mol^{-1} at 25 °C. With saturation of the antibody sites lower values were found. The overall site average was $\Delta H = -15.2 \pm 0.3$ kcal mol^{-1} of site if the site purity was assumed to be 100%. However, spectral data indicated an antibody purity of 87% which would bring the average enthalpy of hapten binding to -17.5 kcal mol^{-1} of site.

Fluorimetric van't Hoff plots gave a maximum enthalpy value of $\Delta H = -11$ kcal mol^{-1} of binding site. If this value is ascribed to the last fraction of the binding sites in the titration, the van't Hoff value could be considered compatible with the calorimetric value. A spread in ΔH values from -23 to -11 kcal mol^{-1} of sites would, in the absence of any entropy compensation, lead to a much broader distribution of binding constants than was indicated by the fluorescence titration experiments. It was therefore suggested that a considerable enthalpy–entropy compensation does take place. Comparison between results of measurements performed at 25 and 5 °C leads to $\Delta C_p = -300 \pm 30$ cal deg^{-1} mol^{-1} of strongest binding sites. The average ΔC_p value, assuming 87% purity of the antibody preparation, was -220 ± 20 cal deg^{-1} mol^{-1} of sites. In accord with the usual assumption, the negative ΔC_p values might reflect a decreased exposure of hydrophobic groups to the water.

As in the case of biotin–avidin binding[61] (see Section 1.3.1.5) it is judged that a considerable fraction of the ΔC_p can be explained as the result of a transfer of the hydrophobic parts of the hapten molecule from water to a hydrophobic site.

In some experiments slightly different haptens were used. ε-DNP-aminocaproate (which carries a full negative charge) and 1-hydroxy-2-DNP azonaphthalene-3,6-disulphonate (in which only the DNP group is common to the other haptens) were shown to bind with nearly the same average ΔH as DNP lysine. It was therefore suggested that electrostatic interactions have little effect on the value of ΔH, which seems to be determined largely by the DNP group itself.

DNP lysine was also reacted with two samples of F_{ab} fragments which were prepared from the same batches of antisera used in the antibody experiments. The shape of the titration curves, as well as derived ΔH values, were in general agreement with those obtained with intact antibodies. The F_{ab} fragment is univalent whereas the antibody molecule is divalent. Thus the results suggest that in hapten antibody binding there is site heterogeneity rather than site interaction.

1.4 PROTEIN–PROTEIN (POLYPEPTIDE) BINDING

Protein–protein binding reactions form a vast field suitable for microcalorimetric investigations such as enzyme–protein inhibitor binding, immunochemical reactions, and self-association processes. As yet, however, only a few studies have been reported in this area.

1.4.1 Binding between different molecules

1.4.1.1 Haemoglobin, haptoglobin

Adams and Weiss[81] measured the combination of haptoglobin and haemoglobin under different sets of conditions. The preparations they used were quite impure, but, as this binding reaction is highly specific, data obtained should still be reasonably valid. Most measurements were performed as calorimetric titrations at 37 °C.

The haemoglobin–haptoglobin coupling reaction is complex. Presumably it includes dissociation of the tetramer haemoglobin molecule ($\alpha_2\beta_2$) into dimers

$$\alpha_2\beta_2 \rightleftharpoons 2\alpha\beta \qquad (1.5)$$

ΔH for reaction (1.5) has been reported to have a small value[82], cf. also the study by Evans *et al.*, Section 1.4.1.3. Further, two coupling reactions seem to compete as shown below where Hp = haptoglobin

$$\mathrm{Hp} + \alpha\beta \rightleftharpoons \mathrm{Hp} - \alpha\beta \qquad (1.6)$$

and

$$\beta\alpha + \mathrm{Hp} - \alpha\beta \rightleftharpoons \beta\alpha - \mathrm{Hp} - \alpha\beta \qquad (1.7)$$

Results of the calorimetric titrations support the view that a haptoglobin molecule associated with one $\alpha\beta$-dimer has a greater affinity for another $\alpha\beta$-dimer than has a haptoglobin molecule with both sites still free.

The enthalpy change per mole of haemoglobin was calculated from the slope of the titration curve, holding haptoglobin constant, and from the maximum heat for a constant amount of haemoglobin. The derived enthalpy value was -70 kcal mol^{-1} at 37 °C. From measurements at 37, 20 and 4 °C an extraordinary large negative ΔC_p value was derived, $\Delta C_p = -2300$ cal mol^{-1} deg^{-1}.

From other studies it has been suggested that hydrophobic binding is involved in the reaction which is in agreement with a negative ΔC_p. However, the strongly exothermic ΔH value suggests that other factors also are involved. Adams and Weiss pointed out that the haemoglobin–haptoglobin reaction appears to be remarkably similar to the biotin–avidin binding reaction. It is interesting to note that recently the same pattern of large negative ΔH and ΔC_p values has been found for this latter reaction[61] (Section 1.3.1.5).

1.4.1.2 Fc fragment of IgG, protein A

It has been shown that the reaction between 'protein A' from *S. aureus* and Fc-fragment from IgG is in several respects similar to an antigen–antibody reaction. In addition to its general suitability for studies of specific protein–protein interactions, it may also be looked upon as a model for antigen–antibody reactions.

Sjöquist and Wadsö[83] measured enthalpy changes for the following reaction in the temperature range 25–43 °C.

$$\text{Fc} + 2\text{A} \rightarrow \text{FcA}_2 \tag{1.8}$$

The enthalpy change at 25 °C was found to be $\Delta H = -19.7$ kcal mol^{-1} of A, and the apparent change in heat capacity, $\Delta C_p = -515$ cal deg^{-1} mol^{-1} of A.

As for many other coupling reactions treated in this review we can note the pattern of large negative ΔH and ΔC_p values. It is particularly interesting to note that this was also found to be the case for binding of DNP-lysin-hapten to antibody molecules (Section 1.3.1.16). It therefore seems possible that the very low ΔH value Steiner and Kitzinger[84] reported for antigen–antibody coupling at 22 °C might turn out to be strongly exothermic if measured at a physiological temperature (cf. discussion by Adams and Weiss[81]).

1.4.1.3 *Association of α and β chains of haemoglobin*

Evans *et al.*[85] have reported results from a microcalorimetric study on the recombination of α and β chains of human haemoglobin

$$4\alpha + \beta_4 \rightarrow (\alpha_2\beta_2)_2 \tag{1.9}$$

A number of samples from different preparations were extensively investigated. ΔH values are highly scattered which seems to be a result of differences in state of protein rather than inadequacy of the calorimetric technique. A minimum ΔH value of about -7 kcal mol^{-1} of dimer was attributed to the recombination process.

1.4.1.4 *S-protein, S-peptide*

Bovine pancreas ribonuclease A (RNaseA) can be split enzymatically into two parts: protein-S and peptide-S. The protein and the peptide remain tightly bound to each other at neutral pH, and the product (called RNaseS) retains enzymatic activity. The protein and peptide can be separated at low pH (loss of activity) and again combined at neutral pH, as shown below.

$$\text{S-peptide} + \text{S-protein} \rightleftharpoons \text{RNaseS}' \tag{1.10}$$

The product formed, RNaseS′, is very similar to, possibly identical with, RNaseS.

Marzatto *et al.*[86] and Hearn *et al.*[87] have reported results from microcalorimetric and equilibrium measurements on reaction (1.10). Hearn *et al.* also made measurements where methionine 13 of the S-peptide was oxidised to the corresponding sulphone, Met(O_2)-13-S-peptide.

The study by Hearn *et al.* was much more extensive and results were documented in more detail than those by Marzotto *et al.* Results of the two studies were not in good agreement and only those data from Hearn *et al.* will be treated here.

The latter investigators made measurements at pH 7 over a wide range of temperatures with S-peptide and with Met(O_2)-13-S-peptide. These measurements were taken with a batch microcalorimeter. In addition flow calorimetric measurements were made on the S-peptide in which concentrations of the reaction components were varied. No trend was found among the ΔH values suggesting that the reaction was complete (and that there was no significant unspecific binding) under the conditions of the batch experiments.

Numerical agreement between batch and flow experiments was reasonably good but not so good as one might have expected (25°, pH 7, 0.3 mol l^{-1} NaCl solution): $\Delta H = -39.8$ kcal mol^{-1} (smoothed out value from the batch experiments) and $\Delta H = -33.6 \pm 0.9$ kcal mol^{-1} (mean value from 14 flow experiments).

The enthalpy values were found to be very dependent on temperature and were therefore fitted to a quadratic function of temperature. Values for ΔG were obtained by integration of the Gibbs–Helmholtz equation. (The integration constant was evaluated by comparison with results from equilibrium measurements.) The derived expression for ΔH led to ΔC_p values, which were found to be large and negative. Both for the S-peptide and the sulphone peptide the ΔC_p values were found to be strongly temperature dependent (more negative at higher temperatures).

The temperature dependence of ΔH was interpreted as being largely the result of thermally-induced conformational transition occurring in both S-protein and RNaseS′. (In separate experiments Hearn *et al.* made some studies of these transitions with a differential scanning microcalorimeter, cf. Tsong *et al.*[88], Section 1.6.1.2).

Substantial differences were observed when the various thermodynamic quantities were compared for the two peptides investigated. No interpretation could be offered for the effect of the oxidised methionine group.

Very recently Rocchi *et al.*[143] have reported from an extensive spectroscopic and calorimetric investigation on the interaction between *S*-protein and *S*-peptide and *S*-peptide analogues. These authors find, contrary to Hearn *et al.*[87], that ΔH of association is a function of the ratio of *S*-protein to *S*-peptide concentration.

1.4.2 Protein self-association

1.4.2.1 α-Chymotrypsin

Shiao and Sturtevant[23] studied the enthalpy of aggregation of α-chymotrypsin in connection with their inhibition studies on this enzyme (see Section 1.3.1.2). Heats of dilution were measured calorimetrically. Results were consistent with the simple reaction scheme

$$(\alpha\text{-CT})_2 \rightleftarrows 2\alpha\text{CT} \qquad (1.11)$$

Assuming this to be the only reaction, K and ΔH values were calculated as $K = (3.5 \pm 0.4) \cdot 10^{-4}$ and $\Delta H = 35.4 \pm 2.3$ kcal mol^{-1} of dimer.

For recent non-calorimetric studies of the thermodynamics of dimerisation of α-chymotrypsin, see Aune *et al.*[89, 90].

1.4.2.2 Hemerythrin

Langerman and Sturtevant[77] in their general calorimetric study on hemerythrin (cf. Sections 1.3.1.12, 1.3.1.14) measured enthalpy changes for dissociation of aggregates

$$(\mathrm{Hr})_8 \rightleftarrows 8\ \mathrm{Hr} \tag{1.12}$$

Hr = metazid-, metaquo-, and methydroxy- hemerythrin. Enthalpy values as well as ΔC_p values were found to be close to zero. Comparison between results obtained in different buffers showed that somewhat less than one proton per aggregate was released during the dissociation process.

1.4.2.3 Virus protein

Stauffer *et al.*[91] have performed microcalorimetric studies on polymerisation and depolymerisation reactions of tobacco mosaic virus protein. The measurements, which were made at room temperature, consisted of changing one or more of the following parameters: pH, ionic strength and concentration. Molecular weight determinations were made in parallel with the calorimetric measurements.

ΔH values for the polymerisation processes arising from a change in concentration and ionic strength agreed with earlier values obtained from equilibrium measurements. However, calorimetric ΔH values for the polymerisation resulting from changes in pH were considerably smaller than corresponding values from equilibrium measurements. This was interpreted as a result of differences in the extent of reaction in the two methods.

It would seem desirable to consider the effect of changes in solvation when enthalpy values of this kind are discussed.

1.5 MISCELLANEOUS PROTEIN REACTIONS

1.5.1 Oxidation of cytochrome c

Watt and Sturtevant[36], using flow microcalorimetry, investigated the oxidation of ferrocytochrome c by ferricyanide ion, and reduction of ferricytochrome c by ascorbate ion. Most measurements were made at 25 °C in the pH range 6–11. Using auxiliary thermochemical data, ΔH for the process

$$\mathrm{Fe^{III}cyt\ c} + 0.5\ \mathrm{H_2} \rightleftarrows \mathrm{Fe^{II}cyt\ c} + \mathrm{H^+} \tag{1.13}$$

was calculated. Identical sigmoidal curves of ΔH versus pH were found for this reaction with data originating from both oxidation and reduction experiments.

The enthalpy value for reaction (1.13) at pH 7 and 25 °C was found to be $\Delta H = -14.1 \pm 1.5$ kcal mol^{-1} in good agreement with the calorimetric value reported by George *et al.*[92], $\Delta H = -14.5$ kcal mol^{-1}. Measurements were also made at 15 °C leading to $\Delta C_p = -60$ cal mol^{-1}.

The small ΔC_p value for reaction (1.13) is consistent with the view that the conformational change, which is believed to accompany the redox process, is not very extensive.

The sigmoidal shape of the ΔH *v.* pH curve was interpreted as a conformational change in Fe^{III}cyt c which takes place with the uptake of, presumably, one proton. The ionising group has a pK of 9.3 and ΔH for the combined conformational change and protonisation reaction was calculated to be *c.* -16 kcal mol^{-1}.

1.5.2 Activation of chymotrypsinogen A

Beres and Sturtevant[37], by microcalorimetric measurements, obtained the enthalpy change of the trypsin catalysed 'rapid activation' of chymotrypsinogen A to form π-chymotrypsin. In these experiments hydrocinnamate, an inhibitor which stops the activation reaction at the π-chymotrypsin stage, was present. Binding of the inhibitor to π-chymotrypsin gave nearly the same enthalpy value as the binding reaction with chymotrypsinogen, $\Delta H = -6.5$ kcal mol^{-1} (pH 7.4, 25 °C). Since the enthalpy of interaction between chymotrypsinogen and the inhibitor was found to be close to zero, it was concluded that the enthalpy change for the activation process must also be close to zero. The value was the same in several buffers having different ΔH_i values and it was thus shown that the extent of protonation of the protein was not affected by the process. On the other hand activation in the absence of inhibitor is accompanied by enthalpy changes which depend linearly of the buffer ionisation enthalpy, which indicates liberation of protons from the protein molecule. From activation studies without inhibitor present the value $\Delta H = -2 \pm 1$ kcal mol^{-1} was calculated for the conversion of π-chymotrypsin to δ-chymotrypsin.

1.5.3 Reconstitution of virus from protein and nucleic acid components

Srinivasan and Laufer[93] have measured by a microcalorimetric technique the coupling process between tobacco mosaic virus protein (TMVP) and the virus RNA (TMV RNA). The heat of a hypothetical reaction between polymerised TMVP (cf. Section 1.4.2.3) and TMV RNA (phosphate buffer, $\mu = 0.01$, pH $= 5.0$) was calculated by application of Hess' law to a series of actual calorimetric experiments. The ΔH value was found to be -0.485 cal g^{-1} of TMVP corresponding to $\Delta H = -2.8$ kcal mol^{-1} of phosphate groups in the RNA chain. The possible nature of the protein-RNA binding process was discussed.

1.6 UNFOLDING PROCESSES IN BIOPOLYMERS

Present understanding of conformational changes of biopolymers in solution has to a large extent been based upon results from equilibrium measurements performed at different temperatures. During recent years, however, there have been developed calorimeters which are suitable for studies of conformational changes. It seems that calorimetry, preferably in conjunction with other techniques, has already provided substantial contributions to our understanding of denaturation processes for proteins and nucleic acids. The calorimetric experiments in this connection are of two kinds: heat capacity measurements at different temperatures and measurements of heat evolution in connection with chemically induced conformational changes. In the former type of experiments temperature scanning heat capacity calorimeters are preferable.

1.6.1 Proteins, thermally induced unfolding processes

1.6.1.1 The chymotrypsinogen family of proteins

Jackson and Brandts have made a calorimetric study of the thermally induced unfolding of chymotrypsinogen which is parallel to earlier equilibrium studies by Brandts[94] and by Lumry and co-workers. Measurements were made with a differential temperature scanning heat capacity calorimeter. The heating rate was 10 deg h^{-1} and protein concentrations were 0.25 %. For comparison an experiment was also run with 1.85 % protein. Measurements were made in the pH range 2–3 and in the temperature range of about 5–65 °C. The ΔH values obtained for the unfolding process varied between +100 kcal mol^{-1} at 42° and pH 2, to +140 kcal mol^{-1} at 54° and pH 3, while

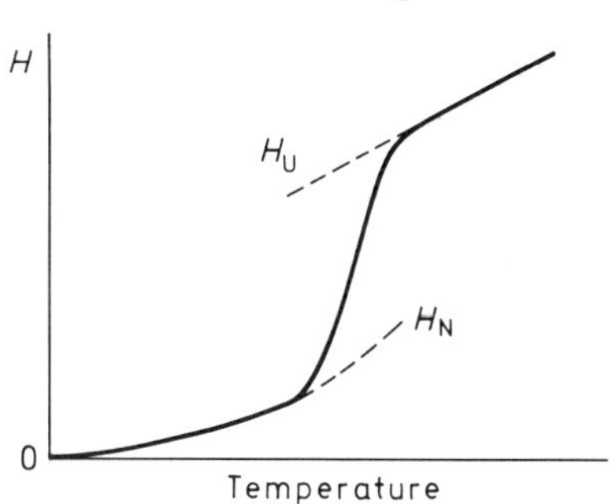

Figure 1.4 Variation of enthalpy with temperature for protein unfolding; N = native; U = unfolded
(Reprinted from Jackson, W. M. and Brandts, J. F., *Biochemistry*, **9**, (1970) 2294. Copyright 1970 by the American Chemical Society. Reprinted by permission of the copyright owner)

ΔC_p values changed from about 3800 to 2800 cal deg^{-1} mol^{-1} over the same range of conditions.

Since calorimetric results were in agreement with earlier estimates based on the two-state analysis of transition curves it was concluded that the chymotrypsinogen denaturation reactions studied obey the two state approximation to a high degree of accuracy. (Jackson and Brandt give in their paper a useful summary of the concept of the two state approximation as used by workers like Lumry, Brandts and Biltonen.) However, it is apparent that this conclusion would not have been substantiated if the calorimetric heat absorption curve had been given the interpretation favoured by Sturtevant and co-workers see Tsong *et al.*[88], Section 1.6.1.2).

In the case of chymotrypsinogen as well as for RNase (Section 1.6.1.12), enthalpy–temperature curves of the type shown in Figure 1.4 are found.

Jackson and Brandts in their treatment define the native state as that distribution of microscopic states which exists exclusively at any temperature below that at which the major unfolding transition begins. Their ΔH value at a certain temperature refers therefore to the difference between the extrapolated curves in Figure 1.4. Sturtevant and co-workers, on the other hand, place primary emphasis on an analysis in which they consider the native state as that in which the experiment starts (i.e. in Figure 1.4 at that point where H arbitrarily is set to zero). Consequently their enthalpy change corresponds to the value for H_D. Clearly, when these two schools refer to calorimetric enthalpies of unfolding they will, as pointed out by Jackson and Brandts, refer to different processes.

Biltonen *et al.*[95] have measured heat capacities for chymotrypsinogen, α-chymotrypsin and dimethionine sulphoxide chymotrypsin in 1 % aqueous solution. The determinations were conducted in a precise 'macrocalorimeter' capable of a precision of 0.01 %. Measurements were made under conditions of pH and temperature where the proteins were in both their folded (denaturated) states. Results from their C_p measurements are summarised in Table 1.4.

Table 1.4 Apparent molar heat capacities of chymotrypsinogen proteins in 1% aqueous solution (cal g^{-1} deg^{-1})

(From Biltonen *et al.*[95]. Copyright 1970 by the American Chemical Society. Reprinted by permission of the copyright owners)

Protein	*Native*, pH 4			Unfolded, pH 2		
	25 °C	40 °C	50 °C	25 °C	40 °C	50 °C
α-Chymotrypsin	0.390 ±0.012	0.414 ±0.020			0.549 ±0.008	
Chymotrypsinogen	0.382 ±0.012		0.431 ±0.020			0.553 ±0.008
Dimethionine sulphoxide	0.382 ±0.008	0.467 ±0.012		0.557 ±0.004	0.545 ±0.008	

It is seen that for all three proteins there is a significant increase in the apparent heat capacity concomitant with the unfolding process. The value for chymotrypsinogen agrees with that determined by Jackson and Brandts[94] using temperature scanning calorimetry. Furthermore, the results are in good agreement with conclusions earlier deduced from the temperature behaviour of the van't Hoff heat for the unfolding process.

It is seen that C_p values in the folded state at 25 °C are the same within uncertainty limits for the three proteins. However, in each case there is found a strong increase in heat capacity with temperature (which is in agreement with data of Jackson and Brandts for chymotrypsinogen). We further note that the temperature dependence of the heat capacity seems to be different for the three compounds.

C_p values for each protein in the unfolded state are identical. This observation, which is in accord with results of the equilibrium studies by Biltonen and Lumry[96], supports the view that the conformations of each species in the unfolded state are thermodynamically identical.

In addition, Biltonen *et al.*[95] studied the unfolding processes as initiated by lowering of the pH, see Section 1.6.2.1. It was concluded that results of the calorimetric measurements strongly support the validity of the two state approximation for the unfolding processes studied.

Crescenzi and Delben[102] have reported from differential scanning calorimetric experiments on chymotrypsinogen in 6–7% solution (see Section 1.6.1.4), $\Delta H = 154 \pm 8$ (pH 3) and $\Delta H = 112 \pm 3$ (pH 2) kcal mol^{-1}.

1.6.1.2 The RNaseA *family*

Tsong *et al.*[88] have made a penetrating calorimetric study of the thermally induced conformational transition of pancreatic bovine ribonuclease A (RNaseA). Some measurements at pH 7 were also made on the RNaseA derivatives RNaseS, RNaseS′ and S-protein (cf. Section 1.4.1.4).

The equipment used was a differential temperature scanning heat capacity calorimeter (being an improved version of the calorimeter earlier used by the Sturtevant group). The heating rate was 18 deg h^{-1} and protein concentrations were 0.5% in experiments in the pH range 0.4–7.8. Concentrations were varied over a range of 0.1–2.7% in experiments performed at pH 2.8. A few additional experiments and comments concerning this study have been reported more recently[87].

ΔH values for the unfolding process were strongly pH dependent: $\Delta H =$ 47.8 kcal mol^{-1} at pH = 0.36 whereas at pH 7.8 ΔH was as high as 105.5 kcal mol^{-1}. ΔC_p values were positive, *c.* 2000 cal mol^{-1} deg^{-1}.

From their results and by comparison with earlier measurements Tsong *et al.* concluded that the thermally induced conformational transition meets the requirements for a two state process at pH values below 2. At higher pH values their analysis was not in accord with the two state model.

As discussed above (Section 1.6.1.1) Tsong *et al.* make a somewhat different analysis of the enthalpy–temperature curve from that favoured by Jackson and Brandts. When Tsong *et al.* apply this latter interpretation of their calorimetric results they do get reasonable agreement with van't Hoff data (i.e. results in accord with the two state model) for experiments up to pH 4. However, at higher pH values the calorimetric ΔH values are still significantly higher than corresponding van't Hoff values.

The RNaseA derivatives all appear to be less stable than the parent compound (lower melting temperatures and lower enthalpies of unfolding). These transition reactions do not appear to meet the requirements for a two state process.

Delben *et al.*[103] have reported differential scanning calorimetric experiments on RNaseA in water [$\Delta H = 99 \pm 8$ kcal mol^{-1}, pH = 6(?)] as well as in the presence of denaturants, see Section 1.6.1.4.

1.6.1.3 Lysozyme

The enthalpy of unfolding of lysozyme has been determined by O'Reilly and Karasz by heat capacity calorimetry and van't Hoff treatment of o.r.d. data.

The calorimeter used was a precision heat capacity single calorimeter of the adiabatic type. Calorimetric and o.r.d. measurements were made on identical protein solutions. The concentrations were rather large (2.5 %) and the pH was low (1.0). Both calorimetric and o.r.d measurements indicated that the process was not completely reversible under the experimental conditions.

The enthalpy change obtained calorimetrically was $\Delta H = 63 \pm 6$ kcal mol^{-1}. The transition temperature was 46 °C. Within uncertainty limits the values agree, which supports the view that the unfolding process can be interpreted in terms of the two state model.

The ΔH value for the unfolding process is in good agreement with a van't Hoff value obtained earlier by Sophianopoulus and Weiss[98], $\Delta H = 66$ kcal mol^{-1} (pH = 1–3). A slightly higher van't Hoff value has recently been derived by McDonald *et al.*[99], $\Delta H = 73$ kcal mol^{-1} (pH 3.3).

A considerably higher value was obtained by Delben and Crescenzi[101] using differential temperature scanning calorimetry, $\Delta H = 138$ kcal mol^{-1} (pH 5.4, *c.* 6 % lysozyme solution). The corresponding van't Hoff value was determined as $\Delta H = 123$ kcal mol^{-1} at low protein concentration[102], see Section 1.6.1.4. Other calorimetric measurements have been conducted on the denaturation of lysozyme in the presence of denaturants, cf. Section 1.6.2.2. However, data from such processes cannot be compared closely with data from thermally induced reactions in the absence of denaturants.

From the calorimetric experiments by O'Reilly and Karasz an approximate ΔC_p value was calculated for the unfolding process, $\Delta C_p = 7000$ cal deg^{-1} mol^{-1}. Compared with data for other protein unfolding processes this value seems too high. Equilibrium studies by Tanford and Aune[100] for the unfolding process in the presence of guanidine hydrochloride indicated a ΔC_p value of about 1000 cal deg^{-1} mol^{-1}.

1.6.1.4 *Chymotrypsinogen,* RNaseA, *lysozyme in more concentrated solutions*

Crescenzi and Delben[102] (cf. Delben *et al.*[103], Delben and Crescenzi[101]) have recently reported extensive temperature scanning calorimetric data on rather concentrated protein solutions: RNaseA (3–14 %), lysozyme (5–10 %) and chymotrypsinogen (6–7 %). Measurements, which were made with a commercial scanning calorimeter (Perkin–Elmer DSC-1B), were performed on the proteins in aqueous media in the absence and in the presence of denaturants. From the data both enthalpies of denaturation and also activation enthalpies were calculated. U.V. and o.r.d. equilibrium measurements were performed on dilute protein solutions at different temperatures in parallel with the calorimetric experiments. From the results of these data van't Hoff enthalpies for the denaturation processes were calculated.

For several of the systems investigated substantial differences were found between values derived calorimetrically and those from equilibrium measurements. However, in most cases the estimated limits of error (± 2 to ± 13 kcal mol^{-1}) included both values and no systematic trend in the differences seemed to be present. It thus appears that useful unfolding enthalpy values can be determined for protein solutions using relatively simple scanning calorimetric equipment. It is necessary, however, to employ more concentrated solutions than those used in the differential temperature scanning studies referred to earlier in this section.

1.6.1.5 Myoglobin

Privalov[104] has reported values for the enthalpy of unfolding for myoglobin, procollagen (see Section 1.6.1.6) and T_2 phage DNA (see Section 1.6.3.7). His technique involves the use of a micro differential scanning calorimeter. At pH 9 and 70 °C he obtained $\Delta H = 110$ kcal mol^{-1} for myoglobin. Cf. the very recent report by Privalov *et al.*[144].

1.6.1.6 Tropocollagen and collagen

The proteins which comprise the fibres in connective tissue are called collagens. They are made up of smaller units, called tropocollagen, having molecular weights of *c.* 300 000. Each such molecule consists of a triple helix composed of polypeptide chains. A characteristic property of these proteins is their very high content of prolin and hydroxyprolins. The helical structure melts (unfolds) at a certain temperature, T_m. It has been found repeatedly that T_m increases with increased prolin content.

The thermochemistry of collagens has been investigated extensively by two groups in the U.S.S.R., each having available a differential temperature scanning calorimeter suitable for investigations of very dilute solutions of biopolymers.

Privalov and Tiktopulo[105] and Privalov[104] have recently reported heat capacity measurements on tropocollagen from a variety of sources (Table 1.5).

Table 1.5 Unfolding of tropocollagens having different prolin contents. Solutions were salt free with pH 3.5. Calorimetric experiments were made on *c.* 0.5 g of 0.04% protein solution. Temperature scanning rate = 0.2 °C per min.

(From Privalov, P. L. and Tiktopulo, E. I.[105] by permission of John Wiley & Sons, Inc.)

Tropocollagen	*Prolin amino acids content per 1000 residues*	T_m/°C	ΔH/cal g^{-1}
Rat	226	40.8	16.8 ± 0.2
Pike	199	30.6	13.6 ± 0.2
Whiting	—	21.5	9.7 ± 0.2
Cod	155	20.0	8.2 ± 0.2

The increase of T_m with increasing content of prolin can be observed as well as the increase in heat of unfolding, ΔH. The value for rat skin tropocollagen corresponds to $\Delta H = 1.53$ kcal mol^{-1} of amino acid residue.

Values given in Table 1.5 were determined on salt free solutions. When measurements were made on rat skin tropocollagen in 0.1 mol l^{-1} NaCl a more complex melting behaviour was observed[106]. The heat is absorbed in two discrete steps with T_m = 32.5 and 36.5 °C. The total heats of melting, however, was the same as those found for salt free solutions. It was found that a change in optical activity takes place only at the higher transition temperature. It was thus postulated that the latter transition represented a helix-coil conformational change whereas the first melting process was of a different nature.

Similar experiments and results have also been reported by Monaselidze and Bakradze[107, 108], cf. Andronikashvili *et al.*[109].

Monaselidze and Bakradze[108] performed additional measurements in which the protein (in 0.1 mol l^{-1} NaCl solution) was left 'at rest' in the calorimeter for different lengths of time before the temperature was increased. If the sample was given one hour of rest, a comparatively small heat absorption peak was observed at the first T_m value. In experiments begun after 12 hours of rest a large peak was found at the lower T_m value whereas only a small quantity of heat was absorbed in the higher transition temperature. The sum of the two Q_m values was constant. Optical activity measurements were not affected by the length of the resting period and gave transition curves with a high T_m value (*c.* 36 °C). It was therefore suggested that ΔH for the helix-coil transition is close to zero.

Measurements were also conducted at different heating rates[108]. Lowering the heating rate (from 0.1 deg min^{-1} to 0.005 deg min^{-1}) resulted in an increase in the fraction of heat absorbed at the lower transition temperature. This temperature was slightly lower than the previous value whereas the higher T_m value did not change. It was therefore concluded that these transition processes depend on kinetic as well as relaxation factors.

Both groups of workers present discussions in their papers of the possible molecular nature of the transition processes; cf. the recent discussion by Cooper[110].

Monaselidze and Bakradze[111] have also reported results from temperature scanning calorimetric experiments on solutions of acid-extracted collagen. These cover a wide range of concentrations. It was found that T_m increases with concentration. At low and at high T_m values sharp transitions were recorded, whereas for intermediate T_m values very broad melting curves were obtained. The maximum width for the melting curve was found at 77% water. This corresponds to a concentration at which (according to earlier studies[112]) all water is bound to the protein.

In a more recent communication Monaselidze *et al.*[113] briefly report similar calorimetric studies, in combination with low angle light scattering and polarisation microscopy studies.

1.6.2 Proteins and polypeptides, chemically induced unfolding processes

1.6.2.1 The chymotrypsinogen family of proteins

Biltonen *et al.*[95] recently reported enthalpy changes accompanying the lowering of pH for solutions of chymotrypsinogen, α-chymotrypsin and dimethylsulphoxide chymotrypsin. Measurements performed by batch and by flow microcalorimetric techniques were conducted in the pH range 4.5–1.2, and at temperatures in the range 25–50 °C. Lowering of the pH resulted in protonation of the protein which was accompanied by unfolding at high enough temperatures. By using the results of earlier equilibrium measurements, it was possible to analyse the derived ΔH values in terms of contributions from protein titration and unfolding. Enthalpy values for the unfolding process were in good agreement with earlier estimates obtained by the van't Hoff method.

The calorimetric titration experiments with chymotrypsinogen at 25 °C in the very low pH region revealed an apparent structural change having a pK of *c.* 2.5. This 'transition' which has a ΔH of about 30 kcal mol^{-1} is not accompanied by any significant change in the u.v. adsorption spectra or the solubility. Hence it is of a nature completely different from the reversible thermal unfolding studied earlier in great detail[114].

A brief report by Shiao and Sturtevant[115] deals with flow microcalorimetric titrations of chymotrypsinogen and chymotrypsin over large pH ranges (pH 11–1.3 and 11–2, respectively). The titration of chymotrypsinogen at pH 2.5 gave the same result as found by Biltonen *et al.*[95].

Both of these calorimetric studies demonstrate convincingly that flow microcalorimetry offers a very convenient and powerful tool for the investigation of chemically induced conformational changes in biopolymers. It seems important, however, to confirm that the conformational changes taking place are rapid enough for the flow rates employed, cf. Lapanje and Wadsö[116].

Skerjanck and Lapanje[117] using a mixing calorimeter studied the enthalpy of urea induced denaturation for chymotrypsinogen A. Results clearly indicated an extensive binding of urea to the protein molecule. ΔH for the overall process in which the urea concentration was increased from 2 mol l^{-1} to 8 mol l^{-1} was $\Delta H = -132$ kcal mol^{-1} (25 °C, pH 6.0). Of this quantity a substantial part is believed to be due to the binding of urea.

1.6.2.2 Lysozyme

Lapanje and Wadsö[116] and Atha and Ackers[118] have recently reported microcalorimetric studies of the denaturation of lysozyme by the guanidine hydrochloride (GuHCl)–HCl system.

Lapanje and Wadsö initiated the unfolding reaction both by lowering the pH (5.2 → 1.4) of lysozyme–GuHCl solutions at 25, 30 and 35 °C and by mixing lysozyme in GuHCl solution at pH 5.2 with stronger GuHCl solutions (25 °C).

The enthalpy change for the overall HCl-induced unfolding reaction at 25 °C and 3 mol l^{-1} GuHCl was calculated to be $\Delta H = 36$ kcal mol^{-1}. A large positive ΔC_p value was found for this process.

From the experiments performed at a constant pH value of 5.2, a protein–GuHCl enthalpy titration curve was constructed. Extrapolation of the linear parts of the curve before and after the unfolding takes place leads to a large exothermic overall enthalpy of unfolding. It is apparent that the main part of this ΔH value is due to solvation effects.

Atha and Ackers determined the enthalpy of unfolding at various GuHCl concentrations by changing the pH from 5.25 to 1.25. In 3 mol l^{-1} GuHCl in which complete transition takes place the observed enthalpy change was 30 ± 3 kcal mol^{-1}. This value is reasonably close to the value determined under similar conditions by Lapanje and Wadsö. Atha and Ackers also determined the corresponding van't Hoff enthalpy by spectral difference measurements and found it to be in excellent agreement with their calorimetric value. The agreement suggests that the unfolding reaction is essentially a two state reaction, cf. Section 1.6.1.3.

1.6.2.3 Poly-L-lysine

Chou and Scheraga have calorimetrically determined the enthalpy change of the helix coil transformation for poly-L-lysine (molecular weight 140 000) in 0.1 mol l^{-1} KCl solution. In the experiments polylysine solutions were brought from pH 11 to pH 7 (or lower) by addition of HCl. The calorimetric measurements, which were made at 25 and 15 °C, were supported by potentiometric titrations and measurements of o.r.d. and far-u.v. spectra. These latter studies showed that poly-L-lysine at pH 11 and 15 °C consists of almost 100 % α-helix. At 25 ° and pH 11, the α-helix content was estimated to be 90 %. The 10 % non-helical portion was believed to be divided about equally 5 % coil and 5 % β-pleated sheet. At both temperatures and pH 7 poly-L-lysine exists as 100 % coil.

The change in enthalpy for the transition process

$$\text{poly-L-lysine (uncharged 100\% helix)} \rightarrow \text{poly-L-lysine (uncharged 100\% coil)} \quad (1.14)$$

was determined to be $\Delta H = 1.21 \pm 0.09$ kcal mol^{-1} of amino acid residue at both 25 and 15 °C. It thus appears that ΔC_p is close to zero. This value of ΔH is nearly the same as that determined earlier by Rialdi and Hermans[120] for poly(L-glutamic acid), $\Delta H = 1.10 \pm 0.2$ kcal mol^{-1}.

In order to correct the experimental ΔH value for charged to uncharged molecule, the protonation of poly-D,L-lysine was measured, $\Delta H = -13.70 \pm 0.09$ kcal mol^{-1} of NH_2 group. (Poly-D,L-lysine exists as 100 % random coil at both pH 11 and pH 7.)

In other calorimetric experiments poly-L-lysine in the metastable 100 % β form (pH 11, 25 °C) was protonated to give the random coil at pH 7. Within uncertainty limits the same value was obtained as in the experiments with the polypeptide as 90 % α-helix. It was therefore concluded that ΔH is close to zero for the $\alpha \rightleftarrows \beta$ conversion.

1.6.2.4 Polypeptides in organic solvents

Giacometti, Turolla and co-workers have studied helix-coil transitions for several synthetic polypeptides in organic solvents. Their method consists of measuring heats of solution of the polymer in organic solvent mixtures at different concentrations. By comparison of heats of solution for concentration ranges in which the polymer exists as helix and coil, enthalpies for the helix-coil transition can be evaluated.

Recently, transition processes have been studied for poly-ε-carbobenzoxy-L-lysine, for a series of its copolymers with phenylalanine[121] and for poly(*N*-γ-carbobenzoxy-L-α,γ-diamino-)butyric acid[122]. Solvent systems used were dichloroethane–dichloroacetic acid.

1.6.3 Conformational transitions in polynucleotides and nucleic acids

1.6.3.1 Poly U, *monomeric adenine derivatives*

Scruggs and Ross[123] have used a temperature scanning calorimeter to observe the melting of structures formed between polyribouridylic acid (poly U) and adenine derivatives (A) in aqueous solution. From u.v. measurements it was shown that the complexes had the composition A(poly U)$_2$, presumably forming a three stranded helix. In separate measurements using a reaction calorimeter they also determined the heat of formation for several of these complexes. Calorimetric results are summarised in Table 1.6.

Table 1.6 Thermochemistry of A(poly U)$_2$ complexes 0.6 mol l^{-1} NaCl, pH 7
(From Scruggs, R. L. and Ross, P. D.[123] by courtesy of Academic Press Inc.)

A	*Complex formation at 20 °C* ΔH_f/ kcal mol^{-1}	*Melting process*		
		ΔH_m/ kcal mol^{-1}	T_m/K	ΔS/cal mol^{-1} deg^{-1}
Adenosine	−12.81	12.8	305.2	41.9
Adenine	−12.82			
Deoxyadenosine	−12.83			
2-Aminoadenosine	−15.82	15.8	322.7	49.0
2,6-Diaminopurine	−15.91			
2-Methylamino-adenosine	−14.45			

From the table it is apparent that the presence or absence of the sugar group or the 2′ OH group does not contribute significantly to ΔH of complex formation. On the other hand reactions between poly U and 2-amino-adenosines and with 2,6-diaminopurine were found to be about 3 kcal mol^{-1} more exothermic than those with adenosine, adenine and deoxyadenosine. It is noteworthy that the bases with the most exothermic ΔH values are capable of forming 3 hydrogen bonds with uracil, while the other compounds can form only two. This difference is about twice as large as would be expected if the effect depended solely on the formation of an extra hydrogen bond.

The values obtained from the temperature scanning calorimeter are in complete agreement with those obtained from the reaction calorimeter which shows that ΔC_p for the process is low (cf. results for poly(dAT) below).

From the results of these studies the authors are led to suggest that the additional favourable enthalpy change accompanying the addition of a second polymer strand to form the 1:2 complex is decisive in overcoming the unfavourably large entropy change accompanying the immobilisation of the monomer species upon incorporation into the 1:1 complex, cf. ΔS values in Table 1.7.

1.6.3.2 Poly C, poly I

Ross and Scruggs[124] used a microcalorimeter to measure the enthalpy change of combination of polyribocytidylic acid (poly C) and polyriboinosinic acid (poly I) to form a two-stranded helix. Measurements were made at 20 and 37 °C, pH 8, and at different salt concentrations (0.06–0.40 mol l^{-1}). Values obtained at the two temperatures were not significantly different: $\Delta H = -5.585 \pm 20$ and -5.593 ± 6 kcal mol^{-1} of base pair, respectively. Thus, ΔC_p for the binding reaction is close to zero.

The invariance of ΔH with changes in salt concentrations for the poly C-poly I system is in sharp contrast with earlier results obtained for the formation of poly (A+U), cf. Section 1.6.3.3.

1.6.3.3 Poly A, poly U

Neuman and Ackermann[125] have investigated the melting of the system polyriboadenylic acid, polyuridylic acid [poly|(A+U)] by temperature scanning calorimetry. Calorimetric work on this system has been reported earlier by several groups; most recently by Krakauer and Sturtevant[126]; cf. also the study by Scruggs and Ross[123] discussed above (Section 1.6.3.1).

The helical structure poly(A+U) formed at moderate salt concentrations and at low temperatures will on heating be converted to single strands

$$\text{poly(A+U)} \rightarrow \text{poly A} + \text{poly U} \tag{1.15}$$

Under the conditions of the experiments poly U forms a random coil whereas poly A forms to some extent a helix. Therefore, measurements were made at different salt concentrations, which caused a change in the T_m value. Extrapolation of the derived melting enthalpies to 95 °C, where poly A is present only as random coil, yields the value $\Delta H = 9.3 \pm 0.5$ kcal mol^{-1} of (A+U). This corresponds to the dissociation process (eqn. 1.15) in which both reaction products are in the form of random coils.

At high ionic strength the curve of C_p *vs.* T shows two separate peaks. They are located at temperatures where u.v. measurements suggest the formation of poly(A+2U) from poly (A+U) and where dissociation of the three-stranded form occurs.

The first (small) peak corresponds to the reaction

$$\text{poly(A+U)} \rightarrow 1/2\ \text{poly(A+2U)} + 1/2\ \text{poly A} \tag{1.16}$$

while the second peak is for the process

$$\text{poly(A+2U)} \rightarrow \text{poly A} + 2\ \text{poly U} \tag{1.17}$$

Enthalpy for the dissociation of the three-stranded helix is given by $\Delta H = 13.5 \pm 0.5$ kcal mol^{-1} of (A+2U).

Neuman and Ackermann have given an analysis of their results in terms of current theoretical approaches.

1.6.3.4 Poly I, poly C

Hinz *et al.* studied the melting of the system polyinosinic acid–polycytidylic acid [poly (I+C)], which forms a two-stranded helix, and of polyinosinic acid which presumably forms a triple-stranded helix, poly (I+I+I). A temperature scanning calorimeter was used in conjunction with spectrophotometric measurements (u.v., i.v.). Experiments were performed at various salt concentrations. In contrast with the case of poly(A+U) (Neuman and Ackermann[125], Section 1.6.3.3) only one transition was observed, corresponding to the reaction,

$$\text{Poly(I+C)} \rightarrow \text{poly I} + \text{poly C} \tag{1.18}$$

The products are believed to be in the form of random coils. The values of ΔH varied from $\Delta H = 6.5 \pm 0.4$ to $\Delta H = 8.4 \pm 0.4$ kcal mol^{-1} of (I+C) for cation concentrations ranging from 0.06 to 1.0 molal, respectively. The respective values of T_m were 51.1 and 73.9 °C. At high concentrations of salt poly I will form a poly (I+I+I) complex. This complex melts at a rather low temperature and has a low transition enthalpy, $\Delta H = 1.9 \pm 0.4$ kcal mol^{-1} of I (1 molal NaCl, $T_m = 43.6$ °C). The enthalpy and entropy values were discussed in terms of hydrogen bonding, contributions from stacking energy and electrostatic effects.

1.6.3.5 d(AT)

The double helix form of the alternating copolymer deoxyadenylic acid–deoxythymidylic acid, dAT, serves as a model for the A–T base pairs in DNA. Scheffler and Sturtevant[128] have reported temperature scanning calorimetric measurements of the helix coil transition of dAT. The enthalpy change was $\Delta H = 7.90 \pm 0.14$ kcal mol^{-1} of base pair ($T_m = 40$ °C, pH = 7.0). The ΔC_p value, derived from the slope of the final portion of the transition curve, was calculated to be $\Delta C_p = 20 \pm 16$ cal deg^{-1} mol^{-1} of base pair.

1.6.3.6 Poly A

Polyriboadenylic acid (poly A) can undergo reversibly a type of helix–coil transition. The coil form is stable at neutral pH, where it predominantly exists as a single-stranded, flexible, rod-like conformation that is stabilised by stacking of the adenine bases. At low pH the single chains associate to form organised linear complexes of varying length with hydrodynamic properties similar to those of un-denatured DNA. The molecular properties of this form have been interpreted in terms of a double-stranded helical model with discontinuities where one strand ends and another begins.

Klump *et al.* have studied the thermally induced double helix-coil transition of poly A in aqueous solution at pH 5.7–4.5 using u.v. measurements and temperature scanning heat capacity calorimetry. From the latter measurements enthalpies of transition were calculated.

By decreasing pH, higher T_m values and higher transition enthalpies were found. For example at pH 5.5 the enthalpy change was found to be 3.4 kcal mol^{-1} of base pair and $T_m = 31.5$ °C, while at pH 4.7 $\Delta H = 5.6$ kcal mol^{-1} of base pair and $T_m = 65.5$ °C. (Polymer concentration 6.8×10^{-3} mol base pair per kg of solution, 0.2 mol l^{-1} citrate, 0.15 mol l^{-1} NaCl, HCl.)

Results were discussed in terms of breaking hydrogen bonds and stacking interactions. It was concluded that the major part of the enthalpy change is due to stacking interactions which are nearly completely eliminated at $T_m =$ 65.5 °C. At lower T_m values a significant degree of stacking still remains after the transition process has taken place.

1.6.3.7 DNA

Rialdi and Profumo[130] used a microcalorimetric technique to determine the enthalpy of unfolding of N4 coliphage DNA, initiated by an increased concentration of urea at 27 °C (pH 3.25, 0.1 mol l^{-1} KCl). In the experiment the urea concentration was increased from 0.8 to 2.2 mol l^{-1}. In separate experiments the melting process was followed spectrophotometrically.

Experiments were also made with previously denaturated DNA. The ΔH values were very similar to those obtained in urea dilution experiments with no DNA present. It was therefore concluded that the interaction between urea and DNA is negligible. The enthalpy change obtained for the transition process, $\Delta H = 9.5 \pm 1.5$ kcal mol^{-1} of base pair, can thus be attributed to the helix–coil transition reaction.

Privalov *et al.*[131], cf. Privalov[104], have measured the enthalpy of melting of T_2 phage DNA at different values of pH and ionic strength. Measurements were made with an automatic temperature scanning calorimeter which required only minute quantities of material, *c.* 0.5 ml of 0.05% aqueous solution of nucleic acid. In separate experiments parallel to the calorimetric measurements, the progress of the unfolding process was followed by u.v. measurements. Substantial differences were found for ΔH and T_m values determined at different conditions of pH and ionic strength. As shown in Figure 1.5, ΔH values were found to be very dependent on pH in the high and low regions. The figure also shows corresponding variations in T_m.

From the calorimetric data determined at different values of T_m and ionic strength Privalov *et al.* were able to calculate two sets of independent ΔG values for the unfolding process. Both sets of data led to the value $\Delta G = 1.2$ kcal mol^{-1} of base pair at conditions close to physiological (pH = 7, ionic strength = 0.2, temperature = 37 °C).

Klump and Ackermann[132] have recently reported a calorimetric study of the helix coil transition for DNA samples of different origin. Calorimetric measurements, which were accompanied by u.v. measurements, showed that for a given set of conditions (pH, salt concentration) transition enthalpies and, in particular, T_m values depend considerably on the base composition of the nucleic acid. Both ΔH and T_m increase with an increased content of guanine–cytosine base pair. (For C–G three hydrogen bonds per base pair can be formed compared to only two for the adenine–thymine pair.)

From studies of two bacterial nucleic acids, two distinct phases of the melting curves were found if the temperature was increased at a slow

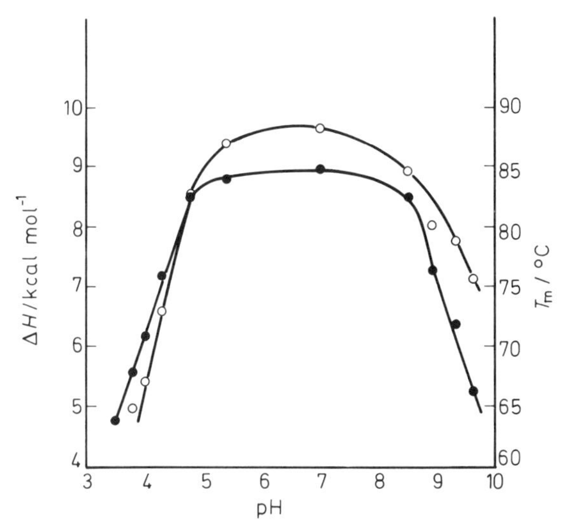

Figure 1.5 Denaturation of T_2 phage DNA ΔH (O) and T_m (●) as a function of pH
(From Privalov *et al.*[131] by permission of John Wiley & Sons, Inc.)

enough rate. The two phases could be distinguished also by the u.v. measurements. It was suggested that the first (smaller) peak on the transition curve is related to the melting of parts of the helix containing a non-random distribution of A–T bases with comparatively weak base pair coupling.

The authors have given an analysis of the calorimetric data in relation to various current theoretical models.

1.7 CALORIMETRIC STUDIES OF THE INTERACTION BETWEEN WATER AND BIOPOLYMERS. HEAT CAPACITY MEASUREMENTS

1.7.1 Degree of hydration of proteins and nucleic acids

A few years ago Mrevlishvili and Privalov[112], cf. Privalov[133], described a calorimetric technique by which the amount of hydrated water in biopolymers can be determined. Using a differential adiabatic heat capacity calorimeter they determined the apparent heat capacity of water below 0°C in biopolymers containing different quantities of water. The heat capacity of dry nucleic acid, for example, was found to be constant over the investigated range of *c.* −30 to +5°C. Addition of moderate quantities of water (*c.* 30%) to the sample increases the heat capacity but it remains constant with temperature, indicating that no melting of ice takes place.

When larger quantities of water are added, definite melting curves are obtained showing that part of the water actually is present as ice. Through analysis of the calorimetric curves it is possible to calculate the amount of hydrate water, i.e. water which does not freeze. Results from measurements on some native and denaturated proteins are shown in Table 1.7. Uncertainties were estimated to be ±0.005 g H_2O per g of dry weight. It is seen

Table 1.7 Hydration of biopolymers
(From Mrevlishvili, G. M. and Privalov, P. L.[112] by courtesy of Nauka Press)

Substance	g of H_2O per g dry weight	
	Native	*Denaturated*
DNA	0.610	0.645
Procollagen	0.465	0.519
Serum albumin	0.315	0.330
Egg albumin	0.323	0.332
Haemoglobin	0.324	0.339

that there is a small but significant increase in hydrate water in going from native to denaturated polymers.

Bakradze *et al.*[134] have continued these studies and they have recently reported a study on transfer RNA. Measurements were made on a new temperature scanning calorimeter capable of precise measurements on mg quantities of material. For a 9.8% t-RNA solution 0.68 g of hydrate water was found per g of nucleic acid. This figure, which corresponds to 13.5 H_2O per nucleotide, is higher than the value found earlier for DNA (see above). For more concentrated RNA solutions a lower degree of hydration was found, e.g. 0.44 g H_2O per gram of RNA in a 60% RNA solution.

Bull and co-workers employed a simple macrocalorimeter for similar types of work. Bull and Breese[135] determined the apparent heat capacity for egg albumin in aqueous solutions at different concentrations. At a protein concentration ⩽70% they found the apparent heat capacity to be 0.396± 0.022 cal g^{-1} deg^{-1} (mean value for the temperature range −3 to +25.5 °C). This value is significantly lower than that reported by Kresheck and Benjamin[136] for dilute solutions at 25 °C, 0.457 cal g^{-1} deg^{-1}.

When less water was present the apparent C_p for the protein was significantly lower, 0.282±0.061 cal g^{-1} deg^{-1} whereas the apparent heat capacity for water associated with the protein was found to be 1.247±0.023 cal g^{-1} deg^{-1}.

From a subsequent study, Chattoraj and Bull[137] report results from measurements on calf thymus DNA in the form of alkali metal salts and the calcium salt. Samples of DNA solutions were super-cooled, e.g. to −6 °C, and the heat required to bring the sample to 25.5 °C was measured. The same solutions were then frozen (dry ice, acetone) after which they were again equilibrated at −6 °C. The heat required to bring the frozen samples to 25.5 °C was determined. The difference between the two heat determinations represents the heat required to melt the ice in the sample and the amount of non-frozen water ('hydrate water') could then be calculated. At −6 °C the amount of hydrate water per mole of nucleotide was 30.4, 24.1, 22.1, 13.6 and 10.7 moles for Li–DNA, Na–DNA, K–DNA, Cs–DNA and Mg–DNA, respectively.

The apparent heat capacities for Na–DNA were found to be 0.371 cal g^{-1} deg^{-1} below and 0.202 cal g^{-1} deg^{-1} above a DNA concentration of 30% (mean value for the temperature range −3 to25.5 °C).

1.7.2 C_p for solid insulin and chymotrypsinogen

Hutchens *et al.* have measured the heat capacities in the temperature interval 11–305 K for crystalline bovine zinc insulin (4% H_2O) and chymotrypsinogen A (10.7%H_2O). C_p was also measured throughout the temperature range for the anhydrous proteins. Comparison with results from earlier entropy determinations for amino acids led to values for mean entropies of formation for the peptide bonds in the anhydrous proteins at 25 °C.

No elevated heat capacities for the hydrated samples in the region of 250–275 K were observed, i.e. there was no evidence of the fusion of ice. By a comparison between C_p values for hydrated and anhydrous protein the apparent heat capacities for water were calculated at different temperatures. For temperatures somewhat above 0 °C the values lie between that for liquid water and the extrapolated value for ice. For the hydrate water in chymotrypsinogen (10.7% water) a value close to that for liquid water was found.

1.7.3 Vaporisation of water from some proteins

Berlin *et al.*[139] used a differential scanning calorimeter to study the binding of water in lyophilised preparation of β-lactoglobulin, bovine serum albumin and calf skin collagen. When the water content was higher than 0.18 g per g of dry protein it was found that the heat of vaporisation of water was 80–125 cal g^{-1} higher than that for liquid water. When less water was absorbed, the ΔH_v for water was normal. It was suggested that at higher moisture levels the solid protein matrix has become swollen thus permitting more H_2O surface contacts.

1.7.4 Water vapour sorption on biopolymers

Bettelheim *et al.*[140] have made calorimetric and gravimetric measurements of the sorption of water vapour at different pressures by polygalacturonic acid, Na-hyaluronate and by the Ca-salt of bovine submaxillary mucin. Prior to the sorption measurements the samples were made anhydrous by an extensive vacuum treatment. Calorimetric heats, as well as enthalpy values calculated from two different isotherms ('isosteric heats'), both showed maxima at a 'monolayer' coverage. (Monolayer means here the amount of water sorbed on 'primary sites'; presumably energetically favourable polar groups of the polymer.)

The sorption processes are not equilibrium processes and only the calorimetric heat values represent true enthalpy values. A comparison between calorimetric and isosteric heats of sorption will indicate the degree of irreversibility of the process.

References

1. Wadsö, I. (1970). *Quart. Rev. Biophysics.*, **3,** 383
2. Westrum, E. (1971). (Ed.). *Bull. Thermodyn. Thermochem.*, No. 14. (Ann Arbor, Michigan: University of Michigan)
3. Reid, D. S., Quickenden, M. A. J. and Franks, F. (1969). *Nature (London)*, **224,** 1293
4. Krishnan, C. V. and Friedman, H. L. (1969). *J. Phys. Chem.*, **73,** 1572
5. Arnett, E. M., Kover, W. B. and Carter, J. V. (1969). *J. Amer. Chem. Soc.*, **91,** 4028
6. Arnett, E. M. and Mckelvey, D. R. (1969). in *Solute–Solvent Interactions*, Ed. by Coetzee, J. F. and Ritchie, C. D., 343. (New York: Marcel Dekker)
7. Konicek, J. and Wadsö, I. (1971). *Acta Chem. Scand.*, **25,** 1541
8. Rüterjans, H., Schreiner, F., Sage, U. and Ackermann, I. (1969). *J. Phys. Chem.*, **73,** 986
9. Kresheck, G. C. and Klotz, J. M. (1969). *Biochemistry*, **8,** 8
10. Spink, C. H. and Auker, M. (1970). *J. Chem. Phys.*, **74,** 1742
11. Corkhill, J. M., Goodman, J. F. and Tate, J. R. (1969). *Trans. Faraday Soc.*, **65,** 1742
12. Franks, F. and Watson, B. (1969). *Trans. Faraday Soc.*, **65,** 2339
13. Franks, F., Quickenden, M. A. J., Reid, D. S. and Watson, B. (1970). *Trans. Faraday Soc.*, **66,** 582
14. Sarma, T. S. and Ahluwalia, J. C. (1970). *J. Phys. Chem.*, **74,** 3547
15. Subramanian, S., Sarma, T. S., Balasubramanian, D. and Ahluwalia, J. C. (1971). *J. Phys. Chem.*, **75,** 815
16. Krishnan, C. V. and Friedman, H. L. (1969). *J. Phys. Chem.*, **73,** 3934
17. Krishnan, C. V. and Friedman, H. L. (1970). *J. Phys. Chem.*, **74,** 3900
18. Krishnan, C. V. and Friedman, H. L. (1970). *J. Phys. Chem.*, **74,** 2356
19. Farquhar, E. L., Downing, M. and Gill, S. J. (1968). *Biochemistry*, **7,** 1224
20. Gill, S. J., Downing, M. and Sheats, G. F. (1967). *Biochemistry*, **6,** 272
21. Stoesser, P. R. and Gill, S. J. (1967). *J. Phys. Chem.*, **71,** 564
22. Binford, J. S., Jr. and Holloway, D. M. (1968). *J. Mol. Biol.*, **31,** 91
23. Shiao, D. D. F. and Sturtevant, J. M. (1969). *Biochemistry*, **8,** 4910
24. Izatt, R. M. and Christensen, J. J. (1968). in Sober, H. A. (Ed.). *Handbook of Biochemistry*, J-49. (Cleveland: The Chemical Rubber Co.)
25. Jencks, W. P. and Regenstein, J. (1968). in Sober, H. A. (Ed.). *Handbook of Biochemistry*, J-151. (Cleveland: The Chemical Rubber Co.)
26. Larson, J. W. and Hepler, L. G. (1969). in *Solute–Solvent Interactions*, Ed. by Coetzee, J. F. and Ritchie, C. D. (Eds.), 1. (New York: Marcel Dekker)
27. Christensen, J. J., Izatt, R. M., Wrathall, D. P. and Hansen, L. (1969). *J. Chem. Soc. A*, 1212
28. Öjelund, G. and Wadsö, I. (1968). *Acta Chem. Scand.*, **22,** 2691
29. Christensen, J. J., Oscarson, J. L. and Izatt, R. M. (1968). *J. Amer. Chem. Soc.*, **90,** 5949
30. Christensen, J. J., Slade, M. D., Smith, D. E., Izatt, R. M. and Tsang, J. (1970). *J. Amer. Chem. Soc.*, **92,** 4164
31. Christensen, J. J., Rytting, J. H. and Izatt, R. M. (1970). *Biochemistry*, **9,** 4907
32. Christensen, J. J., Rytting, J. H. and Izatt, R. M. (1970). *J. Chem. Soc. B*, 1643
33. Christensen, J. J., Rytting, J. H. and Izatt, R. M. (1970). *J. Chem. Soc. B*, 1646
34. Izatt, R. M., Christensen, J. J. and Rytting, J. H. (1971). *Chem. Rev.*, **71,** 439
35. Woolley, E. M., Wilton, R. W. and Hepler, L. G. (1970). *Canad. J. Chem.*, **48,** 3249
36. Watt, G. D. and Sturtevant, J. M. (1969). *Biochemistry*, **8,** 4567
37. Beres, L. and Sturtevant, J. M. (1971). *Biochemistry*, **10,** 2120
38. Leung, C. S. and Grunwald, E. (1970). *J. Phys. Chem.*, **74,** 687

39. Greengard, P., Rudolph, S. A. and Sturtevant, J. M. (1969). *J. Biol. Chem.*, **244,** 4798
40. Rudolph, S. A., Johnson, E. M. and Greengard, P. (1970). *J. Biol. Chem.*, **246,** 1271
41. Strack, E. and Müller, D. M. (1971). *Hoppe-Seyler's Z. Physiol. Chem.*, **352,** 1014
42. Rajender, S., Lumry, R. and Han, M. (1971). *J. Phys. Chem.*, **75,** 1375
43. Letter, J. E. and Bauman, J. E. (1970). *J. Amer. Chem. Soc.*, **92,** 443
44. Meyer, J. L. and Bauman, J. E. (1970). *J. Amer. Chem. Soc.*, **92,** 4210
45. Belaich, J. P. and Sari, J. C. (1969). *Proc. Natl. Acad. Sci.*, **64,** 763
46. Alberty, R. A. (1969). *J. Biol. Chem.*, **244,** 3290
47. Alberty, R. A. (1969). *J. Amer. Chem. Soc.*, **91,** 3899
48. Phillips, R. C., George, P. and Rutman, R. J. (1969). *J. Biol. Chem.*, **244,** 3330
49. Möschler, H. J., Weder, H.-G. and Schwyzer, R. (1971). *Helv. Chim. Acta*, **54,** 1437
50. Früh, P. U., Clerc, J. T. and Simon, W. (1971). *Helv. Chim. Acta*, **54,** 1446
51. Izatt, R. M., Rytting, J. H., Nelson, D. P., Haymore, B. L. and Christensen, J. J. (1969). *Science*, **164,** 443
52. Izatt, R. M., Nelson, D. P., Rytting, J. H., Haymore, B. L. and Christensen, J. J. (1971). *J. Amer. Chem. Soc.*, **93,** 1619
53. Kresheck, G. (1969). *J. Phys. Chem.*, **73,** 2441
54. Kresheck, G. (1970). *J. Chem. Phys.*, **52,** 5966
55. Longo, F. R., Finarelli, J. D. and Schmalzbach, E. (1970). *J. Phys. Chem.*, **74,** 3296
56. Rohdewald, P. and Baumeister, M. (1969). *J. Pharm. Pharmac.*, **21,** 867
57. Papenmeier, G. J. and Campagnoli, J. M. (1969). *J. Amer. Chem. Soc.*, **91,** 6579
58. Canady, W. J. and Laidler, K. J. (1958). *Canad. J. Chem.*, **36,** 1289
59. Vichutinskij, A. A., Zaslowsky, B. J., Platonov, A. L., Timmerman, L. A. and Horlin, A. J. (1969). *Dokl. Acad. Nauk. SSSR*, **189,** 432
60. Bjurulf, C., Laynez, J. and Wadsö, I. (1970). *Eur. J. Biochem.*, **14,** 47
61. Bjurulf, C., Suurkuusk, J. and Wadsö, I. (1971). in Broda, E., Locker, A. and Springer-Lederer, H. (Eds.). *Proceedings of First European Biophysics Congress*, 347. (Wien: Verlag der Wiener Medizinischen Akademie)
62. Shiao, D. D. F. (1970). *Biochemistry*, **9,** 1083
63. Lumry, R. and Rajender, S. (1971). *J. Phys. Chem.*, **75,** 1387
64. Ross, P. D. and Ginsburg, A. (1969). *Biochemistry*, **8,** 4690
65. Hinz, H. J., Shiao, D. D. F. and Sturtevant, J. M. (1971). *Biochemistry*, **10,** 1347
66. Ginsburg, A. and Mehler, A. H. (1966). *Biochemistry*, **5,** 2623
67. Lehrer, G. M. and Barker, R. (1970). *Biochemistry*, **9,** 1533
68. Bolen, D. W., Flögel, M. and Biltonen, R. (1971). *Biochemistry*, **10,** 4136
69. Maurer, W., Haar, W. and Rüterjans, H. (1971). *Proceedings of First European Biophysics Congress*, Ed. by Broda, E., Locker, A. and Springer-Lederer, H., 375. (Wien: Verlag der Wiener Medizinischen Akademie)
70. Velick, S. F., Baggott, J. P. and Sturtevant, J. M. (1971). *Biochemistry*, **10,** 779
71. Wang, J. H., Kwok, S.-C., Wirch, E. and Suzuki, I. (1970). *Biochem. Biophys. Res. Commun.*, **40,** 1340
72. Klopfenstein, W. E. (1969). *Biochim. Biophys. Acta*, **181,** 323
73. Klopfenstein, W. E. (1969). *Biochim. Biophys. Acta*, **187,** 272
74. O'Reilley, R. A., Ohms, J. I. and Mothley, C. H. (1969). *J. Biol. Chem.*, **244,** 1303
75. Lovrien, R. and Anderson, W. (1969). *Arch. Biochem. Biophys.*, **131,** 139
76. Pilcher, G., Jones, M. N., Espada, L. and Skinner, H. A. (1969). *J. Chem. Thermodyn.*, **1,** 381
77. Langerman, N. and Sturtevant, J. M. (1971). *Biochemistry*, **10,** 2809
78. Henkens, R. W., Watt, G. D. and Sturtevant, J. M. (1969). *Biochemistry*, **8,** 1874
79. Keyes, M. H., Falley, M. and Lumry, R. (1971). *J. Amer. Chem. Soc.*, **93,** 2035
80. Barisas, B. G., Sturtevant, J. M. and Singer, S. J. (1971). *Biochemistry*, **10,** 2816
81. Adams, E. C. and Weiss, M. R. (1969). *Biochem. J.*, **115,** 441
82. Kirchner, A. G. and Tanford, C. (1964). *Biochemistry*, **3,** 291
83. Sjöquist, J. and Wadsö, I. (1971). *FEBS Letters*, **14,** 254
84. Steiner, R. F. and Kitzinger, C. (1956). *J. Biol. Chem.*, **222,** 271
85. Evans, W. J., Forlani, L., Brunori, M., Wyman, J. and Antonini, E. (1970). *Biochim. Biophys. Acta*, **214,** 64
86. Marzotto, A., Turolla, A. and Galzigna, L. (1970). *J. Biochem.*, **68,** 833
87. Hearn, R. P., Richards, R. M., Sturtevant, J. M. and Watt, G. D. (1971). *Biochemistry*, **10,** 806

88. Tsong, T. Y., Hearn, R. P., Wrathall, D. P. and Sturtevant, J. M. (1970). *Biochemistry*, **9,** 2666
89. Aune, K. C. and Timasheff, S. N. (1971). *Biochemistry*, **10,** 1609
90. Aune, K. C., Goldsmith, L. C. and Timasheff, S. N. (1971). *Biochemistry*, **10,** 1617
91. Stauffer, H., Srinivasan, S. and Lauffer, M. A. (1970). *Biochemistry*, **9,** 193
92. George, P., Eaton, W. A. and Trachtman, M. (1968). *Fed. Proc.*, **27,** 526
93. Srinivasan, S. and Lauffer, M. A. (1970). *Biochemistry*, **9,** 2173
94. Jackson, W. M. and Brandts, J. F. (1970). *Biochemistry*, **9,** 2294
95. Biltonen, R., Schwartz, A. T. and Wadsö, I. (1971). *Biochemistry*, **10,** 3417
96. Biltonen, R. and Lumry, R. (1971). *J. Amer. Chem. Soc.*, **93,** 232
97. O'Reilly, J. M. and Karasz, F. E. (1970). *Biopolymers*, **9,** 1429
98. Sophianopoulus, A. J. and Weiss, B. J. (1964). *Biochemistry*, **3,** 1920
99. McDonald, C. C., Phillips, W. D. and Glickson, J. D. (1971). *J. Amer. Chem. Soc.*, **93,** 235
100. Tanford, C. and Aune, K. C. (1970). *Biochemistry*, **9,** 206
101. Delben, F. and Crescenzi, V. (1969). *Biochim. Biophys. Acta*, **194,** 615
102. Crescenzi, V. and Delben, F. (1971). *Int. J. Protein Res.*, **3,** 57
103. Delben, F., Crescenzi, V. and Quadrifoglio, F. (1969). *Int. J. Protein Res.*, **1,** 145
104. Privalov, P. L. (1970). *Biofizika*, **15,** 206
105. Privalov, P. L. and Tiktopulo, E. I. (1970). *Biopolymers*, **9,** 127
106. Privalov, P. L. and Tiktopulo, E. I. (1969). *Biofizika*, **14,** 20
107. Monaselidze, D. R. and Bakradze, N. G. (1968). *Dokl. Acad. Nauk. SSSR*, **183,** 1205
108. Monaselidze, D. R. and Bakradze, N. G. (1970). *Soobshch. Acad. Nauk. Gruz. SSR*, **58,** 705
109. Andronikashvili, E. L., Mrevlishvili, G. M., Bakradze, N. G., Majgaladze, G. V., Monaselidze, D. R. and Chanchalashvili, Z. I. (1968). *Dokl. Acad. Nauk. SSSR*, **183,** 212
110. Cooper, A. (1971). *J. Mol. Biol.*, **55,** 123
111. Monaselidze, D. R. and Bakradze, N. G. (1969). *Dokl. Acad. Nauk. SSSR*, **189,** 899
112. Mrevlishvili, G. M. and Privalov, P. L. (1967). *Water in Biological Systems*, Ed. by Kaynshin, L. P., 63. (Moscow: Nauka Press). [Transl. (1969). Consultants Bureau, New York]
113. Monaselidze, D., Shepelevski, A., Bakradze, N., Shaltiko, L. and Frenkel, S. (1970). *Studia Biophysica (Berlin)*, **23,** 169
114. Brandts, J. (1964). *J. Amer. Chem. Soc.*, **86,** 4302
115. Shiao, D. D. F. and Sturtevant, J. M. (1970). *Fed. Proc.*, **29,** 335
116. Lapanje, S. and Wadsö, I. (1971). *Eur. J. Biochem.*, **22,** 345
117. Skerjanc, J. and Lapanje, S. (1969). *Croat. Chem. Acta*, **41,** 111
118. Atha, D. H. and Ackers, G. K. (1971). *J. Biol. Chem.*, **246,** 5845
119. Chou, P. Y. and Scheraga, H. A. (1971). *Biopolymers*, **10,** 657
120. Rialdi, G. and Hermans, Jr., J. (1966). *J. Amer. Chem. Soc.*, **88,** 5719
121. Giacometti, G., Turolla, A. and Boni, R. (1970). *Biopolymers*, **9,** 979
122. Giacometti, G., Turolla, A. and Verdini, A. S. (1971). *J. Amer. Chem. Soc.*, **93,** 3092
123. Scruggs, R. L. and Ross, P. D. (1970). *J. Mol. Biol.*, **47,** 29
124. Ross, P. D. and Scruggs, R. L. (1969). *J. Mol. Biol.*, **45,** 567
125. Neuman, G. and Ackermann, T. (1969). *J. Phys. Chem.*, **73,** 2170
126. Krakauer, H. and Sturtevant, J. M. (1968). *Biopolymers*, **6,** 491
127. Hinz, H.-J., Haar, W. and Ackermann, T. (1970). *Biopolymers*, **9,** 923
128. Scheffler, I. E. and Sturtevant, J. M. (1969). *J. Mol. Biol.*, **42,** 577
129. Klump, H., Ackermann, T. and Neuman, E. (1969). *Biopolymers*, **7,** 423
130. Rialdi, G. and Profumo, P. (1968). *Biopolymers*, **6,** 899
131. Privalov, P. L., Ptitsyn, O. B. and Birshtein, T. M. (1969). *Biopolymers*, **8,** 559
132. Klump, H. and Ackermann, T. (1971). *Biopolymers*, **10,** 513
133. Privalov, P. L. (1967). *Biofizika*, **13,** 163
134. Bakradze, N. G., Monaselidze, D. R., Mrevlishvili, G. M., Bibikova, A. D. and Kisselev, L. L. (1971). *Biochim. Biophys. Acta*, **238,** 161
135. Bull, A. B. and Breese, K. (1968). *Arch. Biochem. Biophys.*, **128,** 497
136. Kresheck, G. C. and Benjamin, L. (1964). *J. Phys. Chem.*, **68,** 2476
137. Chattoraj, D. K. and Bull, H. B. (1971). *J. Colloid Interface Sci.*, **35,** 220
138. Hutchens, J. O., Cole, A. G. and Stout, J. W. (1969). *J. Biol. Chem.*, **244,** 26
139. Berlin, E., Kliman, P. G. and Pallansch, M. J. (1970). *J. Colloid Interface Sci.*, **34,** 488

140. Bettelheim, F. A., Block, A. and Kaufman, L. J. (1970). *Biopolymers,* **9,** 1531
141. Christensen, J. J., Hill, J. O. and Izatt, R. M. (1971). *Science,* **174,** 459
142. Lim, M. C. and Nancollas, G. H. (1971). *Inorg. Chem.,* **10,** 1957
143. Rocchi, R., Borin, G., Marchiori, F., Moroder, L., Peggion, E., Scoffone, E., Crescenzi, V. and Quadrifoglio, F. (1972). *Biochemistry,* in the press
144. Privalov, P. L., Khechinashvili, N. N. and Atanasov, B. P. (1971). *Biopolymers,* **10,** 1865
145. Lovrien, R. and Sturtevant, J. M. (1971). *Biochemistry,* **10,** 3811

2
Thermodynamics of Metal-Complex Formation

R. F. JAMESON
The University, Dundee

2.1 GENERAL INTRODUCTION 45

2.2 DATA AND THEIR INTERPRETATION 46
- 2.2.1 *Introduction* 46
- 2.2.2 *The equilibrium constant and choice of standard states and concentration scales* 47
- 2.2.3 *The equilibrium constant and ionic strengths* 49
- 2.2.4 *Significance of thermodynamic functions* 49

2.3 METHODS 49
- 2.3.1 *Computational techniques* 49
- 2.3.2 *Use of the glass electrode* 50
- 2.3.3 *Other experimental techniques* 50

2.4 SOME SELECTED DATA 51
- 2.4.1 *Introduction and apologia* 51
- 2.4.2 *Precise thermodynamic data* 51
- 2.4.3 *Data of significance or potential significance to bio-systems* 52
- 2.4.4 *Background electrolyte effects* 54
- 2.4.5 *Miscellaneous* 54

2.1 GENERAL INTRODUCTION

The study of metal complexes in solution has had a long history and been the subject of many reviews of which the latest, by Nancollas[1], is particularly valuable. This latter covers the literature up to 1969, and demonstrates that the experimental interest in the thermodynamics of complex formation is shifting from equilibrium studies towards direct calorimetric determinations of metal–ligand interactions. Extensive tabulations of equilibrium constant data (and derived thermodynamic constants) are available in convenient form in two special publications[2] of the Chemical Society (London). The movement towards direct calorimetric measurement of the enthalpies of formation of metal complexes is further revealed in two recent books, both written by practising thermochemists, namely the *Thermochemistry of Transition Metal Complexes*, by Ashcroft and Mortimer[3] and the *Handbook of Metal Ligand Heats*, by Christensen and Izatt[4].

Unfortunately a perusal of these publications soon shows that one is still left with many published data that are obviously unreliable and/or contradictory. Furthermore, the problem of what medium to use and what concentration units to work in is so varied and misunderstood that many data are incompatible for the purposes of comparison.

In view of the recent article by Nancollas[1], this review is of necessity short, and so the opportunity has been taken in Section 2.2 to discuss briefly some of the problems associated with obtaining experimental data (choice of medium, concentration of salt, etc.) and to follow this with a short account of the limitations of the use of these data. Few of the ideas expressed are new (as can be seen by the references used), but many seem to have been overlooked. A section then follows on techniques and finally, a few of the data obtained over the past two years are presented and discussed.

2.2 DATA AND THEIR INTERPRETATION

2.2.1 Introduction

In the past the study of the thermodynamics of complex formation has been confined to equilibrium studies and through these the evaluation of free energy changes (ΔG). More recently, and particularly over the last decade or so with the arrival of commercially available micro- and semimicro-solution calorimeters, the direct evaluation of enthalpy changes (ΔH) on complex formation has become a much more common pursuit. This has led to a tendency to discuss the physical and theoretical properties of complexes in terms of ΔH° and ΔS° correlations instead of, as previously, in terms of linear free energy relationships (LFERs) based on a correlation of ΔG° values with various properties. Although the suggestion by several workers (e.g. by Bell[5]) that ΔG is the more significant function for these purposes has recently been questioned[6], the problem remains as to what type of correlation and interpretation is valid. All too often the *type* of quotient used to define an equilibrium constant (e.g. mole fraction, molarity) and the *units* in which it is measured (e.g. β_6 for the coordination of six monodentate ligands does not have the same units as β_3 for the coordination of three bidentate ligands) is totally disregarded even to the extent of comparing *numerical* values of, say, β_1 and β_2 (it is sobering to note what happens to a lot of these data simply by adopting the SI units of mol m^{-3}!). The effect on the calculation of ΔS° values from ΔG° and ΔH° values was first made clear by Adamson[7] and it follows that the use of ΔS° values in discussing the type (σ or π) of metal–ligand bond must be used with great care[6]. A further problem confronting workers in this field is that of choice of solvent (e.g. aqueous, non-aqueous or mixed), and whether or not a 'background' electrolyte is to be employed. In the latter event, one must also decide what salt is suitable and what is a suitable concentration to use. It is pertinent to begin by discussing some of these problems in more detail.

2.2.2 The equilibrium constant and choice of standard states and concentration scales

When dealing with the thermodynamics of complex formation in solution, the choice of standard state is to a certain extent arbitrary but will depend on concentration terms and hence on the scale adopted. In the same way that the (non-existent) ideal gas is taken as the standard state when discussing free-energy changes in gases, so a hypothetical solution to which 'ideal' properties at unit concentration on the chosen scale can be ascribed is taken as the standard state for electrolyte solutions at the temperature and pressure under consideration.

It is then possible to define an activity of the solute B, according to the concentration scale chosen, namely:

$$\begin{aligned}\bar{G}_B &= \bar{G}_B^{om} + RT\ln \alpha_B(m) \text{ for the molal scale}\\ &= \bar{G}_B^{oc} + RT\ln \alpha_B(c) \text{ for the molar scale} \qquad (2.1)\\ &= \bar{G}_B^{oN} + RT\ln \alpha_B(N) \text{ for the mole fraction scale.}\end{aligned}$$

and note that although for a given solution at fixed T and P, $\bar{G}_B$ is unique, *both the* α_B *and the* $\bar{G}_B^o$ *depend on the scale adopted.*

In order to relate the activity terms defined above to concentration values, an activity coefficient is now defined by the appropriate relationship:

$$\text{activity} = \text{concentration} \times \text{activity coefficient} \qquad (2.2)$$

and note here that an argument can be made out for taking the activity coefficients as dimensionless, whereupon the activities will often have dimensions; the reverse assignment of dimensions where required is quite often a cause of confusion as to the role of units when measuring activity quotients for equilibria. Equations (2.1) and (2.2) can now be used to define the properties of the hypothetical standard states, namely that the limiting value of the activity coefficient on the appropriate scale is unity as the concentration is reduced to zero (for all temperatures and pressures). Finally, it is often convenient to divide the activities, α_B, into components referring to the separate ionic species present.

Thus for an equilibrium between a metal ion M and a ligand L (charges are omitted to maintain clarity and generality) we have

$$M + L \rightleftharpoons ML$$

which gives

$$K_1 = \frac{\alpha_{ML}}{\alpha_M \alpha_L} \qquad (2.3)$$

$$= \frac{[ML]}{[M][L]} \cdot \frac{f_{ML}}{f_M f_L} \qquad (2.4)$$

$$= K_1^c \cdot f_{ML}/f_M f_L \qquad (2.5)$$

where the f_X are the activity coefficients of X and K_1^c is the concentration quotient. This is the factor most often obtained in stability constant work. The activity coefficient quotient is sometimes evaluated by calculation (e.g. by application of equations based on Debye–Hückel theory) or is obtained experimentally by extrapolation to zero concentration or zero ionic strength, thus giving the *true* thermodynamic constants. Most workers, however, merely report values of K^c in defined background ionic media. These results are valid only for that medium and unfortunately the position is not helped by the proliferation of salts and concentrations chosen by different schools. However, provided all the data required are collected for one medium there is, of course, no valid objection to defining a standard state by reference to that medium.

Although most concentration quotients are obtained on a molar scale as far as metal-complex formation is concerned, other scales, especially the molal

Table 2.1 The effect of concentration scales for equilibrium quotients on a series of relevant factors (quotients are quoted in terms of solute B in solvent A)

*Property to be questioned**	*Type of equilibrium quotient*					
	Mole fraction $\frac{n_B}{n_A+n_B}$	*Molar ratio* $\frac{n_B}{n_A}$	*Molarity* $\frac{n_B}{V}$	*Molality* $\frac{n_B}{m_A}$	*Mass fraction* $\frac{m_B}{m_A+m_B}$	*Mass ratio* $\frac{m_B}{m_A}$
Invariant with respect to assignment of mol. wt. of solute?	−	−	−	−	+	+
Invariant with respect to assignment of mol. wt. of (simple) solvent?	−	−	+	+	+	+
Invariant with respect to assignment of mol. wts. of (mixed) solvent?	−	−	+	+	+	+
Invariant with respect to which species is considered to be solvent?	+	−	+	−	+	−
Invariant with respect to assumptions concerning solvation?	−	−	+	−	−	−
Invariant (for a given specimen) with respect to change of temperature?	+	+	−	+	+	+
Invariant (for a given specimen) with respect to change of pressure?	+	+	−	+	+	+
Does log K vary with temperature according to "simple" equations?	+	+	−	+	+	+
Bears simple direct relation to molecular theories?						
(i) Non-ideality	+	−	−	−	−	−
(ii) Properties of ideal dilute solution	−	−	+	−	−	−

*Where − signifies 'no' and + signifies 'yes' in response to the questions in column 1.

one, are often employed for related quantities (e.g. apparent ionic products for water, activity coefficients in dilute electrolyte solutions) and it would not seem that the choice of scale is usually made on particularly valid reasoning. Table 2.1, therefore, is presented in order to indicate briefly the important factors to be kept in mind when making such a choice and to indicate how each scale has its advantages.

2.2.3 The equilibrium constant and ionic strengths

For dilute solutions the Debye–Hückel theory and related theories can often be used to calculate the activity coefficient terms in Equation (2.5) and this method is discussed in some detail in several standard texts, e.g. by Nancollas[8], Rossotti and Rossotti[9] and by Beck[10]. However, when making use of a titration technique, the charge on the entities present often changes considerably during the course of the titration making direct calculation difficult. Hence the practice of recording results obtained in the presence of an 'inert' background ionic medium; sometimes the results are extrapolated to zero ionic strength, but more often than not they are merely reported as concentration quotients in the medium concerned.

An excellent detailed discussion of the use of background electrolytes is to be found in the works already referred to[8,9] but two points should perhaps be emphasised. Firstly, for dilute electrolytes (i.e. *c.* 0.1 mol l^{-1} and below) the mean ionic activity coefficients change rapidly with concentration in most cases (see data in the book by Robinson and Stokes[11]), and hence considerable accuracy in making up solutions is called for. Secondly, the use of very high (e.g. 3 mol l^{-1} $NaClO_4$) concentrations is spreading, and not always for good reason since (i), ionic interactions involving the perchlorate ion become very marked[12] and (ii), the interionic distances in such a medium (*c.* 6 Å) are often very small in comparison with many bulky ligands now being studied, which results in obvious problems of interpretation of data (see also Section 2.4.4, below).

2.2.4 Significance of thermodynamic functions

Far too many workers still assign ΔS° and ΔG° values to reactions without taking into account the problems outlined above (note that ΔH° values, being unitary quantities, do not involve one in this problem) although these points have been well known for some time (see also Rossotti's excellent review[13]). In most instances it would appear that attempts to correlate metal–ligand bonding types (i.e. σ or π) with either ΔH° or ΔS° values should usually be taken *cum grano salis*, as also should too much reliance on differences between 'hard' and 'soft' metal–ligand interactions[14,15]; a large part of the entropy change will originate from ion–solvent interactions[6].

Recently Hancock and Finkelstein[16] have followed up a suggestion due to Erenburg and Peschevitskii[17] in using standard reduction potentials as a basis for correlations of the LFER type, but one is a little sceptical about such sharp distinctions as their results would suggest.

2.3 METHODS

2.3.1 Computational techniques

The use of the computer in the interpretation of data on complex formation is now well established, and it is tempting to predict that the latest paper on this subject by Rossotti *et al*[18] will become a classic of its type. Since this paper contains an exhaustive survey of the literature to date, further elaboration here would be superfluous.

The compilation of IUPAC recommended symbols for complex equilibria by Marcus[19] is of considerable value.

2.3.2 Use of the glass electrode

Bates, in an IUPAC report[20], has given a concise general account of the establishment of pH scales in amphiprotic and mixed solvents, and has also[21] presented standards for a practical pD scale in heavy water (D_2O) at ten temperatures over the range 5–50 °C.

Working in parallel, Prue and Rossotti and co-workers[22] have made the important and significant observation that by careful choice of electronics the precision of the glass electrode in aqueous medium can be held within 0.01 mV. Rather surprisingly, perhaps, this degree of precision is claimed to cover both titration and transfer techniques of calibration, and if this work is confirmed it will mean that in nearly every case the convenient commercially-available high-resistance glass electrode can replace the rather cumbersome and temperamental hydrogen electrode (even for work of the highest precision).

Of relevance to this work is that of Hair[23] on the structure of glass in relation to its electrochemical selectivity.

If the use of mixed solvent systems is contemplated, then the paper by van Veen *et al.*[24] concerning the effect of solutes on liquid junction potentials in mixed solvents should be consulted together with similar work by Arnaud[25] and Jullard[26], although the work is not of the highest precision. Also of interest is the establishment of the standard e.m.f. of Ag/AgCl electrodes in aqueous ethanol solutions[27]. A hydrogen electrode was used with the Ag/AgCl electrodes. The results are consistent with previous work, and comparison with these is made in the paper.

2.3.3 Other experimental techniques

An important development in the basic use of calorimetry that must be noted is the work of Hansen[28], who gives strong arguments for using the reaction of $HClO_4$(aq.) with NaOH(aq.) in place of the generally recommended tris(hydroxymethyl)aminomethane reaction with HCl(aq.) or $HClO_4$(aq.) as a calorimetric standard. Although the work of Eatough[29] is of considerable interest in confirming that titration calorimetry will give precise equilibrium constants in the case of very stable complexes, it would not thereby justify the use of this method in preference to potentiometry.

Ion-selective electrodes are becoming of greater value as their precision increases, and thus the very comprehensive review by Simon[30] is to be welcomed. It would appear that their value will be particularly important for the study of sparingly soluble complexes (cf. for example work on lead(II)[31] and calcium[32] fluoride systems), but see also[56] below.

Although conductiometric work is not widely used in obtaining stability constant data, it can be of value, and the paper by Prue[33] reporting the analysis of precise conductiometric data to abstract information on weak interactions in solutions is well worthy of note.

The use of n.m.r. methods is becoming more widespread although much of the information gained thereby is structural rather than of thermodynamic interest. However, this information can be of immense value for the *interpretation* of thermodynamic data and hence a few papers that illustrate trends in the use of different nuclei are reported here. Firstly the review by Rowe[34] is of wide interest, although concentrating on applications to the study of bio-complexes. Papers on the use of ^{1}H resonance worthy of note include those by Natusch and Porter[35] who discuss applications to bonding and stability problems encountered in a study of cysteine complexes, Grasdalen[36] who discusses solvation of copper(II) acetate, and Rakshys[37] who demonstrates the use of contact shifts in a study of bis(pentane-2,4-dionato) nickel(II) complexes with amines.

Other nuclei that have been used recently include ^{17}O by Chmelnick and Fiat[38] who show that iron(II) and nickel(II) ions in aqueous solution are hexacoordinated. Strouse and Matwiyoff[39] used ^{13}C resonances, reporting chemical shifts in p.p.m. with respect to the ^{13}C in the parent molecule for a series of nickel(II) complexes with amino acids and ethylene diamines. Haque[40] and Buslaev[41] used ^{19}F resonances to study fluoride complexes with gallium(III) and tantalum(V) respectively. Finally, Ovchinnikov *et al.*[42] have used ^{31}P n.m.r. spectroscopy to look at some metal ion complexes of dithiophosphine, and more recently Williams[43] has used ^{31}P chemical shifts in a study of thallium(I) compounds (see also below).

2.4 SOME SELECTED DATA

2.4.1 Introduction and apologia

In a perusal of the literature for 1970 and the first half of 1971, over 190 papers reporting equilibrium data and/or thermodynamic data on complex formation have been noted, and no doubt in due course most of these data will find their way into comprehensive compilations of the type mentioned above[2–4]. For the purposes of this review, however, only work of the following categories has been reported: (i) work of particularly high precision and of general applicability; (ii) work that sheds light on old problems; and (iii) work that constitutes or preludes an exciting new development. Such drastic selection is, of course, highly subjective, if not idiosyncratic, but it is hoped that this Review will be the more readable thereby, and that those whose work is not reported will appreciate the reasons for omission.

2.4.2 Precise thermodynamic data

Of particular value to those workers using calorimetry in aqueous solution is the paper by Grenthe *et al.*[44]. Here are reported full thermodynamic data for the ionisation of water at 5, 20, 25, 35 and 50 °C together with data for the interaction of tris(hydroxymethyl)aminoethane with a proton (see, however, comment on the use of this reagent by Hansen and Lewis[28]). Similarly the calculated entropy values for $Na^+ + OH^-$ given by Larson[45] will be of value; this paper also covers the thermodynamic properties of water. Avedikian[51] has also reported accurate thermodynamic data for the dissociation of water, in both methanol and acetone solutions.

Christensen *et al.*[46] have carried out measurements with the high precision that has come to be expected from his research group on the ionisation of a series of methyl- and ethyl-substituted carboxylic acids. ΔG°, ΔH° and ΔS° values are recorded and correlated at 10, 25 and 40 °C.

2.4.3 Data of significance or potential significance to bio-systems

This section opens with a description of some of the work on lanthanides since the chemistry of these elements is of interest both as marker ions for calcium and as indirect n.m.r. probes (chemical shifts due to paramagnetic ions). Firstly a cautionary tale: Wallace[47] has shown rather clearly the dangers inherent in drawing up correlation schemes in general; he shows that the correlations in ΔG drawn up by Siekierski[48] for lanthanide complexes are almost certainly an artifact of the curvature of the ΔG *v.* atomic number curve.

Williams[49] has continued his work on lanthanide–histidine complexes[50], quoting ΔC_p values with other thermodynamic data for the protonation of these complexes at 25 and 37 °C. The results show a 'gadolinium break' which is interpreted as a change of hydration of the complete complex ion. Another paper worthy of special mention is that of Dellien and Grenthe[51] on the stability of lanthanide complexes of malonic acid and their protonation.

Several other papers present data on lanthanide complexes, and a few are listed here without further comment: Hinchley[52] reports entropy values for lanthanium and yttrium ions; Campbell[53] presents data on tropolonate complexes of the lanthanides and Grenthe and Gardhammar[54] deal with the complexing of europium(III) by iminodiacetic acid.

Thallium(I) has considerable interest as a potential probe for potassium in biological systems, (i) by virtue of its $6s^1p^1 \leftarrow 6s^2$ allowed transition enabling photometric or fluorescence studies, (ii) by n.m.r. based on ^{205}Tl nuclei and (iii) n.m.r. based on chemical shifts (e.g. proton or phosphorus-31 resonances). Thus the paper by Williams and co-workers[43] is of outstanding value. A whole range of ligands, Cl^-, malonate, oxalate, citrate, nitriloacetate, EDTA, phosphate, diphosphate, ribosephosphate, ADP and ATP have been studied and data presented on points (i), (ii) and (iii), above. This work enables the desirability and practicability of substitution of Tl^+ for K^+ in bio-systems to be much more reliably assessed, although the gap in knowledge lies equally in the complexing behaviour of K^+ itself.

The series of cyclic polyethers, the so-called 'crown' compounds, are receiving a lot of interest as shown by the appearance of two first class papers. These compounds are exemplified by dicyclohexyl-18-crown-6 ($C_{20}H_{36}O_6$)(I) which has been studied by Christensen, Izatt *et al.*[55]. They report equilibrium constants, ΔH° values and ΔS° values for Group I and II metal complexes. The data are in the order expected, i.e., $K^+ > Rb^+ > Cs^+ > Na^+ > Li^+$ and $Ba^{2+} > Sr^{2+} > Ca^{2+} > Mg^{2+}$, as required should size of ion be a dominating factor. This work should be read in conjunction with that of Frensdorf[56], who has studied equilibria between several crown compounds and alkali metal ions and silver(I) ions in water and in methanol. The effect of the replacement of O by N or S has also been looked at and the following

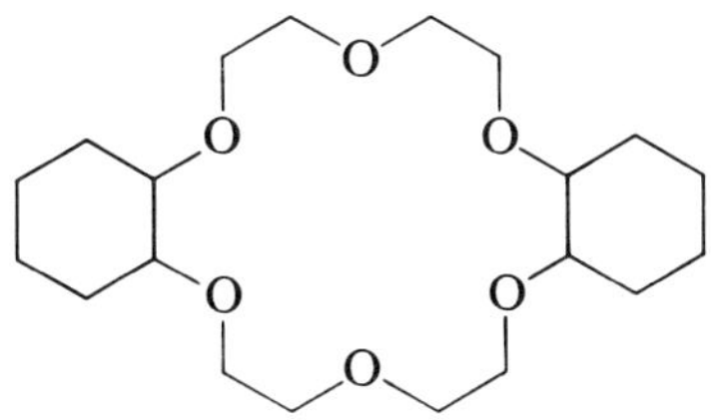

(I)

important points emerge, suggesting that the bonding of the alkali metal ions is very largely electrostatic. Firstly, the constants are larger in methanol than in water; secondly, only for small rings is there good evidence for ML_2 complexes; and thirdly, whereas substitution of N or S for O in the case of silver(I) complexes leads to considerably more effective binding, it has the reverse effect in the case of alkali metal complexes. This work is also important in that it was carried out using cation-sensitive electrodes.

The importance of stereochemical relationships and optical activity in bio-systems makes the paper by Barnes and Pettit[58] of particular interest. They report ΔH values for the formation of copper(II), nickel(II) and zinc(II) complexes of optically active and racemic ligands (α-alanine, valine, proline, asparagine and histidine) finding stereo-selectivity only in the case of histidine. It is at first glance surprising that Cu^{2+} coordinates preferentially with the optically pure species (as was found some time ago also to occur in the case of the α-hydroxyamidines[59]) whereas Ni^{2+} and Zn^{2+} form $[M(+L)(-L)]$ complexes in preference, but reasons for this are discussed. It is also important to note that ΔG values obtained potentiometrically are not sensitive enough to detect selectivity in the case of the histidine complexes[60].

The complexes of copper(II) with the ephedrines have been looked at by Gillard and Wootton[61] and although only approximate thermodynamic data (equilibrium constants for trimerisation) are presented, the paper is important in drawing attention to the dangers inherent in a use of Cotton effects in studying ligand configuration.

An excellent review of metal–porphyrin chemistry by Hambright has just been published[81], and two other papers are worthy of mention here. Burnham reports the interaction of various metal ions with porphyrins[62] and Cole *et al.*[63] have studied the solvation of iron(II) porphyrins with pyridines in benzene, chloroform and carbon tetrachloride. The ΔS° values are remarkably large but vary in the expected order, namely benzene $> CHCl_3 > CCl_4$.

Other work on ligands of biological importance includes studies on metal chelates of arginine[64], complexes with serine-type ligands[65], magnesium complexes of ATP, etc.[66], and beryllium complexes of some antihistamines[77]. It is surprising that in their study of proton and M^{2+} complexes of tryptamine, 5-hydroxytryptamine and 5-hydroxytryptophan, Weber and Simeon[67] used the rather unusual background of 0.37 mol l^{-1} $NaNO_3$. With the increase in the use of nuclear power, transuranic elements may become involved in biosystems; the work of Eberle on aminocarboxylic acid complexes of transuranium(V) elements[68] is perhaps therefore to be included in this section.

Finally, the complexes of the L-ascorbate anion are being intensively studied by Wahlberg *et al.* in a series of papers[69].

2.4.4 Background electrolyte effects

As indicated above (Section 2.2.3) the use of 3 mol l^{-1} perchlorate has been shown by Heck[12] to have a considerable effect on equilibrium constants. Ginstrop[70] using a hydrogen–silver/silver chloride electrode system has produced data, however, that appear to justify, at least in some instances, the use of this medium. This work should be studied in conjunction with that of Ohtaki and Biedermann[71] in which they study the variation of hydrogen ion activity coefficients with various cations, and with that of Lagrange[72] in which a very detailed examination of the applicability of Scatchard's formula is presented.

The effect of ionic strength on outer-sphere complexes, namely cobalt(III) hexammine–chloride complexes, has been studied by Mironov[73], and Andreev and Smirnova[74] have studied the effect on copper(II) chloride and bromide complexes of varying the background electrolyte.

2.4.5 Miscellaneous

Two old problems have been re-investigated; firstly the composition of copper(II) tartrate solutions which have been examined in detail by Bottari and co-workers[75,76]. They explain their e.m.f. measurements by postulating a bewildering array of complexes, protonated and deprotonated (above about pH = 4 these include OH^- complexes[76]). The reason for the confusion in the past is made clear, but the answer proposed embraces so many possibilities that it is not very convincing.

Secondly, new attempts have been made to obtain accurate reduction potentials for gold(I) by use of standard reduction potentials as a basis for the correlations mentioned above[16,17] (Section 2.2.4). Hancock and Finkelstein[16] suggest a potential of 1.73 V but Hawkins *et al.*[78] have, on the basis of equilibrium constants for the reaction between diphenylphosphinobenzene-*m*-sulphonate and gold(I) restored the *status quo* by supporting strongly the old value of 2.12 V. Although the standard reduction potential studies[16,17] are expected to be more reliable, it would seem that the problem remains.

In a study of bridged SCN^- complexes, Purohit and Bjerrum[79] have postulated that in the case of the reaction of *cis*-di-isothiocyanatobis(ethylenediamine)cobalt(III) with silver(I), the Ag^+ ion is 'chelated', (II) to explain the very different equilibrium constant obtained for the *trans*-compound.

To conclude this survey mention must be made of the fascinating water-soluble zinc tetracarbonylferrate(−II)s examined potentiometrically by Galembeck and Krumholz[80]. These are water soluble and yet would seem to contain metal–metal bonds: those reported are $ZnFe(CO)_4$, $[Zn_2Fe(CO)_4]^{2+}$

and $[ZnHFe(CO)_4]^+$. Independent evidence to support (or reject) the assump-

$$(en)_2Co\begin{matrix} N-C-S \\ N-C-S \end{matrix}Ag \qquad (II)$$

tion of such bonds should be readily obtainable—especially for any solid species.

Thanks are due to Dr. P. G. Wright for useful discussions whilst writing Section 2.2, and for suggesting the form of Table 2.1.

References

1. Nancollas, G. H. (1970). *Coord. Chem. Revs.*, **5**, 379
2. Sillen, L. G. and Martell, A. E. (Compilers) (1964). *Stability Constants of Metal-ion Complexes.* (London: The Chemical Society) and Sillen, L. G. and Martell, A. E. (Compilers) (1971). *Stability Constants of Metal-ion Complexes—Suppl. No. 1.* (London: The Chemical Society)
3. Ashcroft, S. J. and Mortimer, C. T. (1970). *Thermochemistry of Transition Metal Complexes.* (London and New York: Academic Press)
4. Christensen, J. J. and Izatt, R. M. (1970). *Handbook of Metal–ligand Heats and Related Thermodynamic Quantities.* (New York: Marcel Dekker)
5. Bell, R. P. (1959). *The Proton in Chemistry,* 71. (London: Methuen)
6. Prue, J. E. (1969). *J. Chem. Educ.*, **46,** 12
7. Adamson, A. W. (1954). *J. Amer. Chem. Soc.*, **76,** 1578
8. Nancollas, G. H. (1966). *Interactions in Electrolytic Solutions.* (London: Elsevier)
9. Rossotti, F. J. C. and Rossotti, H. S. (1961). *The Determination of Stability Constants.* (New York: McGraw Hill)
10. Beck, M. T. (1970). *Chemistry of Complex Equilibria.* (London: Van Nostrand-Reinhold)
11. Robinson, R. A. and Stokes, R. H. (1959). *Electrolyte Solutions,* Appendix 8.9. (London: Butterworths)
12. Heck, L. (1971). *Inorg. Nucl. Chem. Lett.*, **7,** 701, 709
13. Rossotti, F. J. C. (1960). *Thermodynamics of Metal Ion Complex Formation in Solution* in *Modern Coordination Chemistry* (ed. by Lewis, J. and Wilkins, R. G.), 20. (New York: Interscience)
14. Williams, R. J. P. and Hale, J. D. (1966). *Struct. and Bonding,* **1,** 249
15. Ahrland, S. (1968). *Struct. and Bonding,* **5,** 118
16. Hancock, R. D. and Finkelstein, N. P. (1971). *Inorg. Nucl. Chem. Lett.*, **7,** 477
17. Erenburg, A. M. and Peschevitskii, B. I. (1969). *Russ. J. Inorg. Chem.*, **14,** 1429
18. Rossotti, F. J. C., Rossotti, H. S. and Whewell, R. J. (1971). *J. Inorg. Nucl. Chem.*, **33,** 2051
19. Marcus, Y. (1969). *Pure Appl. Chem.*, **18,** 459
20. Bates, R. G. (1969). *Pure Appl. Chem.*, **18,** 421
21. Paabo, M. and Bates, R. G. (1969). *Anal. Chem.*, **41,** 283
22. Henry, R. P., Prue, J. E., Rossotti, F. J. C. and Whewell, R. J. (1971). *Chem. Commun.*, 868
23. Hair, M. L. (1970). *J. Phys. Chem.*, **74,** 1145
24. van Veen, A., Hoefnagel, A. J. and Webster, B. M. (1971). *Rec. Trav. Chim.*, **90,** 289
25. Arnaud, R. (1971). *Bull. Soc. Chim. Fr.*, 2782
26. Jullard, J. (1970). *Bull. Soc. Chim. Fr.*, 2040
27. Sagner, P. and Pechova, I. (1970). *Collect. Czech. Chem. Commun.*, **35,** 334
28. Hansen, L. D. and Lewis, E. A. (1971). *J. Chem. Thermodynamics,* **3,** 35
29. Eatough, D. J. (1970). *Analyt. Chem.*, **42,** 635
30. Simon, W., Wuhrmann, H.-R., Vasak, M., Pioda, L. A. R., Dohner, R. and Stefanac, Z. (1970). *Angew. Chem.*, 445

31. Bond, A. M. (1970). *Inorg. Chem.*, **9**, 1021
32. Elquist, B. (1970). *J. Inorg. Nucl. Chem.*, **32**, 937
33. Hanna, E. M., Pethybridge, A. D. and Prue, J. E. (1971). *Electrochim. Acta,* **16**, 677
34. Rowe, J. J. M., Hinton, J. and Rowe, K. L. (1970). *Chem. Rev.*, **70**, 1
35. Natusch, D. F. S. and Porter, L. J. (1971). *J. Chem. Soc. A,* 2527
36. Grasdalem, H. (1971). *Acta Chem. Scand.*, **25**, 1103
37. Rakshys, J. W. (1970). *Inorg. Chem.*, **9**, 1621
38. Chmelnick, A. M. and Fiat, D. (1971). *J. Amer. Chem. Soc.*, **93**, 2875
39. Strouse, C. E. and Matwiyoff, N. A. (1970). *Chem. Commun.*, 439
40. Haque, R. (1969). *J. Inorg. Nucl. Chem.*, **31**, 3869
41. Buslaev, U. A. and Il'in, E. G. (1970). *Dokl. Akad. Nauk. SSSR,* **190**, 1351
42. Ovchinnikov, I. V., Gainulin, I. E., Garif'yuanov, N. S. and Kozyrev, B. M. (1970). *Dokl. Akad. Nauk. SSSR,* **191**, 395
43. Manners, J. P., Morallee, K. G. and Williams, R. J. P. (1971). *J. Inorg. Nucl. Chem.*, **33**, 2085
44. Grenthe, I., Ots, H. and Ginstrup, O. (1970). *Acta Chem. Scand.*, **24**, 1067
45. Larson, J. W. (1970). *J. Phys. Chem.*, **74**, 685
46. Christensen, J. J., Slade, M. D., Smith, D. E. and Izatt, R. M. (1970). *J. Amer. Chem. Soc.*, **92**, 4164
47. Wallace, R. M. (1971). *Inorg. Nucl. Chem. Lett.*, **7**, 305
48. Siekierski, S. (1970). *J. Inorg. Nucl. Chem.*, **32**, 519
49. Jones, A. D. and Williams, D. R. (1971). *Inorg. Nucl. Chem. Lett.*, **7**, 369
50. Jones, A. D. and Williams, D. R. (1970). *J. Chem. Soc. A,* 1550
51. Dellien, I. and Grenthe, I. (1971). *Acta Chem. Scand.*, **25**, 1387
52. Hinchley, R. I. (1970). *Inorg. Chem.*, **9**, 917
53. Campbell, D. L. (1970). *J. Inorg. Nucl. Chem.*, **32**, 945
54. Grenthe, I. and Gardhammer, G. (1971). *Acta Chem. Scand.*, **25**, 1401
55. Izatt, R. M., Nelson, D. P., Rytting, J. H., Haymore, B. L. and Christensen, J. J. (1971). *J. Amer. Chem. Soc.*, **93**, 1619
56. Frensdorf, H. K. (1971). *J. Amer. Chem. Soc.*, **93**, 600
57. Avedikian, L. (1971). *Bull. Soc. Chim. Fr.*, 2832
58. Barnes, D. S. and Pettit, L. D. (1971). *J. Inorg. Nucl. Chem.*, **33**, 2177
59. Iball, J. and Morgan, C. (1964). *Nature (London),* **202**, 689
60. Morris, P. J. and Martin, R. B. (1970). *J. Inorg. Nucl. Chem.*, **32**, 2891
61. Gillard, R. D. and Wootton, R. (1971). *J. Inorg. Nucl. Chem.*, **33**, 2185
62. Burnham, B. F. (1970). *J. Amer. Chem. Soc.*, **92**, 1547
63. Cole, S. J., Curthoys, G. C. and Magnusson, E. A. (1971). *J. Amer. Chem. Soc.*, **93**, 2153
64. Clark, E. R. (1970). *J. Inorg. Nucl. Chem.*, **32**, 911
65. Letter, J. E. (1970). *J. Amer. Chem. Soc.*, **92**, 437
66. Nørby, J. G. (1970). *Acta Chem. Scand.*, **24**, 3276
67. Weber, D. A. and Simeon, Vl. (1971). *J. Inorg. Nucl. Chem.*, **33**, 2097
68. Eberle, S. H. (1970). *J. Inorg. Nucl. Chem.*, **32**, 109
69. Wahlberg, O. and co-workers (1971). *Acta Chem. Scand.*, **25**, 1000, 1045, 1064, 1079
70. Ginstrup, O. (1970). *Acta Chem. Scand.*, **24**, 875
71. Ohtaki, H. and Biedermann, G. (1971). *Bull. Soc. Chem. Japan,* **44**, 1515
72. Lagrange, J. (1971). *Ann. Chim. (Paris),* **6**, 125
73. Mironow, V. E. (1970). *Russ. J. Phys. Chem.*, **44**, 230
74. Andreev, S. N. and Smirnova, M. F. (1969). *Russ. J. Inorg. Chem.*, **14**, 1737
75. Bottari, E., Liberti, A. and Rufolo, A. (1969). *Inorg. Chim. Acta,* **3**, 201
76. Bottari, E. and Vicedomini, M. (1971). *J. Inorg. Nucl. Chem.*, **33**, 1463
77. Chawla, I. D. (1969). *J. Inorg. Nucl. Chem.*, **31**, 3809
78. Hawkins, C. J., Mønsted, O. and Bjerrum, J. (1970). *Acta Chem. Scand.*, **24**, 1059
79. Purohit, D. N. and Bjerrum, J. (1971). *J. Inorg. Nucl. Chem.*, **33**, 2067
80. Galembeck, F. and Krumholz, P. (1971). *J. Amer. Chem. Soc.*, **93**, 1909
81. Hambright, P. (1971). *Coord. Chem. Revs.*, **6**, 247

3
Thermochemistry of Chemical Compounds

G. PILCHER
University of Manchester

3.1 INTRODUCTION 58

3.2 THERMOCHEMISTRY OF INORGANIC COMPOUNDS 58
  3.2.1 *Techniques of measurement* 59
  3.2.2 *Accuracy of measurement* 59

3.3 ENTHALPIES OF FORMATION OF ORGANIC COMPOUNDS 62
  3.3.1 *The —CH_2— increment in homologous series* 62
  3.3.2 *Hydrocarbons* 62
  3.3.3 *Oxygen compounds* 64
    3.3.3.1 *Acetals* 64
    3.3.3.2 *Aldehydes and ketones* 65
  3.3.4 *Nitrogen compounds* 66
  3.3.5 *Sulphur compounds* 68
  3.3.6 *Halogen compounds* 68
    3.3.6.1 *Fluorine compounds* 69
      (a) *Reaction with alkali metals* 69
      (b) *Combustion calorimetry* 69
      (c) *Miscellaneous procedures* 71
    3.3.6.2 *Miscellaneous organochlorine, bromine and iodine compounds* 72

3.4 ENTHALPIES OF FORMATION OF ORGANOMETALLIC COMPOUNDS 72

3.5 ENTHALPIES OF ORGANIC REACTIONS FROM EQUILIBRIUM STUDIES 75
  3.5.1 *The principles of measurement* 75
  3.5.2 *Hydrocarbons* 75
  3.5.3 *Halogen compounds* 77

3.6 THERMOCHEMISTRY OF INORGANIC COMPOUNDS 78
  3.6.1 *Key compounds* 78
    3.6.1.1 *Hydrogen fluoride and hydrofluoric acid* 79
    3.6.1.2 *Orthophosphoric acid* 79

3.6.2 *Fluorine bomb calorimetry* 80
3.6.2.1 *Fluorides* 80
3.6.2.2 *Oxides* 80
3.6.2.3 *Carbides, nitrides, borides, phosphides, sulphides* 82
3.6.3 *Oxygen static-bomb calorimetry* 83
3.6.4 *Reaction calorimetry studies* 83
3.6.4.1 *Hot-zone calorimetry* 83
3.6.4.2 *Fluorine-flame calorimetry* 85
3.6.4.3 *Miscellaneous reaction calorimetry studies* 85

3.7 DISSOCIATION ENERGIES 87
3.7.1 *Fluorine* 87
3.7.2 *D(Alkyl–H)* 87
3.7.3 *D(Phenyl–H)* 87

3.1 INTRODUCTION

In common with most branches of physical chemistry, interest in the thermochemistry of organic, organometallic and inorganic compounds has increased steadily during the last decade. A reviewer of this subject is considerably assisted by the *Bulletin of Thermodynamics and Thermochemistry* published annually at the University of Michigan, Ann Arbor, Michigan. This Bulletin lists abstracts of work currently in progress submitted by research groups throughout the world: in Bulletin No. 14, 1971, 336 such abstracts are given associated with 960 research workers. The Bulletin also gives an indexed bibliography of work published during the year and the growth of activity in experimental chemical thermodynamics is evidenced by the fact that Bulletin No. 6, 1963 lists 1282 references whereas No. 14, 1971 lists 3666 references. Hence a virtually complete and detailed index of all the work done in this field for the past 8 years is available in convenient form at modest cost.

To review the subject it seems preferable to select some significant advances recently made and examine these in depth, by considering the work which has preceded them and the problems which have arisen. Some aspects, omitted in this review, or only given passing mention, relate to work currently in progress, and it is preferable to defer critical comment until a later date.

All energy quantities will be given in calories defined by 1 cal = 4.184 J. There has recently been a slight increase in the number of authors reporting energy quantities in joules, but the overwhelming majority of authors at present use calories.

3.2 THERMOCHEMISTRY OF ORGANIC COMPOUNDS

Three recent books have reviewed large areas of this subject. '*The Chemical Thermodynamics of Organic Compounds*' by Stull, Westrum and Sinke[1], lists thermodynamic functions for 918 compounds from 298 to 1000 K and values of the enthalpies of formation and entropies at 298 K for about 4500 compounds; the literature was reviewed up to 1966. '*The Thermochemistry of Organic and Organometallic Compounds*' by Cox and Pilcher[2] consists mainly of a critical analysis, giving selected values of the enthalpies of formation of about 3000 compounds at 298 K, published between 1930 and 1967. '*The Thermochemistry of Transition Metal Complexes*' by Ashcroft and Mortimer[3] reviews data for a class of compounds not covered by the other two works. The present review will concentrate on more recent work but naturally much of this work has its basis in the three books mentioned above.

3.2.1 Techniques of measurement

The most widely used calorimetric method of determining enthalpies of formation of organic compounds is by measuring enthalpies of combustion, by static-bomb calorimetry for compounds containing C, H, O, N, and by rotating-bomb calorimetry for compounds containing sulphur, halogens and for metal–organic compounds. The experimental techniques have been described in detail in the two volumes of '*Experimental Thermochemistry*' Volume 1[4]; Volume 2[5]. The experimental methods of thermochemistry are very well established and no fundamental changes have been made in them recently. Although each calorimeter built as a research instrument has some differences in construction, basically all follow the pattern described in '*Experimental Thermochemistry*'. What is new however, is the advent of a commercially available high-precision calorimetry system from L.K.B. Produckter AB, Sweden. Several interchangeable calorimeters can be used in the system, reaction and solution calorimeters of 100 or 25 cm^3 capacity, a titration calorimeter, a sealed bomb reaction calorimeter and a vaporisation calorimeter. The accuracy of results obtained with this equipment is comparable with that from specialised research apparatus and the importance of its appearance on the market lies in the fact that valuable thermochemical data may now be obtained by research workers who have not necessarily been trained in the specialised art of calorimeter construction. Commercially available microcalorimeters are also produced by LKB, namely, a batch and a flow solution microcalorimeter; Beckmann Instruments Inc. produce the Model 190-B calorimeter which is a heat-burst microcalorimeter based on the original design by Benzinger[6]; Seteram, at Lyon (France), produce the Calvet microcalorimeters, a standard model which will operate from room temperature to 200 °C and a high-temperature model which extends the range of operation to 1000 °C. Most of these microcalorimeters are such new additions to the armoury of the thermochemist that their use is still largely in the exploratory stage but we can reasonably expect significant results in the next few years.

3.2.2 Accuracy of measurement

The level of accuracy that may be achieved in bomb calorimetry can be demonstrated by examining the favourable case of the energy of combustion of benzoic acid. Benzoic acid is used as the standard to calibrate bomb calorimeters, and for this purpose the energy of combustion of a particular batch is measured in a standardising laboratory by means of an electrically calibrated calorimeter. A large number of measurements have been made of the energy of combustion of benzoic acid under standard conditions (i.e. 30 atm pressure of oxygen, 25 °C, the mass of benzoic acid and of the water added to the bomb to be three times the internal volume of the bomb in litres). A list of values is given in Table 3.1. These results demonstrate the precision with which an enthalpy of combustion can be determined by various

Table 3.1 The energy of combustion of benzoic acid

Reference	$-\Delta u_c/\mathrm{J g^{-1}}$
Jessup and Green[7]	26 432 ± 3
Jessup[8]	26 434 ± 2
Prosen and Rossini[9]	26 435 ± 4
Challoner, Gundry and Meetham[10]	26 436 ± 4
Coops and Adriaanse[11]	26 435 ± 6
Gundry and Meetham[12]	26 436 ± 3
Meetham and Nicholls[13]	26 432 ± 2
Hu, Yen and Geng[14]	26 433 ± 5
Peters and Tappe[15]	26 433 ± 3
Churney and Armstrong[16]	26 434 ± 3
Mosselman and Dekker[17]	26 433 ± 2
Gundry, Harrop, Head and Lewis[18]	26 434 ± 1

workers using quite different apparatus in many parts of the world, provided that the compound is easy to purify, is not hygroscopic and burns easily. There are few substances however, for which such an impressive table could be given and some recent studies have been directed to the general problem of obtaining *accurate* results.

Good, in a series of investigations of the enthalpies of combustion of hydrocarbons, has considered whether it is better to base the result on the mass of sample or the mass of CO_2 produced. This problem is discussed by Good and Smith[19] who concluded that for a highly-purified hydrocarbon sample where precautions were taken to exclude moisture, there was no significant difference in the two methods of evaluating results. However, the mass of CO_2 is more difficult to measure precisely than the mass of sample, so that the precision of the result based on CO_2 is slightly worse than that based on sample mass. If one suspects that a particular sample does contain water, then the result should be based on the CO_2 produced, and Good's work gives us confidence in the accuracy of results obtained in this way. Moreover, measuring both the mass of sample and mass of CO_2 produced can provide evidence of the completeness of combustion.

An important test of the accuracy of rotating-bomb calorimetry for organochlorine and organobromine compounds has been made by Laynez *et al*[20]. The enthalpy change for the reaction

$$\mathrm{Tris(c) + HX(g) \rightarrow TrisHX(c)} \text{ with } \mathrm{X = Cl, Br}^*$$

*Tris = tris(hydroxymethyl)aminomethane, $(HOCH_2)_3CNH_2$.

was derived (a) from rotating-bomb calorimetry by measuring the enthalpies of combustion of Tris(c), Tris·HCl(c) and Tris·HBr(c), and (b) from reaction calorimetry using the LKB reaction calorimeter by measuring the enthalpies of solution of Tris(c) and Tris·HCl(c) in aqueous HCl and Tris(c) and Tris·HBr(c) in aqueous HBr. The comparison is given in Table 3.2. The results from the two methods agree to within the limits of experimental error demonstrating the absence of any serious systematic error in the rotating-bomb technique for organochlorine and organobromine compounds.

Irving and Wadsö[21] proposed the enthalpy of solution of Tris(c) in 0.1 mol l^{-1} HCl as a suitable test reaction for solution calorimetry. This suggestion has met with approval in several laboratories and the National Bureau of Standards, Washington now produce a standard sample of Tris (acidometric titration 99.94 ± 0.01 mol %) for this purpose. There have been many investigations of this enthalpy of solution but two recent ones are outstanding. Hill *et al.*[22] have measured this enthalpy of solution at 0, 15, 25 and 50 °C using the LKB reaction calorimeter. They also examined the endothermic solution of Tris(c) in NaOH at 25 °C and found that this depended markedly

Table 3.2 Comparison of rotating-bomb and reaction calorimetry

Tris(c) + HX(g) → Tris · HX(c)

X	ΔH_r/kcal mol^{-1} *Rotating-bomb calorimetry*	*Reaction calorimetry*
Cl	-30.29 ± 0.46	-29.96 ± 0.01
Br	-33.05 ± 0.39	-32.93 ± 0.01

on the concentration of NaOH. From their results they also derived the aqueous enthalpy of protonation of Tris.

In an international collaborative effort between Gunn, Lawrence Radiation Laboratory, University of California, Livermore, Watson and Mackle of Queen's University, Belfast, Gundry and Head of the National Physical Laboratory, Teddington, and Månsson and Sunner of Lund, Sweden, the enthalpy of solution of Tris(c) in 0.1 mol l^{-1} HCl has been determined in terms of the energy of combustion of benzoic acid (Gunn *et al.*[23]). Firstly the enthalpy of reaction of $H_2SO_4 \cdot 8H_2O$ with $2.5(NaOH \cdot 10H_2O)$ was measured in three different rotating-bomb calorimeters each of which had been calibrated using benzoic acid; the results obtained were, at Belfast 36038 ± 5 cal mol^{-1}, at NPL 36044 ± 7 cal mol^{-1} and at Lund 36039 ± 11 cal mol^{-1}. Then Gunn[24] using a rocking sealed-bomb reaction calorimeter compared the enthalpy of this neutralisation reaction with the enthalpy of solution of Tris(c) in 0.1 mol l^{-1} HCl, obtaining a result for the latter in excellent agreement with that of Hill *et al.*[22].

Hansen and Lewis[25] have concluded that the reaction of aqueous Tris with aqueous $HClO_4$ is less satisfactory as a standard reaction for titration calorimetry than the reaction of aqueous NaOH with aqueous $HClO_4$. Absorption of CO_2 would affect the former reaction but not the latter.

3.3 ENTHALPIES OF FORMATION OF ORGANIC COMPOUNDS

3.3.1 The —CH_2— increment in homologous series

Sunner[26] has pointed out that the best values for the —CH_2— increment in the enthalpies of formation of gaseous homologous series are, in kcal mol^{-1}. n-alkanes 4.91 ± 0.03, n-alkanols 4.84 ± 0.04, alkanones 5.04 ± 0.10, thia-alkanes 5.01 ± 0.02. The constancy of this —CH_2— increment is a necessary feature of most practicable bond-energy schemes, so that the above figures could indicate that existing schemes are imperfect. The experimental proof of constancy is difficult, as the higher members of a homologous series are not easy to purify, their enthalpies of combustion are large and the low volatility of these compounds makes measurement of the enthalpies of vaporisation difficult.

3.3.2 Hydrocarbons

In a recent series of papers, Good has published the enthalpies of formation of a large number of hydrocarbons, determined by bomb calorimetry. Good's values for the isomeric pentanes[27] agree to within the limits of experimental error with values obtained from flame-calorimetry by Pilcher and Chadwick[28]. Good has also reported on the enthalpies of formation of n-propylcyclo-hexane and six methylethylcyclohexane isomers[29], and on n-propylcyclo-pentane and five methylethylcyclopentane isomers[30]. The enthalpies of formation of eleven nonanes were reported by Good[31] and of six hydro-carbons, (including benzene and toluene) by Good and Smith[19]. Good has indicated in the *Bulletin of Thermodynamics and Thermochemistry*, No. 14, 1971 that more work of this type is in progress. This large extension in our knowledge of the enthalpies of formation of hydrocarbons is important because only highly purified samples are examined and the measurements are made by a skilled experimentalist using one apparatus; i.e. the measure of consistency in the results is much better than previously. Where Good's measurements overlap with earlier ones carried out at the National Bureau of Standards, the agreement is, in general, excellent.

Boyd and co-workers[32] have made an interesting study of the strain energies in fused-ring compounds, the bicyclo[*n*,*m*,0]hydrocarbons. The enthalpies of formation were determined by aneroid bomb-calorimetry and the enthalpies of vaporisation were determined from vapour-pressure–temperature measurements. Strain energy may be derived by subtracting from the observed enthalpy of formation of the gaseous compound, a cal-culated value for an unstrained structure derived using a bond energy scheme; hence the numerical value of the strain energy depends on the parameters of the bond energy scheme. Boyd used the scheme of Schleyer *et al.*[33] which was derived for discussion of the enthalpy of formation of adamantane. The strain energies are given in Table 3.3.

The idea that the strain can be approximated by the sum of the strain energies of the separate rings (column headed Σ Rings) was also suggested by Cox and Pilcher[2]. This approach is only moderately successful. Boyd,

Table 3.3 Strain in Bicyclo[n,m,0] hydrocarbons

n	m			ΔH_f^0(g)	*Strain* (kcal mol^{-1})	*Σ Rings* (kcal mol^{-1})	ΔH_f^0(calc)
1	1	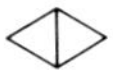	*cis*	51.9	65.9	56	
2	1	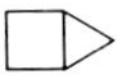	*cis*	37.0	56.1	55	
3	1	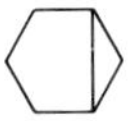	*cis*	9.3	33.5	35	10.3
4	1	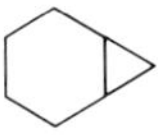	*cis*	0.4	29.6	29	2.2
5	1	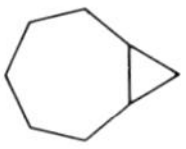	*cis*	−3.8	30.5	35	−3.8
6	1	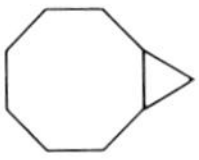	*cis*	−7.6	31.8	39	−5.0
4	2	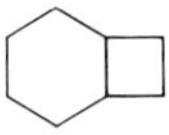	*cis*	−6.1	28.2	28	−5.1
3	3	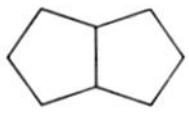	*cis*	−22.3	12.0	14	−22.1
			trans	−15.9	18.4	14	−9.1
4	3	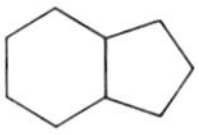	*cis*	−30.4	8.9	8	−31.0
			trans	−31.4	7.9	8	−30.5
5	3	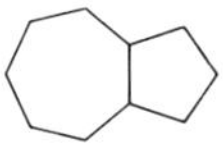	*cis*	−31.1	13.4	14	−32.0
			trans	−31.4	13.1	14	−31.8
4	4	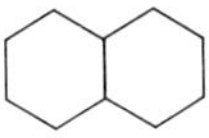	*cis*	−40.4	4.1	2	−39.7
			trans	−43.5	1.0	2	−43.8

however, has achieved excellent results by calculating molecular geometries and strain energies using transferable valence force potential functions for intramolecular distortion and minimising the energy of distortion. The calculations are elaborate, and calculated ΔH_f^0 values are given in the last column of Table 3.3. Boyd has successfully calculated the strain energy in all these compounds to within 1 kcal except for *trans*-bicyclo[3,3,0] octane for which the discrepancy is 6 kcal.

3.3.3 Oxygen compounds

In this group of compounds, the most extensive recent advances have been made with acetals and with aldehydes and ketones.

3.3.3.1 Acetals

The enthalpies of formation of the simplest acetals, dimethoxymethane and 1,1-dimethoxyethane were measured using flame calorimetry by Pilcher and Fletcher[34]. Månsson[35] using bomb calorimetry has determined the enthalpies of formation of some straight chain oxa-compounds of formula, $C_2H_5(OCH_2)_nCH_3$ with $n = 2,3,4,5$. Månsson *et al.*[36] have measured the enthalpies of combustion of trioxan and tetroxan, cyclic compounds containing the —O—CH_2—O— group, and Pihlaja and Heikkila[37] report enthalpies of combustion of dioxolan and some derivatives. These recent results may be analysed using the Allen bond-energy scheme[38] which leads to the following expression for the enthalpy of atomisation,

$$\Delta H_a(C_nH_{2n+2}O_m) \quad (2n+2)B_{CH}+2mB_{CO}+(n-1-m)B_{CC}+b_1\Gamma_{CCC} +b_2\Gamma_{CCO}+b_3\Gamma_{COC}+b_4\Gamma_{OCO}+C_1\Delta_{CCC}+C_2\Delta_{CCO}+C_3\Delta_{COO}-[S]$$

where the general term Γ_{ABC} is the contribution of the neighbouring bond interaction A—B—C, Δ_{ABC} is the contribution of non-bonded trios attached to carbon, C—C(A)(B) : the coefficients b and C are the numbers of such contributions within the molecule and [S] is the destabilising energy effect due to strain or steric hindrance, which is taken as zero in the following analysis.

To calculate the enthalpies of atomisation the following values were used: $\Delta H_f^0(\text{C,g}) = 170.9$, $\Delta H_f^0(\text{H,g}) = 52.10$, $\Delta H_f^0(\text{O,g}) = 59.56$ kcal mol^{-1}. The following parameter values were used: B_{CH} 99.30, B_{CC} 78.84, B_{CO} 78.15, Γ_{CCC} 2.58, Γ_{CCO} 5.66, Γ_{COC} 5.90, Γ_{OCO} 12.96, Δ_{COO} -3.65 kcal mol^{-1}. The terms Γ_{OCO} and Δ_{COO} were chosen to give the best agreement between observed and estimated ΔH_a values for acetals, the remaining terms were derived from data on alkanes, alcohols and ethers by Skinner and Pilcher[39]. It is apparent from Table 3.4 that apart from the cyclic compounds, the experimental values are in accord with the Allen scheme; also we would expect the 5-membered dioxolan ring to show strain energy. The strain in trioxan and tetroxan may arise because there are bonds of high polarity within the ring.

Månsson[35] also measured the enthalpy of formation of 3,-6-dioxaoctane [-97.51 ± 0.24 kcal mol^{-1}] and application of the Allen scheme with the above parameters indicates an apparent strain energy in this molecule of 2.33 kcal mol^{-1}. From this enthalpy of formation it is possible to derive the enthalpy of the isomerisation,

$$\text{—CH}_2\text{—O—CH}_2\text{—CH}_2\text{—O— (g)} \rightarrow \text{—CH}_2\text{—O—CH}_2\text{—O—CH}_2\text{— (g)}$$

as $\Delta H = -6.55$ kcal; in accord with the enthalpy of isomerisation,

$$\text{1,4-dioxane(g)} \rightarrow \text{1,3-dioxane(g)}$$

$\Delta H = -8.20$ kcal. Månsson raises the question whether it is correct to regard for example, 1,4-dioxane as strained, or 1,-3-dioxane as being stabilised, with respect to the value we would expect if the molecule behaved 'normally'. From this work it is evident that caution must be exercised when applying any bond energy scheme to predict ΔH_f^0(g) values for new families

Table 3.4 Enthalpies of atomisation of acetals

Compound	ΔH_f^0(g)/kcal mol^{-1}	ΔH_a(obs)/ kcal mol^{-1}	ΔH_a(est)/ kcal mol^{-1}	δ
$CH_2(OCH_3)_2$	-83.21 ± 0.19	1131.83	1131.76	-0.07
$CH_3CH(OCH_3)_2$	-93.15 ± 0.20	1416.87	1416.87	0.00
$C_2H_5(OCH_2)_2CH_3$	-99.14 ± 0.19	1697.96	1697.96	0.00
$C_2H_5(OCH_2)_3CH_3$	-138.89 ± 0.24	2072.37	2071.72	-0.65
$C_2H_5(OCH_2)_4CH_3$	-177.10 ± 0.33	2445.24	2445.48	$+0.24$
$C_2H_5(OCH_2)_5CH_3$	-216.52 ± 0.45	2819.32	2819.26	-0.06
1,3-Dioxolan	-72.10 ± 0.53	1016.52	1023.32	$+6.80$
2-Methyl-1,3-dioxolan	-84.14 ± 0.74	1303.66	1308.23	$+4.57$
trans-2,4-Dimethyl-1,3-dioxolan	-90.96 ± 0.88	1585.58	1592.68	$+7.10$
Trioxan	-111.32 ± 0.12	1115.30	1121.28	$+5.98$
Tetroxan	-148.24 ± 0.16	1486.88	1495.04	$+8.16$

of compounds even when the formal structure is similar to those involved in deriving the bond energy parameters.

3.3.3.2 *Aldehydes and ketones*

Until 1970, literature values for the enthalpy of formation of formaldehyde ranged from -23 to -29 kcal mol^{-1}. Two recent determinations are in close accord. Fletcher and Pilcher[40] measured the enthalpy of combustion by flame calorimetry and obtained $\Delta H_f^0(CH_2O,g) = -25.95 \pm 0.11$ kcal mol^{-1}. Birley and Skinner[41] measured the enthalpy of hydrolysis

$$CH_2(OCH_3)_2(l) + aq. = CH_2O(aq.) + 2CH_3OH(aq.)$$

and measured the enthalpy of solution of formaldehyde gas in the same acid medium used for the hydrolysis and derived $\Delta H_f^0(CH_2O,g) = -25.90 \pm 0.21$ kcal mol^{-1}. Birley and Skinner[41] also measured the enthalpy of hydrolysis of 1,1-dimethoxyethane and derived ΔH_f^0(acetaldehyde, 1) $= -46.60 \pm 0.50$ kcal mol^{-1} in reasonable agreement with the only previous determination of this enthalpy of formation from the enthalpy of hydrogenation by Kistiakowsky *et al.*[42], $\Delta H_f^0(CH_3CHO, l) = -45.88 \pm 0.13$ kcal mol^{-1}. Fletcher and Pilcher[40] also measured the enthalpy of combustion of glyoxal and derived $\Delta H_f^0(g) = -50.66 \pm 0.19$ kcal mol^{-1}; a value quite different from that generally listed in compilations of data.

The enthalpy of hydrolysis of ketene was measured by Rice and Greenberg[43] in 1934, from which the value $\Delta H_f^0(g) = -14.23$ kcal mol^{-1} was derived. New measurements by Nuttall *et al.*[44] for the reaction

$$CH_2CO(g) + NaOH{\cdot}500H_2O(l) \rightarrow CH_3CO_2Na(aq{\cdot}NaOH)$$

gave $\Delta H_r^0 = -49.75 \pm 0.38$ kcal mol^{-1}, leading to $\Delta H_f^0(CH_2CO, g) = -11.4 \pm 0.4$ kcal mol^{-1}. The older value is quoted in most compilations, including that by Cox and Pilcher[2] and must now be discarded.

A marked increase in our knowledge of the enthalpies of formation of aliphatic ketones in recent years is due to work at the National Physical Laboratory by Harrop *et al.*[45] on straight-chain ketones and at Lund by Sellers[46] on branched-chain ketones. Sellers correlated the enthalpies of formation of the aliphatic ketones by the Allen scheme, using the equation:

$$-\Delta H_f^0(C_nH_{2n}O, g)/\text{kcal mol}^{-1} = 41.98 + 2.38n + 2.67b_3 - 0.92C_4 - 1.48C_4'$$

where b_3 is the number of bond interactions of type (1), C_4 the number of bond interactions of type (2) and C_4' the number of bond interactions of type (3).

C—C—C (1)

C C / C / C (2)

C CO / C / C (3)

The comparison between observed and calculated enthalpies of formation is given in Table 3.5. The correlation of the experimental values with the Allen scheme is very good, provided that the discrepancy for 2,2,4,4-Tetramethylpentan-3-one can be ascribed to steric hindrance of the bulky t-butyl groups. The intervention of a —CH_2— group between one of the t-butyls and the carbonyl group, as in 2,2,5,5-Tetramethylhexan-3-one, completely removes this strain energy.

3.3.4 Nitrogen compounds

There has not been extensive activity in the thermochemistry of this class of compounds in recent years but some investigations have produced interesting conclusions.

Scott *et al.*[47] have reported on the thermodynamic properties of pyrrole, including a new measurement of ΔH_c^0 and the gaseous enthalpy of formation at 298 K. The latter differs by *c.* 6 kcal mol^{-1} from the previously accepted value and makes for a corresponding change in the apparent delocalisation energy in pyrrole. This can be seen from the calculated enthalpies of hydrogenation given in Table 3.6.

Since there is no exceptional strain in pyrrolidine, these enthalpies of hydrogenation indicate that stabilisation in pyrrole due to delocalisation of the π-electrons is more effective energetically by *c.* 10 kcal than in furan or thiophene.

Good[48] has reported the enthalpies of combustion and vaporisation of ethylenediamine, propane-1,2-diamine, butane-1,2-diamine, 2-Methylpropane-1,2-diamine and isobutylamine, and has discussed the C—N bond-energy terms in these compounds. Wadsö[49] has measured calorimetrically

Table 3.5 Enthalpies of formation of gaseous aliphatic ketones

	$-\Delta H_f^0(g)$/kcal mol^{-1}		
	Exp.	*Calc.*	δ
Propan-2-one	51.90 ± 0.12	51.79	−0.11
Butan-2-one	57.02 ± 0.20	56.84	−0.18
Pentan-2-one	61.91 ± 0.26	61.88	−0.03
3-Methylbutan-2-one	62.76 ± 0.21	63.07	+0.31
Pentan-3-one	61.65 ± 0.20	61.88	+0.23
Hexan-2-one	66.87 ± 0.24	66.93	+0.06
3,3-Dimethylbutan-2-one	69.47 ± 0.22	69.59	+0.12
Hexan-3-one	66.50 ± 0.21	66.93	+0.43
2-Methylpentan-3-one	68.38 ± 0.22	68.12	−0.26
2,2-Dimethylpentan-3-one	75.00 ± 0.32	74.63	−0.37
2,4-Dimethylpentan-3-one	74.41 ± 0.27	74.36	−0.05
2,2,4-Trimethylpentan-3-one	80.85 ± 0.28	80.87	+0.02
Nonan-5-one	82.44 ± 0.32	82.07	−0.37
2,6-Dimethylheptan-4-one	85.48 ± 0.27	85.58	+0.10
2,2,5,5-Tetramethylhexan-3-one	94.15 ± 0.55	94.13	−0.02
Undecan-6-one	92.59 ± 0.47	92.17	−0.42
2,2,4,4-Tetramethylpentan-3-one	82.65 ± 0.27	87.39	+4.74

Table 3.6 Enthalpies of hydrogenation of some unsaturated 5-membered rings

Reaction			ΔH_r^0/kcal mol^{-1}
Cyclopentadiene(g) + $2H_2$	→	Cyclopentane(g)	−50.38 ± 0.36
Furan(g) + $2H_2$	→	Tetrahydrofuran(g)	−35.72 ± 0.23
Thiophene(g) + $2H_2$	→	Thiacyclopentane(g)	−35.66 ± 0.35
Pyrrole(g) + $2H_2$	→	Pyrrolidine(g)	−26.68 ± 0.21

the enthalpies of vaporisation of 15 amines and correlated ΔH (vaporisation) with the boiling point by a linear equation; the Wadsö equations are extremely useful for estimating ΔH (vaporisation) from the boiling point.

An interesting study by Lebedeva *et al.*[50] on the enthalpies of combustion and vaporisation of some cyanoketones, gave the results listed in Table 3.7. The deviations from values calculated using a bond energy scheme are ascribed to repulsion between the polar cyano and carbonyl groups; this repulsion effectively drops to zero when these groups are separated by two CH_2 groups.

There has been some recent interest in nitro compounds, hexanitroethane[51], *trans*-2,2′,4,4′,6,6′-hexanitrostilbene[52]. Thermochemical data on nitroso compounds are rare, but Hamilton and Fagley[53] report the enthalpies of combustion of three nitrosonaphthols. The enthalpies of combustion of some simple triazoles were measured by Denault *et al.*[54], and one of the most complex heterocyclic compounds studied thermochemically is porphin, $C_{20}H_{14}N_4$, by Longo *et al.*[55] who calculated 419.0 kcal difference between the observed ΔH_f^0 in the gas state (-264.6 kcal mol^{-1}) and that obtained from a bond energy scheme, assuming a localised bond structure, ($+154.5$ kcal mol^{-1}). Longo *et al.* rationalised 75 % of this difference as due to stabilisation

Table 3.7 Enthalpies of formation of cyanoketones

	ΔH_f^0(g)/kcal mol^{-1}		
	Obs.	*Calc.*	δ
C_6H_5COCN	+28.09	+20.18	7.91
$C_6H_5COCH_2CN$	+16.76	+11.67	5.09
$C_6H_5COCH_2CH_2CN$	+7.22	+7.16	−0.06

by electron delocalisation but detailed discussion of such a complex molecule may be too ambitious at the present time.

3.3.5 Sulphur compounds

The major recent activity has been from Mackle and co-workers in Belfast on the enthalpies of formation of sulphones, alkyl sulphites and alkyl sulphates. Mackle and McNally[56] report the gaseous enthalpies of formation of α-butadiene sulphone, α and β-isoprene sulphones and deduce that the α-isomers are slightly more stable thermochemically by *c.* 1 kcal. Mackle *et al.*[57] investigated eleven $\alpha\beta$-unsaturated sulphones and came to the definite conclusion that there is no thermochemical evidence for conjugation between the sulphone group and the unsaturated double bonds adjacent to it. In the following paper Mackle and Steele[58] investigated some $\beta\gamma$ and $\gamma\delta$ unsaturated sulphones and showed that the $\beta\gamma$ unsaturated sulphones are thermochemically more stable than the $\alpha\beta$-unsaturated sulphones by *c.* 2 kcal mol^{-1}, possibly because the intervening —CH_2— group reduces the inductive withdrawal of electrons from the double bond by the —SO_2R group. This study was carried further by examining some aryl-propargyl and propadiene sulphones[59], again finding the $\beta\gamma$-isomers to be more stable than the $\alpha\beta$-isomers. Mackle and Steele[60] report the enthalpies of combustion of dimethyl, diethyl, di-n-propyl and di-n-butyl sulphites and sulphates. The gaseous enthalpies of formation were in accord with a group bond energy scheme and the mean dissociation energies of the RO—S bonds were derived; those in the sulphates were *c.* 10 kcal less than those in the sulphites.

3.3.6 Halogen compounds

The major effort in recent years has been concentrated on fluorine compounds; it is convenient to consider these separately from the other halogen compounds.

3.3.6.1 Fluorine compounds

There are some methods of general application that may be applied to determine the enthalpies of formation of these compounds and some special methods for particular compounds.

(a) *Reaction with alkali metals*—This technique is applicable to gaseous perfluorocompounds and gaseous perfluoro-chloro compounds, and has been developed by Kolesov and Skuratov. The general reaction may be written:

$$C_nF_{2n+2-a}Cl_a(g) + (2n+2)Na(c) \rightarrow nC(am) + (2n+2-a)NaF(c) + aNaCl(c)$$

The formidable experimental difficulties in measuring the enthalpy of such a reaction makes one cautious in fully accepting the results, but the success of this work is of great credit to these experimentalists. The reaction was carried out in a sealed bomb after calibration of the calorimeter by the combustion of benzoic acid; the pure alkali metal was placed in a crucible and the gaseous compound admitted to the bomb under pressure. The amount of reaction was determined by analysing for residual sodium (by adding water and measuring the volume of hydrogen produced), by determining the amount of fluoride produced and by measuring the quantity of substance placed in the bomb. The carbon produced was in an amorphous strained state and its strain energy found by measuring its enthalpy of combustion in oxygen; typically, strain energies of 4–5 kcal mol^{-1} were found. A list of compounds studied with this technique is given in Table 3.8.

Table 3.8 Enthalpies of formation of some perfluoro- and perfluoro chloro-compounds

Compounds	ΔH^0_f/kcal mol^{-1}	
Carbon tetrafluoride	-226.1 ± 2.2	Vorob'ev and Skuratov[61]
Trifluorochloromethane	-177.3 ± 1.1	Kolesov *et al.*[62]
Difluorodichloromethane	-114.8 ± 1.3	Kolesov *et al.*[62]
Tetrafluoroethylene	-158.8 ± 1.1	Kolesov *et al.*[63]
Trifluorochloroethylene	-135.5 ± 1.3	Kolesov *et al.*[64]
1,2-Dichlorotetrafluoroethane	-225.1 ± 1.2	Kolesov *et al.*[65]
1,1,2-Trichlorotrifluoroethane	-181.4 ± 1.8	Kolesov *et al.*[65]
Octafluoropropane	-427.0 ± 1.2	Kolesov *et al.*[66]
Octafluorocyclobutane	-369.5 ± 2.6	Kolesov *et al.*[67]

Apart from CF_4, for which there is a precise determination, $\Delta H^0_f(g) = -223.05 \pm 0.20$ kcal mol^{-1} [68], there is no other enthalpy of formation in this list which has been determined by another route with greater certainty.

(b) *Combustion calorimetry*—The combustion of an organofluorine compound can result in formation of both CF_4 and $HF{\cdot}nH_2O$ according to the equation,

$$C_aH_bO_cF_h + \left(\frac{4a+b-2c-h}{4}\right)O_2 + \left(\frac{(h-4z)(1+2n)-b}{2}\right)H_2O(l)$$
$$= (a-z)CO_2(g) + zCF_4(g) + (h-4z)HF{\cdot}nH_2O(l)$$

The proportion of CF_4–HF in the final product depends on the H : F ratio in the substance or mixture burned. The analysis is done either by measuring the total CO_2 produced and deducing the CF_4 by difference or by measuring the total fluoride produced. The method can be applied to perfluorocompounds if these are mixed with an auxiliary substance to promote combustion. Generally these measurements are made in a rotating-bomb calorimeter; the bomb is lined with platinum to avoid corrosion by the hydrofluoric acid. To deduce the enthalpy of formation from the enthalpy of combustion the enthalpies of formation of CF_4 and particularly of $HF{\cdot}nH_2O$ are crucially important; these are discussed later.

Static-bomb combustion measurements have been made by Kolesov *et al.*[69] on 1,1-difluoroethane ($\Delta H_f^0(g) = -118.8 \pm 2.0$ kcal mol^{-1}) and on 1,1,1-trifluoropropene ($\Delta H_f^0(g) = -146.9 \pm 1.6$ kcal mol^{-1})[70]. In these measurements the gaseous fluoride was admitted to the bomb with oxygen under pressure, the mixture ignited and a complete analysis of the products made to determine the extent of reaction.

A recent investigation using rotating bomb calorimetry by Cox *et al.*[71] on the enthalpies of formation of pentafluorobenzene derivatives has demonstrated that quite large strain energies, due to electrostatic repulsion between the polar C—F bonds, exist in these molecules. The results are summarised in Table 3.9.

Table 3.9 Strain energies in C_6H_5X compounds

X	$-\Delta H_f^0(C_6F_5X$, g/kcal mol^{-1}) *Experimental*	*calculated*	ΔH(*destabilisation*)
CH_3	201.55	226.00	24.5
H	192.61	218.14	25.5
Cl	193.59	225.81	32.2
OH	228.75	260.97	32.2
CO_2H	274.47	307.89	33.4
F	228.49	265.73	37.2

The calculated values were obtained from the relation,

$$\Delta H_f^0(C_6F_5X,g) = \Delta H_f^0(C_6H_5X,g) + 5\Delta H_f^0(C_6H_5F,g) - 5\Delta H_f^0(C_6H_6,g)$$

which is derived on the assumption of constant difference between the C—H and C—F bond energy terms. It is noteworthy that in hexafluorobenzene the destabilisation energy is of the same order of magnitude as the conventional resonance energy in benzene. A calculation of the electrostatic repulsion energy between the dipoles depends critically upon assumptions made about the location of the dipoles; such calculations indicate that two-thirds of the repulsion arises within the C_6F_5 group and one-third from the interaction of this group with the remaining C—X dipole, so that, approximately,

$$\Delta H(\text{destab.}, C_6F_5X) = \tfrac{2}{3}\Delta H(\text{destab.}, C_6F_6) + \tfrac{1}{3}\Delta H(\text{destab.}, C_6X_6)$$

Applied to CF_5Cl, this relation gives ΔH(destab.) = 30.7 kcal mol^{-1}, compared with the observed value 32.2 kcal mol^{-1}.

Baroody *et al.*[72] have measured the enthalpies of combustion of 1,1,1-

trinitro-4,4-bis(difluoroamino)pentane and 2-fluoro-2,2-dinitroethanol by use of diethyl oxalate and diethyl phthalate as moderators for the violent combustion reactions. Good and Smith[73] have measured the enthalpy of combustion of 1,1-bis(difluoroamino)heptane and deduced the N—F bond energy term.

(c) *Miscellaneous procedures*—$\Delta H_f^0(CF_4,g)$ can be connected with $\Delta H_f^0(HF{\cdot}nH_2O)$ through the determination of enthalpies of reactions such as

$$\text{(i)}\quad CF_4(g)+4Na(c) \rightarrow C(gr)+4NaF(c)$$

$$\text{(ii)}\quad C_2F_4(\text{polymer})+2F_2(g) \rightarrow 2CF_4(g)$$

$$\text{(iii)}\quad CF_4(g)+82H_2O(l) \rightarrow CO_2(g)+4[HF{\cdot}20H_2O](l)$$

In reaction (i), ΔH_f^0(NaF, c) depends on ΔH_f^0(HF·aq.). In reaction (ii), the enthalpy of formation of polytetrafluoroethylene is determined from the enthalpy of combustion in oxygen in which one of the products is HF·aq. The enthalpy of reaction (ii) has been measured directly by Domalski and Armstrong[74] and by Wood *et al.*[75]. CF_4 is resistant to hydrolysis, but the enthalpy of reaction (iii) can be deduced indirectly from bomb calorimeter measurements on mixtures of different F:H ratios. Cox *et al.*[76] obtained the enthalpy of reaction (iii), $\Delta H_r^0 = -41.38 \pm 0.32$ kcal mol^{-1} from combustion of mixtures of perfluorobicyclohexyl and benzoic acid. The principle of this indirect determination can be seen most simply by considering the extremes. If the combustion mixture is chosen so that the fluorocarbon produces no CF_4, then it reacts according to the equation

$$C_aF_b+\left(a-\frac{b}{4}\right)O_2+\frac{b(1+2n)}{2}H_2O(l) \rightarrow aCO_2(g)+b[HF{\cdot}nH_2O]$$

whereas if by burning a mixture of different composition CF_4 is produced according to the reaction,

$$C_aF_b+\left(a-\frac{b}{4}\right)O_2+\frac{(b-4)(1+2n)}{2}H_2O(l) \rightarrow (a-1)CO_2(g)+ CF_4(g)+(b-4)[HF{\cdot}nH_2O]$$

then by subtraction the enthalpy of hydrolysis is obtained,

$$CF_4(g)+2(1+2n)H_2O(l) \rightarrow CO_2(g)+4[HF{\cdot}nH_2O](l).$$

As long as there is uncertainty about both the enthalpies of formation of CF_4 and HF·aq., these interconnections cause a complex situation. A procedure for determining $\Delta H_f^0(CF_4, g)$ independent of ΔH_f^0(HF·aq.) is provided by the combustion of graphite in fluorine and very careful measurements were made by Greenberg and Hubbard[68]. A two-compartment bomb was used to prevent contact of the fluorine with the graphite before initiating the reaction. A small quantity of silicon powder was used as initiator. The correction for the energy effect due to the expansion of the fluorine from the storage compartment to the reaction compartment was measured in separate experiments. Graphite was used from several sources, with the result, $\Delta H_f^0(CF_4, g) = -223.05 \pm 0.20$ kcal mol^{-1}.

The use of NF_3 as a fluorinating agent is shown by the calorimetric determination of the enthalpy of formation of trifluoroacetonitrile by Walker *et al.*[77]. From the enthalpy of reaction:

$$CF_3CN(g) + \tfrac{5}{3}NF_3(g) \rightarrow CF_4(g) + \tfrac{4}{3}N_2(g)$$

$\Delta H_r^0 = -274.80 \pm 0.22$ kcal mol^{-1}, this gives $\Delta H_f^0(CF_3CN, g) = -118.9 \pm 0.5$ kcal mol^{-1}. In the same work, a study of the equilibrium over the temperature range 865–925 K,

$$2CF_3CN(g) \leftrightarrow C_2F_6(g) + C_2N_2(g)$$

gave $\Delta H_r^0 = -10.54 \pm 0.14$ leading to $\Delta H_f^0(CF_3CN, g) = -118.4 \pm 1.0$ kcal mol^{-1}, in agreement with the calorimetric value.

3.3.6.2 *Miscellaneous organochlorine, bromine and iodine compounds*

Kolesov *et al.*[78], using rotating-bomb calorimetry, have carefully redetermined the enthalpy of formation of chlorobenzene(l) and obtained the value $\Delta H_f^0 = +2.58 \pm 0.16$ kcal mol^{-1}, in good agreement with earlier measurements. With the same technique, Tomareva *et al.*[79] have determined the heats of formation of *cis*,1-2-dimethyl-3,3-dichlorocyclopropane(l), $\Delta H_f^0 = -20.76 \pm 0.21$ kcal mol^{-1}, and of the *trans*-isomer, $\Delta H_f^0(l) = -21.57 \pm 0.26$ kcal mol^{-1}. Hu and Sinke[80] have measured enthalpies of combustion of 11 chloro compounds, from which enthalpies of formation were derived, including ΔH_f^0(1,1,1-trichloroethane, l) $= -40.87 \pm 0.15$ kcal mol^{-1}; $\Delta H_f^0(CCl_4, l) = -30.69 \pm 0.14$ kcal mol^{-1}; $\Delta H_f^0(CHCl_3, l) = -32.10 \pm 0.20$ kcal mol^{-1}; $\Delta H_f^0(CH_2Cl_2, l) = -29.70 \pm 0.02$ kcal mol^{-1}; the latter two values are consistent with earlier measurements.

Carpenter *et al.*[81] have determined the enthalpy of formation of bromotrinitromethane, using an auxiliary substance to moderate the violent combustion. Anthoney *et al.*[82] have determined the enthalpy of hydrolysis of carbonyl bromide from which they derive $\Delta H_f^0(COBr_2, l) = -34.7 \pm 0.2$ kcal mol^{-1} and $\Delta H_f^0(COBr_2, g) = -27.3 \pm 0.4$ kcal mol^{-1}. The latter value differs significantly from that (-20.1 ± 0.1) selected by Cox and Pilcher[2] which was based on available data on the thermal equilibrium of dissociation of $COBr_2$; the new value measured by calorimetry is to be preferred. Devore and O'Neal[83] re-examined the enthalpies of hydrolysis of acetyl chloride, bromide and iodide with good agreement with the earlier study by Carson and Skinner[84].

3.4 ENTHALPIES OF FORMATION OF ORGANOMETALLIC COMPOUNDS

In view of the general interest in the chemistry of organometallic compounds in recent years, it is surprising that this has not been matched by a corresponding activity in studies of their thermochemistry. There are, in fact, very few additions to be made to the comprehensive list of thermochemical data on organometallics prepared seven years ago by Skinner[85]. The experimental difficulties with organometallics are often formidable, but bearing in mind that the classic work by Good has clearly shown the power of rotating-bomb calorimetry applied to studies on (a) lead tetramethyl[86] and tetraethyl[87], (b) hexamethyldisiloxane[88] and (c) dimanganese decacarbonyl[89]; it remains puzzling that Good's methods have not been exploited further by other workers.

Recent work has been mainly concerned with the compounds of the Group IV elements. An interesting paper by Potzinger and Lampe[90], who measured the appearance potentials of ions produced by electron impact on some silicon alkyls, demonstrates the inconsistency of available enthalpies of formation, as derived by static-bomb combustion measurements. The principle of the method used by Potzinger and Lampe is shown by consideration of the appearance potentials of SiH_2^+ formed from SiH_4 and Si_2H_6 by the reactions:

$$SiH_4 + e^- \rightarrow SiH_2^+ + H_2 + 2e^-$$
$$Si_2H_6 + e^- \rightarrow SiH_2^+ + SiH_4 + 2e^-$$

The appearance potentials combined with enthalpies of formation of SiH_4 and Si_2H_6 determined by Gunn and Green[91] lead to the same value of $\Delta H_f^0(SiH_2^+)$ and thus showed the internal consistency of Gunn's ΔH_f^0 values for silanes. Using this derived value for $\Delta H_f^0(SiH_2^+)$ and the appearance potential for the process

$$CH_3SiH_3 + e^- \rightarrow SiH_2^+ + CH_4 + 2e^-$$

a value for $\Delta H_f^0(CH_3SiH_3, g)$ was obtained. Proceeding in this way, Potzinger and Lampe were able to deduce the following enthalpies of formation, in kcal mol^{-1}, CH_3SiH_3, +1.0; $(CH_3)_2SiH_2$, −7.8; $(CH_3)_3SiH$, −18.0; $(CH_3)_4Si$, −33.0. The authors give ±3 kcal as a reasonable estimate of the uncertainty and showed that their values are in accord with the Allen bond energy scheme. These values are quite different from those derived from static-bomb measurements, and the latter are *not* in accord with any bond energy scheme. It would appear that only by employing the techniques developed by Good will accurate calorimetric values of the enthalpies of formation of organosilicon compounds be obtained.

Carson, at Leeds, has applied rotating-bomb calorimetry to measure the enthalpies of formation of organogermanium compounds. The aneroid rotating-bomb calorimeter was described by Adams *et al.*[92] who determined the enthalpy of formation of tetraphenylgermanium, using the technique introduced by Bills and Cotton[93] for tetraethylgermanium. The tetraphenylgermanium was burnt in 40 atm of oxygen in a bomb containing aqueous KOH solution to act as solvent for the germania produced on combustion. Comparison experiments were made by burning benzoic acid to produce the same mass of CO_2 as in the germanium tetraphenyl combustions, at the same time dissolving GeO_2(c, hexagonal) in the bomb solution. Hence the enthalpy of solution of the CO_2 and GeO_2 in the KOH is corrected for, and the combination of the two experiments yields the enthalpy of the reaction:

$$GePh_4(c) + 30O_2(g) \rightarrow GeO_2(c, hexagonal) + 24CO_2(g) + 10H_2O(l)$$

leading to $\Delta H_f^0(\text{GePh}_4, c) = +67.2 \pm 3.3$ kcal mol^{-1}. Carson *et al.*[94] with the same technique have determined $\Delta H_f^0(\text{GeBenzyl}_4, c) + 52.5 \pm 2.5$ kcal mol^{-1} and $\Delta H_f^0(\text{Ge}_2\text{Ph}_6, c) = +106.7 \pm 2.5$ kcal mol^{-1}.

A particular problem with organogermanium compounds is that germanium oxide exists in three modifications; the enthalpies of transition are large[95], GeO_2(c, hexagonal) → GeO_2(c, tetragonal)$\Delta H = -6.08 \pm 0.34$ kcal mol^{-1}, GeO_2(amorphous) → GeO_2(c, hexagonal)$\Delta H = -3.75 \pm 0.14$ kcal mol^{-1}. Static-bomb measurements thus suffer from the inherent uncertainty as to the form of the solid GeO_2 produced. Shaulov *et al.*[96] measured the enthalpies of combustion of GeMe_4 and GeEt_4 using the static-bomb technique, but Carson *et al.*[97], by fitting a bond energy scheme to selected alkylgermaniums, concluded that $\Delta H_f^0(\text{GeMe}_4, g)$ should be *c.* -32.5 ± 4 kcal mol^{-1}, whereas the value obtained by Shaulov *et al.* is -20.7 ± 2 kcal mol^{-1}. The probability is that the Shaulov static-bomb measurements are inadequate for alkylgermaniums.

The enthalpies of combustion of organotin compounds can be measured with a static bomb because conditions can be found for essentially complete combustion. Results for tin alkyls from static-bomb calorimetry are internally consistent with respect to bond energy schemes.

Adams *et al.*[98] have measured $\Delta H_f^0(\text{SnPh}_4, c) = +98.0 \pm 1.3$ kcal mol^{-1} in close agreement with Pope and Skinner[99] who obtained $+98.54 \pm 0.87$ kcal mol^{-1}. Carson *et al.*[97] measured hexaphenyldistannane $\Delta H_f^0(c) = +157.8 \pm 2$ kcal mol^{-1}, confirming the earlier value of Lautsch *et al.*[100] but in disagreement with Telnoi and Rabinovich[101]. Keiser and Kana'an[102] have remeasured ΔH(sublimation) for SnPh_4(c), 38.5 ± 1 kcal mol^{-1}; a value more acceptable than the previous value of 15.9 kcal mol^{-1} [103].

For a compound MR_n (where n R-groups are attached to a central atom M), the *mean* bond dissociation energy, $\bar{D}$(M—R) is defined as $1/n$ of ΔH_r^0 for the process,

$$\text{MR}_n(g) \rightarrow \text{M}(g) + n\text{R}(g)$$

i.e. $\bar{D}(\text{M—R}) = 1/n[\Delta H_f^0(\text{M}, g) + n\Delta H_f^0(\text{R}, g) - \Delta H_f^0(\text{MR}_n, g)]$.
From their work, and the best available subsidiary data, Carson *et al.*[97] gave the values for $\bar{D}$(M—R) for the Group IV elements listed in Table 3.10.

The abnormally small increase in $\bar{D}$(M—R) in the case of germanium in going from R = Et to R = Me is in accordance with the suggestion that $\Delta H_f^0(\text{GeMe}_4, g)$ may be in error. Certain trends however are apparent, (i) $\bar{D}$(M—Me) > $\bar{D}$(M—Et), (ii) $\bar{D}$(M—Ph) > $\bar{D}$(M—Me) by *c.* 13 kcal mol^{-1}, and (iii) the $\bar{D}$(M—R) values fall progressively as Group IV is descended.

Table 3.10 Mean bond dissociation energies (kcal mol^{-1})

R	$\bar{D}$(C—R)	$\bar{D}$(Ge—R)	$\bar{D}$(Sn—R)	$\bar{D}$(Pb—R)
Me	86	61	52	37
Et	82	59	47	32
Pr^n		57	46	
Pr^i		—	42	
Bu		—	47	
Bz		45	—	
Ph		77	64	

3.5 ENTHALPIES OF ORGANIC REACTIONS FROM EQUILIBRIUM STUDIES

3.5.1 The principles of measurement

The basic principles of determining enthalpies of reaction from equilibrium constants as a function of temperature are well known. The most common procedure, particularly for organic compounds, is to use the *second-law method*, which is simply the application of the Van't Hoff equation,

$$\frac{\mathrm{d}\ln K_p}{\mathrm{d}T} = \frac{\Delta H^0_{\mathrm{r},T}}{RT^2}$$

If the simplest assumption is made, that $\Delta H^0_{\mathrm{r},T}$ is temperature independent, i.e. $\Delta C^0_p = 0$, then the equation may be integrated to give,

$$\log K_p = \frac{-\Delta H^0_{\mathrm{r},T}}{2.303RT} + \text{constant}$$

This is often reasonable over a narrow temperature range, then $\Delta H^0_{\mathrm{r},T}$ is assigned to the mid-temperature of the range, and correction is then made to 298 K. If ΔC^0_p terms are known, or can be reasonably estimated, then they can be included in an expression for $\Delta H^0_{\mathrm{r},T}$ before integration.

The *third-law method* involves knowledge of the equilibrium constant at a single temperature T_1, then

$$\Delta H^0_{\mathrm{r},298} = -RT_1 \ln K_p + T_1\left\{\sum_{\text{reactants}} -\frac{(G^0 - H^0_{298})}{T_1} - \sum_{\text{products}} -\frac{(G^0 - H^0_{298})}{T_1}\right\}$$

The Gibbs energy function for ideal gases may be calculated by statistical mechanics, but requires information on the molecular geometry, the fundamental vibration frequencies and barriers to internal rotation. The accurate calculation is therefore restricted to simple molecules, but Benson[104] has developed group methods to estimate entropies, enabling approximate applications of the third-law method to be made more generally. This is useful because good agreement between enthalpy of a reaction as determined by the second-law and third-law methods from the same experimental data provides evidence that equilibrium was attained, and adds considerably to confidence in the final result.

3.5.2 Hydrocarbons

In recent years, Benson and co-workers have studied the equilibria between the geometrical and positional isomers of olefins; these isomerisations are catalysed by nitric oxide or iodine. The first study was by Benson and Bose[105] and Golden *et al.*[106] for the equilibria,

$$\text{But-1-ene(g)} \rightarrow \textit{trans}\text{-but-2-ene(g)} \tag{3.1}$$
$$\textit{cis}\text{-But-2-ene(g)} \rightarrow \textit{trans}\text{-but-2-ene(g)} \tag{3.2}$$

These reactions were catalysed by iodine atoms generated photochemically and thermally, the reaction was quenched by expansion of the mixture and analysis made by g.l.c. The enthalpies of reaction were calculated using the second-law method. These enthalpies of reaction may also be deduced from the enthalpies of combustion[107] and from the enthalpies of hydrogenation of these compounds[108]; the comparison is made in Table 3.11.

The values from the three different methods overlap within the limits of error but those from equilibrium studies have the advantage over the others

Table 3.11 Comparison of value of enthalpy of reaction for isomerism of butenes

	ΔH_r/kcal mol^{-1}		
	Combustion	*Hydrogenation*	*Equilibria*
Reaction (1)	-2.43 ± 0.35	-2.70 ± 0.26	-2.80 ± 0.20
Reaction (2)	-0.75 ± 0.40	-0.95 ± 0.26	-1.20 ± 0.15

because the latter involve the difference between two larger measured quantities, and are thus subject to greater experimental uncertainty.

Egger and Benson extended these studies to pentenes[109] and obtained

	$\Delta H^0_{r\,(298)}$/kcal mol^{-1}
pent-1-ene(g) $\leftrightarrow$ *trans*-pent-2-ene(g)	-2.60 ± 0.20
cis-pent-2-ene(g) $\leftrightarrow$ *trans*-pent-2-ene(g)	-0.90 ± 0.15

and to the equilibrium[110]

$$cis\text{-penta-1,3-diene(g)} \leftrightarrow trans\text{-penta-1,3-diene(g)}$$

obtaining $\Delta H^0_{r(298)} = -1.01 \pm 0.18$ kcal mol^{-1}. The value for this reaction, derived from enthalpies of combustion by Fraser and Prosen[111], was -1.66 ± 0.27 kcal mol^{-1}.

Furyama *et al.*[112] studied the gaseous equilibria,

$$I_2 + \text{cyclopentene} \leftrightarrow \text{cyclopentadiene} + 2HI \quad (3.3)$$

$$I_2 + \text{cyclopentane} \leftrightarrow \text{cyclopentene} + 2HI \quad (3.4)$$

The equilibrium constants were determined spectrophotometrically. Reaction (3.3) reached equilibrium very rapidly, but more than a day was required for reaction (3.4) to reach equilibrium at 318 °C. The enthalpy of reaction (3.4) was $\Delta H^0_{r(298)} = +24.40$ kcal mol^{-1} (third-law method) leading to ΔH^0_f(cyclopentene, g) $= +8.2$ kcal mol^{-1}, in excellent agreement with that derived from enthalpies of hydrogenation[113], $+8.23 \pm 0.22$ kcal mol^{-1}. The enthalpy of reaction (3.3) was $\Delta H^0_{r(298)} = +21.4 \pm 0.3$ kcal mol^{-1} (second-law method) which gave ΔH^0_f(cyclopentadiene, g) $= +31.8 \pm 0.3$ kcal mol^{-1}, again in excellent agreement with that derived from enthalpies of hydrogenation[114], $+31.94 \pm 0.28$ kcal mol^{-1}.

The degree of agreement with some of the most accurate calorimetric measurements of enthalpies of reaction gives confidence in accepting Benson's values.

3.5.3 Halogen compounds

Benson and co-workers have studied several equilibria involving substitution reactions of iodine. The reaction of propene,

$$\text{Propene(g)} + I_2\text{(g)} \leftrightarrow \text{Allyl iodide(g)} + \text{HI(g)}$$

was studied by Rodgers *et al.*[115], giving $\Delta H^0_{r(298)} = -8.0 \pm 0.6$ kcal mol^{-1} (second law) and -9.5 ± 1.0 kcal mol^{-1} (third law). The second law value corresponds to ΔH^0_f(allyl iodide, g) $= +21.5 \pm 0.7$ kcal mol^{-1}.

Walsh *et al.*[116] studied the iodination of toluene,

$$C_6H_5CH_3 + I_2 \leftrightarrow C_6H_5CH_2I + HI$$

and obtained $\Delta H^0_{r(298)} = +9.86 \pm 0.32$ kcal mol^{-1}, whence ΔH^0_f($C_6H_5CH_2$ I, g) $= +30.43 \pm 0.32$ kcal mol^{-1}. This result is to be preferred to the earlier values determined by Gellner and Skinner[117] and by Ashcroft *et al.*[118] by reaction calorimetric studies.

The enthalpies of bromination[119] and iodination of acetone have been determined from equilibrium studies,

$$CH_3COCH_3\text{(g)} + Br_2\text{(g)} \leftrightarrow CH_3COCH_2Br\text{(g)} + HBr\text{(g)}$$

$$CH_3COCH_3\text{(g)} + I_2\text{(g)} \leftrightarrow CH_3COCH_2I\text{(g)} + HI\text{(g)}$$

and the following results were obtained ΔH^0_f(CH_3COCH_2Br, g) $= -4.3 \pm 2.0$ kcal mol^{-1}, and ΔH^0_f(CH_3COCH_2I, g) $= -31.0 \pm 2.0$ kcal mol^{-1}, these enthalpies of formation have not been determined previously.

Solley and Benson[121] have studied the equilibrium

$$C_6H_5CHO + I_2 \leftrightarrow C_6H_5COI + HI$$

in the gas phase spectrophotometrically over the range 240–341 °C; after estimating S^0[PhCOI – PhCHO] = 12.5 cal mol^{-1} deg^{-1} from group methods and applying the third-law method $\Delta H^0_{r(298)} = +3.0 \pm 1.0$ kcal mol^{-1}. From their results they suggest ΔH^0_f(PhCHO, g) $= -8.9 \pm 1.0$ kcal mol^{-1} and ΔH^0_f(PhCOI, g) $= +2.6 \pm 1.0$ kcal mol^{-1}.

Norén and Sunner[122] have made a very careful study of the equilibrium,

$$\text{2-Chloropropane(g)} \leftrightarrow \text{Propene(g)} + \text{HCl(g)}$$

in the range 444–574 K. Small samples were withdrawn from the reaction mixture and analysed by g.l.c., corrections were made for side products and each experiment was continued until a constant equilibrium constant was obtained. From this work, ΔH^0_f(chloropropane, g) $= -35.03 \pm 0.21$ kcal mol^{-1} in close agreement with previous equilibrium studies by Kabo and Andreevski[123] and Howlett[124], but in disagreement with the value -32.95 ± 0.22 kcal mol^{-1} derived from the enthalpy of hydrogenation of 2-chloropropane to propane and HCl [125].

Lord and Pritchard[126] studied the equilibrium

$$COCl_2 \leftrightarrow CO + Cl_2$$

in the range 635–760 K. This equilibrium was also studied by Bodenstein and Plaut[127] 40 years earlier and the two sets of results plotted by the second-law method lie on the same line giving $\Delta H^0_{r(298)} = +27.0 \pm 0.4$ kcal mol^{-1}; curvature of the line detracts from the accuracy of this result. The third-law method gave $\Delta H^0_{r(298)} = +25.95 \pm 0.10$ kcal mol^{-1}, leading to $\Delta H^0_f(COCl_2, g) = -52.37 \pm 0.14$ kcal mol^{-1}. Tachoire[128] measured the energy liberated in the photochemically induced combination of CO and Cl_2 in a microcalorimeter and obtained $\Delta H^0_f(COCl_2, g) = -53.5 \pm 0.3$ kcal mol^{-1}. An accurate measurement of the enthalpy of hydrolysis of phosgene could establish the enthalpy of formation more reliably.

Lord and Pritchard[129] also studied the equilibrium,

$$CO_2 + CCl_4 \leftrightarrow 2COCl_2$$

from 628 to 718 K. They obtained $\Delta H^0_{r(298)} = +15.5 \pm 2.6$ kcal mol^{-1} (second law method) and $\Delta H^0_{r(298)} = 16.8 \pm 0.5$ kcal mol^{-1} (third law method). The experimental results showed considerable scatter but no systematic trends; the scatter may have resulted from the difficulties in analysing the corrosive gases Cl_2 and $COCl_2$ by g.l.c. Accepting the third law values as most reliable, the equilibrium results lead to $\Delta H^0_f(CCl_4, g) = -27.5 \pm 0.7$ kcal mol^{-1}. The extreme limits of the equilibrium measurements and Tachoire's value of $\Delta H^0_f(COCl_2, g)$ place $\Delta H^0_f(CCl_4, g)$ between -26.1 and -29.5 kcal mol^{-1}, more negative than the most recent value from rotating-bomb combustion, -22.94 ± 0.15 kcal mol^{-1} [80]. This recent work makes one less confident in the generally accepted value, $\Delta H^0_f(CCl_4, g) = -25.2 \pm 1.5$ kcal mol^{-1}, but further investigations are required before we would wish to change this selected value.

3.6 THERMOCHEMISTRY OF INORGANIC COMPOUNDS

The inorganic thermochemist uses many calorimetric methods in common with the organic thermochemist, such as combustion calorimetry and reaction calorimetry, but the number of subsidiary data required in inorganic thermochemistry is much larger. To obtain the enthalpies of formation of organic compounds containing C, H, O and N from enthalpies of combustion, only $\Delta H^0_f(H_2O, l)$ and $\Delta H^0_f(CO_2, g)$ are required, and when this is extended to sulphur and halogen compounds, then additionally the enthalpies of formation of sulphuric acid and the halogen acids are needed. The subsidiary data for inorganic themochemistry can generally be taken from compilations such as National Bureau of Standards, *Technical Note*, 270-3, 4. Occasionally, a particular enthalpy of formation is required so often that it becomes of key importance and for consistency it is best if one particular value is commonly accepted. In this connection the work of CODATA in producing tables of internationally agreed key values is welcomed. The first CODATA list has been published in *J. Chem. Thermodynamics* (1971)[130] and further sets of agreed values will be forthcoming.

3.6.1 Key compounds

The enthalpies of formation of key compounds of importance in inorganic thermochemistry already listed by CODATA are in kcal mol^{-1}; H_2O, (1) =

-68.315 ± 0.010; $HCl \cdot \infty H_2O = -39.333 \pm 0.33$; $HBr \cdot \infty H_2O = -29.039 \pm 0.041$; $HI . \infty H_2O = -13.60 \pm 0.20$; $CO(g) = -26.417 \pm 0.041$; $CO_2(g) = -94.051 \pm 0.031$; $HCl(g) = -22.063 \pm 0.031$; $HBr(g) = -8.695 \pm 0.041$, $HI(g) = +6.30 \pm 0.19$.

3.6.1.1 *Hydrogen fluoride and hydrofluoric acid*

Since it was realised in the early 1960s that the values for $\Delta H_f^0(HF, g)$ and $\Delta H_f^0(HF \cdot aq)$ recommended in the National Bureau of Standards, *Circular* 500 were probably in error, there has been a surfeit of work on this subject which has led to correction of the previous values. The most convincing determination of $\Delta H_f^0(HF, g)$ is that by Feder *et al.*[131] based on the following series of reactions:

		ΔH_{298}^0/kcal mol^{-1}
$SiO_2(\alpha$, quartz$) + 2F_2(g)$	$\rightarrow SiF_4(g) + O_2(g)$	-168.26 ± 0.28
SiO_2(crystabolite)	$\rightarrow SiO_2(\alpha$-quartz$)$	-0.35 ± 0.05
$2H_2(g) + O_2(g)$	$\rightarrow 2H_2O(g)$	-115.60 ± 0.02
$SiF_4(g) + 2H_2O(g)$	$\rightarrow SiO_2$(crystabolite)$ + 4HF(g)$	$+24.53 \pm 0.36$
$\therefore 2F_2(g) + 2H_2(g)$	$\rightarrow 4HF(g)$	-259.68 ± 0.48

hence, $\Delta H_f^0(HF, g) = -64.92 \pm 0.12$ kcal mol^{-1}. This value should not be in error by more than the error limits quoted.

It would appear that the most reliable value at present for $\Delta H_f^0(HF \cdot nH_2O)$ is to be derived from the enthalpy of hydrolysis of CF_4,

$$CF_4(g) + 82H_2O(l) = CO_2 + 4[HF \cdot 20H_2O]$$

$\Delta H_r^0 = -41.38 \pm 0.32$ kcal mol^{-1}, obtained by Cox *et al.*[76] indirectly from combustions of perfluorobicyclohexyl and benzoic acid mixtures. Together with the direct determination by Greenberg and Hubbard[68] of $\Delta H_f^0(CF_4, g) = -223.05 \pm 0.20$ kcal mol^{-1}, this gives $\Delta H_f^0(HF \cdot 20H_2O) = -76.75 \pm 0.15$ kcal mol^{-1}. To derive $\Delta H_f^0(HF \cdot nH_2O)$ it is necessary to apply the enthalpies of dilution as listed in NBS *Tech. Note* 270-3. In view of the importance of $\Delta H_f^0(HF \cdot nH_2O)$ it is surprising that the enthalpies of dilution have not been recently re-examined. There are several other measurements which are in accord with the values selected above; however, several determinations of the enthalpy of solution of HF(g) in water are not in good agreement. Thus it seems that further work will be necessary in order to reduce the error limits to acceptable magnitude for the requirements of the CODATA programme.

3.6.1.2 *Orthophosphoric acid*

An excellent investigation by Head and Lewis[132] has established $\Delta H_f^0(H_3PO_4 \cdot 40H_2O, l)298\ K = -309.35 \pm 0.38$ kcal mol^{-1}. White phosphorus, encapsulated in polythene, was burned in *c.* 10 atm of oxygen in a rotating-bomb calorimeter at 50 °C. The products were completely hydrolysed to orthophosphoric acid with a solution of perchloric acid in the bomb. This value is in agreement with that obtained by Birley and Skinner[133] of $\Delta H_f^0(H_3PO_4 \cdot 40H_2O) = -309.66 \pm 0.35$ kcal mol^{-1}, obtained from the enthalpy of hydrolysis of PCl_5(c).

3.6.2 Fluorine bomb calorimetry

In Chapter 6, *Experimental Thermochemistry*, Vol. 2 (1962), Hubbard stated that fluorine bomb calorimetry was then in its infancy; under Hubbard's guidance this technique has been developed into a very powerful one. The experimental procedure is superficially analogous to static-bomb combustion calorimetry, the same type of calorimeter and temperature measurements are made and the calorimeter is calibrated by use of benzoic acid. For combustion in fluorine, a nickel bomb pretreated with fluorine is used. This precaution prevents further corrosion in subsequent combustion experiments. Highly purified fluorine is used; there are problems in selecting materials to support the sample, especially as the sample holder will reach a high temperature. For substances spontaneously inflammable in fluorine, Nuttall *et al.*[134] constructed a two compartment bomb; in one section fluorine can be contained at high pressure and to initiate the reaction it is then admitted to the other section containing the sample. The versatility of the technique can be seen by considering some of its major applications.

3.6.2.1 Fluorides

The enthalpies of formation of fluorides can be found by direct synthesis, e.g.

$$W(c) + 3F_2(g) \rightarrow WF_6(g)$$

or for lower fluorides by reaction with fluorine, e.g.

$$PF_3(g) + F_2(g) \rightarrow PF_5(g)$$

A list of the enthalpies of formation of fluorides measured by fluorine bomb calorimetry is given in Table 3.12. The importance of the values listed lies not only in the large number but also in their accuracy; they supersede all previous values and from some of these the deviations are large. The *Bulletin of Thermodynamics and Thermochemistry* No. 14 (1971) states that further work, particularly on the rare earth fluorides, is in progress.

3.6.2.2 Oxides

Fluorine bomb calorimetry can be used to determine the enthalpies of formation of oxides, especially when gaseous products are formed. Wise *et al.*[142] determined the enthalpy of formation of silica by the reaction,

$$SiO_2(c) + 2F_2(g) \rightarrow SiF_4(g) + O_2(g)$$

and obtained $\Delta H_f^0(SiO_2, \alpha\text{–c}) = -217.72 \pm 0.34$ kcal mol^{-1} and $\Delta H_f^0(SiO_2$, vitreous$) = -215.94 \pm 0.31$ kcal mol^{-1}. This is in agreement with the value by Good[158] obtained by rotating-bomb calorimetry; silicon mixed with a polymeric fluorine compound was burned in oxygen and the products dissolved in HF·aq. to produce H_2SiF_6, aq: Good obtained $\Delta H_f^0(SiO_2, \alpha\text{–c}) = -217.5 \pm 0.5$ kcal mol^{-1}. The previous best value, obtained by combustion of silicon in oxygen, was -209.33 ± 0.25 kcal mol^{-1} [159], and was in error because of incomplete combustion of the silicon sample.

Table 3.12 Enthalpies of formation of fluorides

	ΔH_f^0/kcal mol^{-1}	*Reference*
BeF_2(am)	-244.32 ± 0.23	Churney and Armstrong[135]
MgF_2(c)	-268.7 ± 0.3	Rudzitis *et al.*[136]
ZnF_2(c)	-182.7 ± 0.4	Rudzitis *et al.*[137]
CdF_2(c)	-167.39 ± 0.23	Rudzitis *et al.*[138]
BF_3(g)	-271.65 ± 0.22	Johnson *et al.*[139]
AlF_3(c)	-360.7 ± 0.3	Rudzitis *et al.*[140]
YF_3(c)	-410.7 ± 0.8	Rudzitis *et al.*[141]
CF_4(g)	-223.05 ± 0.20	Greenberg and Hubbard[68]
SiF_4(g)	-385.98 ± 0.19	Wise *et al.*[142]
GeF_4(g)	-284.57 ± 0.21	O'Hare *et al.*[143]
	-284.42 ± 0.38	Adams *et al.*[144]
	-284.37 ± 0.15	Gross *et al.*[95]
GeF_2(c)	-157.4 ± 1.0	Adams *et al.*[145]
TiF_4(c)	-394.19 ± 0.35	Greenberg *et al.*[146]
ZrF_4(c)	-456.80 ± 0.25	Greenberg *et al.*[147]
HfF_4(c)	-461.40 ± 0.85	Greenberg *et al.*[146]
PF_5(g)	-380.8 ± 0.3	O'Hare and Hubbard[148]
PF_3(g)	-151.99 ± 0.16	Rudzitis *et al.*[149]
AsF_5(g)	-295.59 ± 0.19	O'Hare and Hubbard[150]
NbF_5(c)	-433.50 ± 0.15	Greenberg *et al.*[151]
TaF_5(c)	-454.97 ± 0.19	Greenberg *et al.*[151]
SF_6(g)	-291.77 ± 0.24	O'Hare *et al.*[152]
SeF_6(g)	-266.95 ± 0.14	O'Hare *et al.*[152]
TeF_6(g)	-327.20 ± 0.56	O'Hare *et al.*[152]
MoF_6(g)	-372.35 ± 0.22	Settle *et al.*[153]
WF_6(g)	-411.5 ± 0.4	O'Hare and Hubbard[154]
NiF_2(c)	-157.2 ± 0.4	Rudzitis *et al.*[155]
RuF_5(c)	-213.41 ± 0.35	Porte *et al.*[156]
UF_6(c)	-522.64 ± 0.43	Settle *et al.*[157]

The enthalpy of formation of boric oxide has been recently determined by Johnson and Hubbard[160]; from the reaction:

$$B_2O_3(c) + 3F_2(g) \rightarrow 2BF_3(g) + \tfrac{3}{2}O_2(g)$$

they obtained $\Delta H_f^0(B_2O_3, c) = -304.48 \pm 0.59$ kcal mol^{-1} in agreement with the value obtained by Good and Månsson[161] -304.10 ± 0.40 kcal mol^{-1}, derived from rotating-bomb calorimetry; crystalline boron mixed with a polymeric fluorine compound was burned in oxygen and the products dissolved in aqueous HF.

An elegant application of combustion in fluorine was made by Gross *et al.*[95], who used Pyrex reaction vessels charged with 5 atm of fluorine. The enthalpy of formation of $GeF_4(g)$ was measured directly, and the enthalpies of formation of the crystal forms of germanium dioxide were determined from measurements on the reaction,

$$GeO_2(c)+2F_2(g) \rightarrow GeF_4(g)+O_2(g)$$

namely, $\Delta H_f^0(GeO_2, \text{c–hex}) = -132.58 \pm 0.19$ kcal mol^{-1}; $\Delta H_f^0(GeO_2, \text{c–tetrag.}) = -138.66 \pm 0.31$ kcal mol^{-1}; the remarkably large difference in enthalpy between two crystal forms is noteworthy.

3.6.2.3 *Carbides, nitrides, borides, phosphides, sulphides*

Fluorine bomb calorimetry is now being extended to determine the enthalpies of formation of compounds which pose difficulties for study by the more conventional routes. To illustrate the versatility and the potential of fluorine bomb calorimetry, we will consider some simple examples of its application.

Greenberg *et al.*[162] have determined the enthalpy of formation of silicon carbide from measurements on the reaction in a bomb.

$$SiC(c)+4F_2(g) \rightarrow SiF_4(g)+CF_4(g)$$

They obtained $\Delta H_f^0(\text{SiC}, \alpha\text{-hexagonal}) = -17.23 \pm 0.46$ kcal mol^{-1}; $\Delta H_f^0(\text{SiC}, \beta\text{-cubic}) = -17.49 \pm 0.43$ kcal mol^{-1}.

Wise *et al.*[163] determined the enthalpy of formation of boron nitride by the reaction

$$BN(c)+\tfrac{3}{2}F_2(g) \rightarrow BF_3(g)+\tfrac{1}{2}N_2(g)$$

obtaining $\Delta H_f^0(\text{BN, c}) = -59.97 \pm 0.37$ kcal mol^{-1}. The enthalpy of formation of boron phosphide was similarly determined by Gross *et al.*[164] from,

$$BP(c)+4F_2(g) = BF_3(g)+PF_5(g)$$

and $\Delta H_f^0(\text{BP, c}) = -27.6 \pm 1.1$ kcal mol^{-1}.

Germanium sulphide was fluorinated by Adams *et al.*[165]

$$GeS(c)+5F_2(g) \rightarrow GeF_4(g)+SF_6(g)$$

to give $\Delta H_f^0(\text{GeS, c}) = -18.2 \pm 1.0$ kcal mol^{-1}. O'Hare *et al.*[166] have similarly determined the enthalpy of formation of molybdenum disulphide:

$$MoS_2(c)+9F_2(g) = MoF_6(g)+2SF_6(g)$$

obtaining $\Delta H_f^0(\text{MoS}_2\text{, c}) = -65.8 \pm 1.2$ kcal mol^{-1}.

In all these applications, the products of the reaction are gaseous and complete combustion was achieved. A more conventional approach using oxygen bomb calorimetry would result in solid products, and as is apparent from the static-bomb combustion of silicon, complete oxidation in such cases can be difficult to achieve.

O'Hare *et al.*[167] have determined the enthalpy of formation of thiazyltrifluoride by the reaction

$$NSF_3(g)+\tfrac{3}{2}F_2(g) \rightarrow \tfrac{1}{2}N_2(g)+SF_6(g)$$

The thiazyltrifluoride was doped with PF_3 to accelerate the reaction and the two-compartment bomb was used to hold each reactant separately before mixing. The value obtained was $\Delta H_f^0(NSF_3, g) = -85.2 \pm 0.5$ kcal mol^{-1}. By combination with the appearance potentials of SF_3^+ from SF_4 and NSF_3 reported by Glemser *et al.*[168], the enthalpy of formation of SF_4, g was calculated to be -161.8 ± 1.1 kcal mol^{-1}; literature values range from -162 to -208 kcal mol^{-1} so that a direct determination of $\Delta H_f^0(SF_4, g)$ is still needed. Assuming $D(SF_3\text{–}F) = \bar{D}(SF_4) = 76 \pm 1$ kcal mol^{-1}, then $D(N{\equiv}SF_3) = 93 \pm 2$ kcal mol^{-1}; also estimated was $D(N{\equiv}SF) = 71 \pm 5$ kcal mol^{-1}.

3.6.3 Oxygen static-bomb calorimetry

The static bomb can be used to measure the enthalpies of combustion of metals, metal carbides and nitrides; much work of this type has been carried out by Holley and co-workers at Los Alamos. There are however, severe experimental difficulties to be overcome in order to obtain reliable enthalpies of formation. Many carbides and nitrides are difficult to ignite and auxiliary substances to promote combustion are required, also it is difficult to establish that complete combustion occurred, particularly when a solid product is formed. In some cases it is possible to determine the enthalpy of formation of the oxide from the difference in the enthalpies of solution of the metal and the oxide in acid; e.g. neodynium oxide, $\Delta H_f^0(Nd_2O_3, c) = -432.15 \pm 0.24$ kcal mol^{-1} from the enthalpy of combustion[169], $\Delta H_f^0(Nd_2O_3, c) = -431.95 \pm 0.51$ kcal mol^{-1} from enthalpies of solution in hydrochloric acid[170]. The solution measurements are less precise, but the excellent agreement in this case removes the possibility of systematic error in the combustion value. Problems may also arise concerning the crystalline form of the oxide produced and its stoichiometry. Enthalpies of formation of non-stoichiometric carbides have been measured using this technique. A representative list of recent measurements is given in Table 3.13.

3.6.4 Reaction calorimetry studies

In this Section we consider some recent developments in technique which seem potentially very useful and also some interesting measurements recently made by conventional methods.

3.6.4.1 Hot-zone calorimetry

In hot-zone calorimetry a reaction is carried out in a calorimeter at room temperature by heating the reactants to a high temperature in an isolated section within the calorimeter. The enthalpy of reaction is found at 25 °C, but the energy input to provide the hot-zone may be a large fraction of the total heat. The first approaches to accurate measurements by this type of calorimetry were made at the National Bureau of Standards and are described by Evans in Chapter 13 of *Experimental Thermochemistry*, Vol. 2. There have been two recent advances in hot-zone calorimetry.

Hajiev[185] measured the enthalpy of formation of lead selenide. The two components were sealed in a small ampoule and placed in a micro-furnace situated inside a calorimetric combustion bomb. A small electric motor was used inside the bomb to swing the ampoule so that the molten mixture could be agitated to produce a homogeneous alloy. The calorimeter was calibrated in the usual way by combustion of benzoic acid. For a measurement, the electrical energy to heat the furnace and drive the motor was *c.* 85% of the total energy. Several compounds have been examined by this method, Hajiev reports the following $\Delta H_f^0(c)$ values, in kcal mol^{-1}, SnSe -21.7 ± 0.5; InSb -7.78 ± 0.08; InAs -13.9 ± 0.8; GaSb -10.7 ± 0.6; PbSe -25.1 ± 0.3. These results compare favourably with others reported in the literature and

Table 3.13 Enthalpies of formation of oxides, carbides and nitrides

	ΔH_f^0/kcal mol^{-1}	*Reference*
ZrO_2	-263.1 ± 0.5	Huber *et al.*[171]
	-263.04 ± 0.16	Kornilov *et al.*[172]
HfO_2	-273.6 ± 0.3	Huber and Holley[173]
Ce_2O_3	-429.31 ± 0.68	Baker and Holley[174]
CeO_2	-260.6 ± 0.2	Baker *et al.*[175]
EuO	-141.3 ± 1.3	Huber and Holley[176]
Dy_2O_3	-445.3 ± 1.0(comb.)	Huber *et al.*[177]
	-445.5 ± 1.6(soln.)	Huber *et al.*[177]
U_3O_8	-854.4 ± 0.6	Huber and Holley[178]
UO_2	-259.3 ± 0.2	Huber and Holley[178]
NpO_2	-256.7 ± 0.6	Huber and Holley[179]
PuO_2	-252.35 ± 0.17	Johnson *et al.*[180]
$ThC_{1.00 \pm 0.01}$	-29.6 ± 1.1	Huber *et al.*[181]
$ThC_{1.91 \pm 0.01}$	-29.9 ± 1.1	Huber *et al.*[181]
$CeC_{1.5}$	-21.1 ± 0.7	Baker *et al.*[175]
CeC_2	-23.2 ± 1.3	Baker *et al.*[175]
ZnC	-49.5 ± 0.6	Baker *et al.*[182]
U_2C_3	-43.4 ± 1.8	Huber *et al*[183]
$PuC_{0.878 \pm 0.003}$	-11.41 ± 0.68	Johnson *et al.*[184]
PuN	-71.51 ± 0.62	Johnson *et al.*[180]

this technique appears to be a promising method for determining the enthalpies of formation of semiconducting compounds.

Nuñez *et al.*[186] described a reaction calorimeter in which a hydrogen in oxygen flame is used to heat a hot-zone vessel. The calorimeter was calibrated by burning hydrogen in oxygen. The energy required to heat the hot-zone vessel is measured by collecting the water produced by the hydrogen flame; this energy input is *c.* 90% of the total. The enthalpies of reduction by hydrogen of cupric and cuprous oxides were measured together with the enthalpies of oxidation of copper and cuprous oxide to give ΔH_f^0(CuO, granular) -38.65 ± 0.30 kcal mol^{-1}; ΔH_f^0(CuO, fine powder) -37.54 ± 0.20 kcal mol^{-1}; $\Delta H_f^0(Cu_2O, c)$ -41.39 ± 0.32 kcal mol^{-1}. The enthalpies of formation of these oxides are influenced by their previous history, state of sub-division and stoichiometry. A further study by Espada *et al.*[187] on the reduction of the lead oxides gave ΔH_f^0(PbO, yellow) -52.12 ± 0.15 kcal mol^{-1}; $\Delta H_f^0(PbO_2, c)$ -65.6 ± 0.7 kcal mol^{-1}; $\Delta H_f^0(Pb_3O_4, c)$ -171.77 ± 1.50 kcal mol^{-1}. Here the main source of uncertainty arises from the imprecise stoichiometry of these compounds. This technique of hot-zone reduction of oxides promises to be useful in those cases where formation of the oxide by combustion of the metal in oxygen is difficult.

3.6.4.2 Fluorine-flame calorimetry

The fluorine-flame calorimeter has been developed at the Nation Bureau of Standards by Armstrong and co-workers as described in *Experimental Thermochemistry*, Vol. 2 (1962). The experimental difficulties appear to be even more severe than those associated with fluorine bomb calorimetry, because as well as corrosion problems there are the analytical problems of determining the extent of reaction and any side reactions. King and Armstrong[188] described a fluorine-flame calorimeter in which the gaseous products from the flame are dissolved in water held in a reaction vessel within the calorimeter. They measured the enthalpies of the reactions,

$$OF_2(g)+2H_2(g)+99H_2O(l) \rightarrow 2[HF{\cdot}50H_2O](l)$$

$$F_2(g)+H_2(g)+100H_2O(l) \rightarrow 2[HF{\cdot}50H_2O](l)$$

and obtained $\Delta H_f^0(HF{\cdot}50H_2O, l) = -76.68 \pm 0.09$ kcal mol^{-1} and $\Delta H_f^0(F_2O, g) = +5.86 \pm 0.38$ kcal mol^{-1}.

Armstrong and King[189], and Armstrong[190] report the enthalpies of formation of the chlorine fluorides determined from the enthalpy of reaction;

$$ClF_n(g)+\left(\frac{n+1}{2}\right)H_2(g)+mH_2O(l) \rightarrow [HCl+nHF+mH_2O](l)$$

and obtained, $\Delta H_f^0(ClF, g) = -14.36 \pm 0.79$ kcal mol^{-1}; $\Delta H_f^0(ClF_3, g) = -38.36 \pm 0.50$ kcal mol^{-1}; $\Delta H_f^0(ClF_5, g) = -54.52 \pm 1.0$ kcal mol^{-1}. The value for $\Delta H_f^0(ClF, g)$ has consequences in determining the dissociation energy of fluorine; these are discussed later. The enthalpy of formation of liquid chlorine pentafluoride was also determined by Bisbee *et al.*[191] from its reaction with hydrogen and with ammonia

$$ClF_5(l)+8NH_3(g) = NH_4Cl(c)+5NH_4F(c)+N_2(g)$$

carried out in a Monel-metal bomb. The mean value obtained was $\Delta H_f^0(ClF_5, l) = -60.9 \pm 4.5$ kcal mol^{-1}, in accord with Armstrong's value but with much larger uncertainty limits.

3.6.4.3 Miscellaneous reaction calorimetry studies

In this Section we consider some recent studies made by well-established methods which have given results of special interest.

Irving and co-workers have continued studies on the enthalpies of formation of metal acetonylacetonates by measuring their enthalpies of hydrolysis in acid media. Hill and Irving[192] hydrolysed $Fe(C_5H_7O_2)_3$ in aqueous HCl and $Mn(C_5H_7O_2)_3$ in a $FeCl_2$/HCl aqueous mixture. Irving and Walter[193] hydrolysed $Ga(C_5H_7O_2)_3$ in aqueous perchloric acid. By considering the reaction,

$$M(g) + 3C_5H_7\dot{O}_2(g) \rightarrow M(C_5H_7O_2)_3(g)$$

it is possible to calculate an average (M—O) bond dissociation energy, thus, Mn—O = 44, Fe—O = 47, Cr—O = 56, Al—O = 64, Ga—O = 48 kcal mol^{-1}.

Finch *et al.*[194] measured the enthalpy of reaction

$$P_2I_4(c) + I_2(nCS_2) \rightarrow 2PI_3(nCS_2)$$

and derived $\Delta H_f^0(P_2I_4, c) = -27.0 \pm 1.2$ kcal mol^{-1}. Assuming that the mean bond dissociation energy in PI_3, 50.2 kcal mol^{-1} is equal to the mean (P—I) bond energy in P_2I_4, then E(P—P) was calculated as 55.5 ± 2 kcal mol^{-1}.

In calorimetric studies of chlorination Gross *et al.*[195] measured the enthalpy of formation of silicon tetrachloride by reacting silicon with chlorine gas in the presence of an excess of liquid chlorine. The reaction was carried out in a glass reaction vessel, and was initiated by a small piece of titanium attached to the silicon. Separate experiments were made to correct for the evaporation of liquid chloride and for the mixing at the end of the experiment of the silicon tetrachloride and chlorine. A precise value was obtained, $\Delta H_f^0(SiCl_4, l) = -165.49 \pm 0.16$ kcal mol^{-1}; this supersedes previous values which were discordant. With the same experimental method, Gross and Hayman[196] measured the enthalpy of formation of aluminium chloride, $\Delta H_f^0(AlCl_3, c) = -168.80 \pm 0.16$ kcal mol^{-1}.

A recent notable example of investigation of rare-gas compounds was by Pepekin *et al.*[197] who measured the enthalpy of combustion of xenon difluoride, the sample was enclosed in a polythene bag, and the combustion yielded as $HF{\cdot}30H_2O(l)$. They obtained the value $\Delta H_f^0(XeF_2, c) = -41.5 \pm 0.6$ kcal mol^{-1}, and from $\Delta H_{sub} = 12.3 \pm 0.2$, they derived $\Delta H_f^0(XeF_2, g) = -29.2 \pm 0.8$ kcal mol^{-1}. O'Hare *et al.*[198] have measured the enthalpy of formation of aqueous xenon trioxide by measuring the enthalpy of reaction:

$$XeO_3(aq.) + 9I^-(aq.) + 6H^+(aq.) \rightarrow Xe(g) + 3I_3^-(aq.) + 3H_2O(l)$$

They obtained $\Delta H_f^0(XeO_3, aq.) = 99.94 \pm 0.24$ kcal mol^{-1}. By estimating the standard partial molar entropy $S^0(XeO_3, aq.) = 44 \pm 4$ cal deg^{-1} mol^{-1}, the standard electrode potential for the following half-reaction is calculated

$$Xe(g) + 3H_2O(l) \rightarrow XeO_3(aq.) + 6H^+ + 6e^-$$

$E^0 = 2.10 \pm 0.01$ V. For the corresponding reaction in alkaline solution,

$$Xe(g) + 7OH^-(aq.) \rightarrow HXeO_4^-(aq.) + 3H_2O(l) + 6e^-$$

O'Hare calculated $E^0 = 1.24 \pm 0.01$ V. XeO_3 is one of the strongest oxidants in aqueous media and, in fact, aqueous solutions of XeO_3 are thermodynamically unstable with respect to oxidation of water. This reaction does not occur indicating a high activation barrier possibly because there are no stable intermediate oxidation states between the free elements and XeO_3.

3.7 DISSOCIATION ENERGIES

A large number of dissociation energies have been measured in recent years, determined by the conventional methods, from kinetic studies of organic reactions, spectroscopy and high-temperature mass spectrometer studies. There are some recent measurements of special interest to the inorganic and organic thermochemist.

3.7.1 Fluorine

At present it is generally accepted that the dissociation energy of fluorine is in the region of 40 kcal mol^{-1}; e.g. from $D(H—F) = 136.10$ kcal mol^{-1}, (Johns and Barrow[199]) and $\Delta H_f^0(HF, g) = -64.92$ kcal mol^{-1}, the value $D(F_2) = 38.2$ kcal mol^{-1} is obtained. This is in excellent accord with thermal equilibrium studies of the dissociation of fluorine by Doescher[200] and Wise[201] which were subjected to careful third law analysis by Stamper and Barrow[202].

There are, however, recent indications that this dissociation energy may be lower, perhaps as low as 30 kcal mol^{-1}. A photo-ionisation study of fluorine by Dibeler *et al.*[203] indicated $D(F_2) = 30.9 \pm 0.7$ kcal mol^{-1}. Dibeler also reported $D(H—F)$ and hence calculated $\Delta H_f^0(HF, g) = -65.3$ kcal mol^{-1} which is not very different from the accepted value. Furthermore, the dissociation energy of fluorine can be derived from $D(Cl—F)$ and $\Delta H_f^0(ClF, g)$, using Armstrong's value for the latter; Dibeler *et al.*[204] obtain $D(F_2) = 32.0 \pm 1.6$ kcal mol^{-1}. This recent work cannot be regarded as definitive, but if proved correct it would mean that the careful analysis of Johns and Barrow on $D(H—F)$ and also the thermal dissociation equilibrium studies on fluorine must contain surprisingly large errors.

3.7.2 *D*(Alkyl–H)

Chupka[205] applied the photo-ionisation technique to obtain $D(CH_3—H) = 104.8 \pm 0.1$ kcal mol^{-1} at 298 K: the important advance here is the reduction of the uncertainty by an order of magnitude from the value previously derived from kinetic studies. Chupka and Lifshitz[206] from the photo-ionisation of methyl radicals obtained $D(CH_2—H) = 109.0 \pm 1.0$ kcal mol^{-1} at 298 K; again a considerable improvement in precision over previous values.

3.7.3 *D*(Phenyl–H)

$D(C_6H_5—H)$ has been the subject of recent controversy, but the situation has now been clarified. Kerr[207] reviewed work prior to 1966 and selected $D(C_6H_5—H) = 104 \pm 2$ kcal mol^{-1}. Rodgers *et al.*[208] studied the kinetics of the thermal equilibrium,

$$C_6H_5I + I\cdot \leftrightarrow C_6H_5\cdot + I_2$$

from which they derived ΔH_r^0, combination with the available $\Delta H_f^0(C_6H_5I, g)$

yields $\Delta H_f^0(C_6H_5, g) = +78.6 \pm 1.6$, corresponding to $D(C_6H_5—H) = 110.9 \pm 1.6$ kcal mol^{-1}. This higher value for $D(C_6H_5—H)$ has now been confirmed by Chamberlain and Whittle[209] who studied hydrogen abstraction from benzene by CF_3 radicals at high temperature. Above 140 °C genuine hydrogen abstraction occurs,

$$CF_3\cdot + C_6H_6 \rightarrow CF_3H + C_6H_5\cdot$$

and from the difference in activation energies the value, $\Delta H_r^0 = 3.8$ kcal mol^{-1} at 298 K is obtained. Since $\Delta H_r^0 = D(C_6H_5—H) - D(CF_3—H)$ and $D(CF_3—H)$ is well established, 106.4 ± 0.5 kcal mol^{-1}, then $D(C_6H_5—H) = 110.2$ kcal mol^{-1}. This result, in contrast to the value by Rodgers *et al.*[208] is independent of subsidiary thermochemical data. Chamberlain and Whittle suggest the best value is a weighted mean, $D(C_6H_5—H) = 110.0 \pm 2.0$ kcal mol^{-1}.

References

1. Stull, D. R., Westrum, E. F. and Sinke, G. C. (1969). *The Chemical Thermodynamics of Organic Compounds.* (New York: John Wiley and Sons, Inc.)
2. Cox, J. D. and Pilcher, G. (1970). *Thermochemistry of Organic and Organometallic Compounds.* (London: Academic Press Inc.)
3. Ashcroft, S. J. and Mortimer, C. T. (1970). *Thermochemistry of Transition Metal Complexes.* (London: Academic Press Inc.)
4. Rossini, F. D., editor, (1956). *Experimental Thermochemistry, Volume 1.* (New York: Interscience Publishers Inc.)
5. Skinner, H. A., editor, (1962). *Experimental Thermochemistry, Volume 2.* (New York: Interscience Publishers Inc.)
6. Kitzinger, C. and Benzinger, T. H. (1955). *Z. Naturforsch.*, **106,** 375
7. Jessup, R. S. and Green, C. B. (1934). *J. Res. Nat. Bur. Stand.*, **13,** 469
8. Jessup, R. S. (1942). *J. Res. Nat. Bur. Stand.*, **29,** 247
9. Prosen, E. J. and Rossini, F. D. (1944). *J. Res. Nat. Bur. Stand.*, **33,** 439
10. Challoner, A. R., Gundry, H. A. and Meetham, A. R. (1955). *Phil. Trans. Roy. Soc. London,* **A247,** 553
11. Coops, J., Adriaanse, N. and Van Nes., K. (1956). *Rec. Trav. Chim.*, **75,** 237
12. Gundry, H. A. and Meetham, A. R. (1958). *Trans. Faraday Soc.*, **54,** 664
13. Meetham, A. R. and Nicholls, J. A. (1960). *Proc. Roy. Soc. London,* **A256,** 384
14. Hu, J-H, Yen, H-K and Geng, Y-L. (1966). *Acta Chimica Sinica,* **32,** 242
15. Peters, H. and Tappe, E. (1967). *Mber. Deut. Akad. Wiss. Berlin,* **9,** 828
16. Churney, K. L. and Armstrong, G. T. (1968). *J. Res. Nat. Bur. Stand.*, **72A,** 453
17. Mosselman, C. and Dekker, H. (1969). *Rev. Trav. Chim.*, **88,** 161
18. Gundry, H. A., Harrop, D., Head, A. J. and Lewis, G. B. (1969). *J. Chem. Thermodynamics,* **1,** 321
19. Good, W. D. and Smith, N. K. (1969). *J. Chem. Eng. Data,* **14,** 102
20. Laynez, J., Ringner, B. and Sunner, S. (1970). *J. Chem. Thermodynamics,* **2,** 603
21. Irving, R. J. and Wadsö, I. (1964). *Acta. Chem. Scand.*, **18,** 195
22. Hill, J. O.,Öjelund, G. and Wadsö, I. (1969). *J. Chem. Thermodynamics,* **1,** 111
23. Gunn, S. R., Watson, J. A., Mackle, H., Gundry, H. A., Head, A. J., Månsson, M. and Sunner, S. (1970). *J. Chem. Thermodynamics,* **2,** 549
24. Gunn, S. R. (1970). *J. Chem. Thermodynamics,* **2,** 535
25. Hansen, L. D. and Lewis, E. A. (1971). *J. Chem. Thermodynamics,* **3,** 35
26. Sunner, S. (1971). *Huffman Memorial Lecture, 25th U.S. Calorimetry Conference: Bulletin of Thermodynamics and Thermochemistry, (1971),* **14.** (University of Michigan Press)
27. Good, W. D. (1970). *J. Chem. Thermodynamics,* **2,** 237
28. Pilcher, G. and Chadwick, J. D. M. (1967). *Trans. Faraday Soc.*, **63,** 2357

29. Good, W. D. (1970). *J. Chem. Thermodynamics,* **2,** 399
30. Good, W. D. (1971). *J. Chem. Thermodynamics,* **3,** 97
31. Good, W. D. (1969). *J. Chem. Eng. Data,* **14,** 231
32. Chang, S., McNally, D., Tehrany, S. S., Hickey, M. J. and Boyd, R. H. (1970). *J. Amer. Chem. Soc.,* **92,** 3109
33. Schleyer, P. R., Williams, J. E. and Blanchard, K. R. (1970). *J. Amer. Chem. Soc.,* **92** 2377,
34. Pilcher, G. and Fletcher, R. A. (1969). *Trans. Faraday Soc.,* **66,** 794
35. Månsson, M. (1969). *J. Chem. Thermodynamics,* **1,** 141
36. Månsson, M., Morawetz, E., Nakase, Y. and Sunner, S. (1969). *Acta Chem. Scand.,* **23,** 56
37. Pihlaja, K. and Heikkila, J. (1969). *Acta. Chem. Scand.,* **23,** 1053
38. Allen, T. L. (1959). *J. Chem. Phys.,* **31,** 1039
39. Skinner, H. A. and Pilcher, G. (1963). *Quart Rev. Chem. Soc.,* **17,** 264
40. Fletcher, R. A. and Pilcher, G. (1970). *Trans. Faraday Soc.,* **66,** 794
41. Birley, G. I. and Skinner, H. A. (1970). *Trans. Faraday Soc.,* **66,** 791
42. Dolliver, M. A., Gresham, T. L., Kistiakowsky, G. B., Smith, E. A. and Vaughan, W. E. (1938). *J. Amer. Chem. Soc.,* **60,** 440
43. Rice, F. O. and Greenberg, J. (1934). *J. Amer. Chem. Soc.,* **56,** 2268
44. Nuttall, R. L., Laufer, A. H. and Kilday, M. V. (1971). *J. Chem. Thermodynamics,* **3,** 167
45. Harrop, D., Head, A. J. and Lewis, G. B. (1970). *J. Chem. Thermodynamics,* **2,** 203
46. Sellers, P. (1970). *J. Chem. Thermodynamics,* **2,** 211
47. Scott, D. W., Berg, W. T., Hossenlopp, I. A., Hubbard, W. N., Messerly, J. F., Todd, S. S., Douslin, D. R., McCullough, J. P. and Waddington, G. (1967). *J. Phys. Chem.,* **71,** 2263
48. Good, W. D. (1970). *J. Chem. Eng. Data,* **15,** 150
49. Wadsö, I. (1969). *Acta Chem. Scand.,* **23,** 2061
50. Lebedeva, N. D., Dneprovskii, A. S. and Katin, Yu, A. (1969). *Russ. J. Phys. Chem.,* **43,** 770
51. Pepekin, V. I., Miroshnichenko, E. A., Lebedev, Yu. A. and Apin, A. Y. (1968). *Russ. J. Phys. Chem.,* **42,** 1583
52. Marantz, S. and Armstrong, G. T. (1968). *J. Chem. Eng. Data,* **13,** 118, 455
53. Hamilton, J. V. and Fagley, T. F. (1968). *J. Chem. Eng. Data,* **13,** 523
54. Denault, G. C., Marx, P. C. and Takimoto, H. H. (1968). *J. Chem. Eng. Data,* **13,** 514
55. Longo, F. R., Finarelli, J. D. and Adler, A. D. (1970). *J. Phys. Chem.,* **74,** 3296
56. Mackle, H. and McNally, D. V. (1969). *Trans. Faraday Soc.,* **65,** 1738
57. Mackle, H., McNally, D. V. and Steele, W. V. (1969). *Trans. Faraday Soc.,* **65,** 2060
58. Mackle, H. and Steele, W. V. (1969). *Trans. Faraday Soc.,* **65,** 2069
59. Mackle, H. and Steele, W. V. (1969). *Trans. Faraday Soc.,* **65,** 2073
60. Mackle, H. and Steele, W. V. (1969). *Trans. Faraday Soc.,* **65,** 2053
61. Vorob'ev, A. F. and Skuratov, S. M. (1960). *Zh. Neorg. Khim.,* **5,** 1398
62. Kolesov, V. P., Zenkov, I. D. and Skuratov, S. M. (1963). *Russ. J. Phys. Chem.,* **37,** 378
63. Kolesov, V. P., Zenkov, I. D. and Skuratov, S. M. (1962). *Russ. J. Phys. Chem.,* **36,** 45
64. Kolesov, V. P., Zenkov, I. D. and Skuratov, S. M. (1963). *Russ. J. Phys. Chem.,* **37,** 115
65. Kolesov, V. P., Talakin, O. G. and Skuratov, S. M. (1968). *Russ. J. Phys. Chem.,* **42,** 1617
66. Kolesov, V. P., Talakin, O. G. and Skuratov, S. M. (1967). *Vestnik. Moscow Univ. (Khim.),* **5,** 60
67. Kolesov, V. P., Talakin, O. G. and Skuratov, S. M. (1968). *Russ. J. Phys. Chem.,* **42,** 1218
68. Greenberg, E. and Hubbard, W. N. (1968). *J. Phys. Chem.,* **72,** 222
69. Kolesov, V. P., Shetekher, S. N., Martynov, A. M. and Skuratov, S. M. (1968). *Russ. J. Phys. Chem.,* **42,** 975
70. Kolesov, V. P., Martynov, A. M. and Skuratov, S. M. (1967). *Russ. J. Phys. Chem.,* **41,** 482
71. Cox, J. D., Gundry, H. A., Harrop, D. and Head, A. J. (1969). *J. Chem. Thermodynamics,* **1,** 77
72. Baroody, E. E., Carpenter, G. A., Robb, R. A. and Zimmer, M. F. (1968). *J. Chem. Eng. Data,* **13,** 215
73. Good, W. D. and Smith, N. K. (1970). *J. Chemical Eng. Data,* **15,** 147

74. Domalski, E. S. and Armstrong, G. T. (1967). *J. Res. Nat. Bur. Stand.*, **71A,** 105
75. Wood, J. L., Lagow, R. J. and Margrave, J. L. (1967). *J. Chem. Eng. Data,* **12,** 255
76. Cox, J. D., Gundry, H. A. and Head, A. J. (1965). *Trans. Faraday Soc.,* **61,** 1594
77. Walker, L. C., Sinke, G. C., Perette, D. J. and Janz, G. J. (1970). *J. Amer. Chem. Soc.,* **92,** 4525
78. Kolesov, V. P., Tomareva, E. M., Skuratov, A. M. and Alekhina, S. P. (1967). *Russ. J. Phys. Chem.,* **41,** 817
79. Tomareva, E. M., Kolesov, V. P., Nefedov, O. M. and Skuratov, S. M. (1968). *Russ. J.* Phys. Chem., **42,** 1379
80. Hu, A. T. and Sinke, G. C. (1969). *J. Chem. Thermodynamics,* **1,** 507
81. Carpenter, G. A., Zimmer, M. F., Baroody, E. E. and Robb, R. A. (1970). *J. Chem. Eng. Data,* **15,** 553
82. Anthoney, M. E., Finch, A. and Gardner, P. J. (1970). *J. Chem. Thermodynamics,* **2,** 697
83. Devore, J. A. and O'Neal, H. E. (1969). *J. Phys. Chem.,* **73,** 2644
84. Carson, A. S. and Skinner, H. A. (1949). *J. Chem. Soc.,* 936
85. Skinner, H. A. (1964) in *Advances in Organometallic Chemistry, Vol. 2.* (New York: Academic Press Inc.)
86. Good, W. D., Scott, D. W., Lacina, J. L. and McCullough, J. P. (1959). *J. Phys. Chem.,* **63,** 1139
87. Good, W. D., Scott, D. W. and Waddington, G. (1956). *J. Phys. Chem.,* **60,** 1090
88. Good, W. D., Lacina, J. L., De Prater, B. L. and McCullough, J. P. (1964). *J. Phys. Chem.,* **68,** 579
89. Good, W. D., Fairbrother, D. M. and Waddington, G. (1958). *J. Phys. Chem.,* **62,** 853
90. Potzinger, P. and Lampe, F. W. (1970). *J. Phys. Chem.,* **74,** 719
91. Gunn, S. R. and Green, L. G. (1961). *J. Phys. Chem.,* **65,** 779
92. Adams, G. P., Carson, A. S. and Laye, P. G. (1969). *Trans. Faraday Soc.,* **65,** 113
93. Bills, J. L. and Cotton, F. A. (1964). *J. Phys. Chem.,* **68,** 806
94. Carson, A. S., Carson, E. M., Laye, P. G., Spencer, O. A. and Steele, W. V. (1970). *Trans. Faraday Soc.,* **66,** 2459
95. Gross, P., Hayman, C. and Bingham, J. T. (1966). *Trans. Faraday Soc.,* **62,** 2338
96. Shaulov, Yu. Kh., Fedorov, A. K. and Genchel, V. G. (1969). *Russ. J. Phys. Chem.,* **43,** 744
97. Carson, A. S., Laye, P. G., Spencer, J. A. and Steele, W. V. (1970). *J. Chem. Thermodynamics,* **2,** 659
98. Adams, G. P., Carson. A. S. and Laye, P. G. (1969). *J. Chem. Thermodynamics,* **1,** 393
99. Pope, A. E. and Skinner, H. A. (1964). *Trans. Faraday Soc.,* **60,** 1402
100. Lautsch, W. F., Trober, A., Zimmer, W., Mehner, L., Linck, W., Lehmann, H.-M., Brandenberger, H., Korner, H., Metschker, H-J., Wagner, K. and Kaden, R. (1963). *Z. Chem.,* **3,** 415
101. Tel'noi, V. I. and Rabinovich, I. B. (1966). *Russ. J. Phys. Chem.,* **40,** 1556
102. Keiser, D. and Kana'an, A. S. (1969). *J. Phys. Chem.,* **73,** 4264
103. Carson, A. S., Cooper, R. and Stranks, D. R. (1962). *Trans. Faraday Soc.,* **58,** 2125
104. Benson, S. W. (1968). *Thermochemical Kinetics.* (New York: John Wiley and Sons)
105. Benson, S. W. and Bose, A. N. (1963). *J. Amer. Chem. Soc.,* **85,** 1385
106. Golden, D. M., Egger, K. W. and Benson, S. W. (1964). *J. Amer. Chem. Soc.,* **86,** 5411
107. Prosen, E. J., Maron, F. W. and Rossini, F. D. (1951). *J. Res. Nat. Bur. Stand.,* **46,** 106
108. Kistiakowsky, G. B., Ruhoff, J. R., Smith, H. A. and Vaughan, W. E. (1935). *J. Amer. Chem. Soc.,* **57,** 876
109. Egger, K. W. and Benson, S. W. (1966). *J. Amer. Chem. Soc.,* **88,** 236
110. Egger, K. W. and Benson, S. W. (1965). *J. Amer. Chem. Soc.,* **87,** 3311
111. Fraser, F. M. and Prosen, E. J. (1955). *J. Res. Nat. Bur. Stand.,* **54,** 143
112. Furuyama, S., Golden, D. M. and Benson, S. W. (1970). *J. Chem. Thermodynamics,* **2,** 161
113. Dolliver, M. A., Gresham, T. L., Kistiakowsky, G. B. and Vaughan, W. E. (1937). *J. Amer. Chem. Soc.,* **59,** 831
114. Kistiakowsky, G. B., Ruhoff, J. R., Smith, H. A. and Vaughan. W. E. (1936). *J. Amer. Chem. Soc.,* **58,** 146
115. Rodgers, A. S., Golden, D. M. and Benson, S. W. (1966). *J. Amer. Chem. Soc.,* **88,** 3194
116. Walsh, R., Golden, D. M. and Benson, S. W. (1966). *J. Amer. Chem. Soc.,* **88,** 650
117. Gellner, O. H. and Skinner, H. A. (1949). *J. Chem. Soc.,* 1145

118. Ashcroft, S. J., Carson, A. S. and Pedley, J. P. (1963). *Trans. Faraday Soc.*, **59**, 2713
119. King, K. D., Golden, D. M. and Benson, S. W. (1971). *J. Chem. Thermodynamics*, **3**, 129
120. Solley, R. K., Golden, D. M. and Benson, S. W. (1970). *J. Amer. Chem. Soc.*, **92**, 4653
121. Solley, R. K. and Benson, S. W. (1971). *J. Chem. Thermodynamics*, **3**, 203
122. Norén, I. and Sunner, S. (1970). *J. Chem. Thermodynamics*, **2**, 597
123. Kabo, G. Ya. and Andreevski, D. N. (1963). *Neftekhimiya*, **3**, 764
124. Howlett, K. E. (1955). *J. Chem. Soc.*, 1784
125. Davies, J. V., Lacher, J. R. and Park, J. D. (1965). *Trans. Faraday Soc.*, **61**, 2413
126. Lord, A. and Pritchard, H. O. (1970). *J. Chem. Thermodynamics*, **2**, 187
127. Bodenstein, M. and Plaut, H. (1924). *Z. Phys. Chem.*, **110**, 399
128. Tachoire, H. (1967). *Coll. Int. Centre Natl. Rech. Sci. (Paris)*, **156**, 329
129. Lord, A. and Pritchard, H. O. (1969). *J. Chem. Thermodynamics*, **1**, 495
130. CODATA, (1971). *J. Chem. Thermodynamics*, **3**, 1
131. Feder, H. M., Hubbard, W. N., Wise, S. S. and Margrave, J. L. (1963). *J. Phys. Chem.*, **67**, 1148
132. Head, A. J. and Lewis, G. B. (1970). *J. Chem. Thermodynamics*, **2**, 701
133. Birley, G. I. and Skinner, H. A. (1968). *Trans. Faraday Soc.*, **64**, 3232
134. Nuttall, R. L., Wise, S. and Hubbard, W. N. (1961). *Rev. Sci. Instrum.*, **32**, 1402
135. Churney, K. L. and Armstrong, G. T. (1969). *J. Res. Nat. Bur. Stand.*, **73A**, 281
136. Rudzitis, E., Feder, H. M. and Hubbard, W. N. (1964). *J. Phys. Chem.*, **68**, 2978
137. Rudzitis, E., Terry, R., Feder, H. M. and Hubbard, W. N. (1964). *J. Phys. Chem.*, **68**, 617
138. Rudzitis, E., Feder, H. M. and Hubbard, W. N. (1963). *J. Phys. Chem.*, **67**, 2388
139. Johnson, G. K., Feder, H. M. and Hubbard, W. N. (1966). *J. Phys. Chem.*, **70**, 1
140. Rudzitis, E., Feder, H. M. and Hubbard, W. N. (1967). *Inorg. Chem.*, **6**, 1716
141. Rudzitis, E., Feder, H. M. and Hubbard, W. N. (1965). *J. Phys. Chem.*, **69**, 2305
142. Wise, S. S., Margrave, J. L., Feder, H. M. and Hubbard, W. N. (1963). *J. Phys. Chem.*, **67**, 815
143. O'Hare, P. A. G., Johnson, J., Klamecki, B., Mulvihill, M. and Hubbard, W. N. (1969). *J. Chem. Thermodynamics*, **1**, 177
144. Adams, G. P., Charlu, V. and Margrave, J. L. (1970). *J. Chem. Eng. Data*, **15**, 42
145. Adams, G. P., Margrave, J. L. and Wilson, P. W. (1970). *J. Chem. Thermodynamics*, **2**, 741
146. Greenberg, E., Settle, J. L. and Hubbard, W. N. (1962). *J. Phys. Chem.*, **66**, 1345
147. Greenberg, E., Settle, J. L., Feder, H. M. and Hubbard, W. N. (1961). *J. Phys. Chem.*, **65**, 1168
148. O'Hare, P. A. G. and Hubbard, W. N. (1966). *Trans. Faraday Soc.*, **62**, 2709
149. Rudzitis, E., Van Deventer, E. H. and Hubbard, W. N. (1970). *J. Chem. Thermodynamics*, **2**, 221
150. O'Hare, P. A. G. and Hubbard, W. N. (1965). *J. Phys. Chem.*, **69**, 4358
151. Greenberg, E., Natke, C. A. and Hubbard, W. N. (1965). *J. Phys. Chem.*, **69**, 2089
152. O'Hare, P. A. G., Settle, J. K. and Hubbard, W. N. (1966). *Trans. Faraday Soc.*, **62**, 558
153. Settle, J. K., Feder, H. M. and Hubbard, W. N. (1961). *J. Phys. Chem.*, **65**, 1337
154. O'Hare, P. A. G. and Hubbard, W. N. (1966). *J. Phys. Chem.*, **70**, 3358
155. Rudzitis, E., Van Deventer, E. H. and Hubbard, W. N. (1967). *J. Chem. Eng. Data*, **12**, 133
156. Porte, H. A., Greenberg, E. and Hubbard, W. N. (1965). *J. Phys. Chem.*, **69**, 2308
157. Settle, J. K., Feder, H. M. and Hubbard, W. N. (1963). *J. Phys. Chem.*, **67**, 1892
158. Good, W. D. (1962). *J. Phys. Chem.*, **66**, 380
159. Humphrey, G. L. and King, E. G. (1952). *J. Amer. Chem. Soc.*, **74**, 2041
160. Johnson, G. K. and Hubbard, W. N. (1969). *J. Chem. Thermodynamics*, **1**, 459
161. Good, W. D. and Månsson, M. (1966). *J. Phys. Chem.*, **70**, 97
162. Greenberg, E., Natke, C. A. and Hubbard, W. N. (1970). *J. Chem. Thermodynamics*, **2**, 193
163. Wise, S. S., Margrave, J. L., Feder, H. M. and Hubbard, W. N. (1966). *J. Phys. Chem.*, **70**, 7
164. Gross, P., Hayman, C. and Stuart, M. C. (1969). *Trans. Faraday Soc.*, **65**, 2628
165. Adams, G. P., Margrave, J. L. and Wilson, P. W. (1970). *J. Chem. Thermodynamics*, **2**, 591
166. O'Hare, P. A. G., Benn, E., Cheng, F. Yu and Kuzmycz, G. (1970). *J. Chem. Thermodynamics*, **2**, 797

167. O'Hare, P. A. G., Hubbard, W. N., Glemser, O. and Wegener, J. (1970). *J. Chem. Thermodynamics,* **2,** 71
168. Glemser, O., Müller, A., Bohler, D. and Krebs, B. (1968). *Z. Anorg. Allgem. Chem.,* **357,** 184
169. Huber, E. J. and Holley, C. E. (1952). *J. Amer. Chem. Soc.,* **74,** 5530
170. Fitzgibbon, G. E., Pavone, D. and Holley, C. E. (1968). *J. Chem. Eng. Data,* **13,** 547
171. Huber, E. J., Head, E. L. and Holley, C. E. (1964). *J. Phys. Chem.,* **68,** 3040
172. Kornilov, A. N., Ushakova, I. M. and Skuratov, S. M. (1968). *Russ. J. Phys. Chem.,* **42,** 817
173. Huber, E. J. and Holley, C. E. (1968). *J. Chem. Eng. Data,* **13,** 252
174. Baker, F. B. and Holley, C. E. (1968). *J. Chem. Eng. Data,* **13,** 405
175. Baker, F. B., Huber, E. J. and Holley, C. E. (1971). *J. Chem. Thermodynamics,* **3,** 77
176. Huber, E. J. and Holley, C. E. (1970). *J. Chem. Thermodynamics,* **2,** 896
177. Huber, E. J., Fitzgibbon, G. E. and Holley, C. E. (1971). *J. Chem. Thermodynamics,* **3,** 643
178. Huber, E. J. and Holley, C. E. (1969). *J. Chem. Thermodynamics,* **1,** 267
179. Huber, E. J. and Holley, C. E. (1968). *J. Chem. Eng. Data,* **13,** 545
180. Johnson, G. K., Van Deventer, E. H., Kruger, O. L. and Hubbard, W. N. (1969). *J. Chem. Thermodynamics,* **1,** 89
181. Huber, E. J., Holley, C. E. and Krikorian, N. H. (1968). *J. Chem. Eng. Data,* **13,** 253
182. Baker, F. B., Storms, E. K. and Holley, C. E. (1969). *J. Chem. Eng. Data,* **14,** 244
183. Huber, E. J., Holley, C. E. and Witterman, W. G. (1969). *J. Chem. Thermodynamics,* **1,** 579
184. Johnson, G. K., Van Deventer, E. H., Kruger, O. L. and Hubbard, W. N. (1970). *J. Chem. Thermodynamics,* **2,** 617
185. Hajiev, S. H. (1970). *J. Chem. Thermodynamics,* **2,** 765
186. Nuñez, L., Pilcher, G. and Skinner, H. A. (1969). *J. Chem. Thermodynamics,* **1,** 31
187. Espada, L., Pilcher, G. and Skinner, H. A. (1970). *J. Chem. Thermodynamics,* **2,** 647
188. King, R. C. and Armstrong, G. T. (1968). *J. Res. Nat. Bur. Stand.,* **72A,** 113
189. King, R. C. and Armstrong, G. T. (1970). *J. Res. Nat. Bur. Stand.,* **74A,** 769
190. Armstrong, G. T. (1971). *Colloques Internationeaux du Centre National de la Recherche Scientifique Thermochemie, Marseilles*
191. Bisbee, W. R., Hamilton, J. V., Gerhauser, J. M. and Rushworth, R. (1968). *J. Chem. Eng. Data,* **13,** 382
192. Hill, J. O. and Irving, R. J. (1968). *J. Chem. Soc. A,* 1052, 3116
193. Irving, R. J. and Walter, G. W. (1969). *J. Chem. Soc. A,* 2690
194. Finch, A., Gardner, P. J. and Sen Gupta, K. K. (1969). *J. Chem. Soc. A,* 2958
195. Gross, P., Hayman, C. and Mwroka, S. (1969). *Trans. Faraday Soc.,* **65,** 2856
196. Gross, P. and Hayman, C. (1970). *Trans. Faraday Soc.,* **66,** 30
197. Pepekin, V. I., Lebedev, A. Yu. and Apin, A. Ya. (1969). *Russ. J. Phys. Chem.,* **43,** 869
198. O'Hare, P. A. G., Johnson, G. K. and Appelman, E. H. (1970). *Inorg. Chem.,* **9,** 332
199. Johns, J. W. C. and Barrow, R. F. (1959). *Proc. Roy. Soc. London,* **A251,** 504
200. Doescher, R. N. (1952). *J. Chem. Phys.,* **20,** 330
201. Wise, H. (1954). *J. Phys. Chem.,* **58,** 389
202. Stamper, J. G. and Barrow, R. F. (1958). *Trans. Faraday Soc.,* **54,** 1592
203. Dibeler, V. H., Walker, J. A. and McCulloh, K. E. (1969). *J. Chem. Phys.,* **51,** 4230
204. Dibeler, V. H., Walker, J. A. and McCulloh, K. E. (1970). *J. Chem. Phys.,* **53,** 4414
205. Chupka, W. A. (1968). *J. Chem. Phys.,* **48,** 2337
206. Chupka, W. A. and Lifshitz, C. (1968). *J. Chem. Phys.,* **48,** 1109
207. Kerr, J. A. (1966). *Chem. Rev.,* **66,** 465
208. Rodgers, A. S., Golden, D. M. and Benson, S. W. (1967). *J. Amer. Chem. Soc.,* **89,** 4578
209. Chamberlain, G. A. and Whittle, E. (1971). *Trans. Faraday Soc.,* **67,** 2077

4
Critically Evaluated Tables of Thermodynamic Data

B. J. ZWOLINSKI and JING CHAO
Thermodynamics Research Centre, Texas A & M University

4.1	INTRODUCTION	94
4.2	EARLY HISTORY AND PERSPECTIVES	95
	4.2.1 *Early history*	95
	4.2.2 *The U.S. National Standard Reference Data System*	95
	4.2.3 *International co-operation*	96
4.3	SOURCES OF THERMODYNAMIC DATA	97
	4.3.1 *Early compilations*	97
	4.3.1.1 *Comprehensive compilations*	97
	4.3.1.2 *Work on special properties*	97
	4.3.2 *Recent continuing critical data compilations*	98
	4.3.2.1 *Selected values of properties of hydrocarbons and related compounds*	98
	4.3.2.2 *Selected values of properties of chemical compounds*	99
	4.3.2.3 *Selected values for the thermodynamic properties of metals and alloys*	100
	4.3.2.4 *Thermodynamic properties of chemical substances*	101
	4.3.2.5 *Phase diagrams for ceramists*	102
	4.3.2.6 *Constitution of binary alloys*	103
	4.3.2.7 *JANAF thermochemical tables*	103
	4.3.2.8 *Chemical thermodynamics in non-ferrous metallurgy*	105
	4.3.2.9 *Thermochemistry for steelmaking*	105
	4.3.2.10 *Thermodynamic functions of gases*	106
	4.3.2.11 *Selected values of chemical thermodynamic properties*	106
	4.3.2.12 *Thermodynamic constants of substances*	107
	4.3.2.13 *High-temperature behaviour of inorganic salts*	109
	4.3.2.14 *IUPAC thermodynamic tables project*	110

4.3.3 *Selected recent books on thermodynamic properties of chemical substances* 110
4.3.3.1 *General* 111
4.3.3.2 *Elements* 111
4.3.3.3 *Inorganic substances* 112
4.3.3.4 *Organic substances* 113
4.3.3.5 *Organometallic compounds* 114
4.3.4 *Other pertinent sources of information* 114

4.4 CRITICAL EVALUATION OF THERMOCHEMICAL DATA 114
4.4.1 *Thermochemical input data from an ideal experimental programme* 114
4.4.2 *Imperfections of thermochemical input data from the open literature* 115
4.4.3 *Evaluation and selection of best values* 115

4.5 USE OF ELECTRONIC COMPUTER FOR HANDLING THERMODYNAMIC DATA 116

ACKNOWLEDGEMENT 118

4.1 INTRODUCTION

The critical evaluation and compilation of thermodynamic data are rapidly becoming one of the more important activities in science and technology today. In this age of knowledge explosion and information pollution, whenever an investigator needs to know the present status of his own specialised field, he usually has to face two problems; namely, (a) he will be unable to search and digest all of the literature relevant to his topic of interest because of its magnitude, and (b) he most probably will not possess the expertise nor have the necessary facilities to reduce the enormous amount of literature in a given field into manageable, understandable, and useful formats.

About 150 years ago, scientists had only about 50 scientific journals; 100 years ago, about 500; 50 years ago, about 5000; and today, of the order of 50 000 journals (Rossini[1]). More than one million scientific and technical papers appear in publication each year in journals, bulletins, and related documents. It is thus impractical and much too costly to have each user develop his own individual set of tables containing selected numerical best values of physical and chemical properties. An effective solution to this problem of providing comprehensive critical tables is to have these tables generated by groups of competent, experienced experts or compilers who are adequately supported on a continuing basis. Once such critically evaluated tables of standard reference data are available, then scientists in research laboratories, whether in education, governmental, or industrial organisations, will be able to devote maximum efforts to research and development. It is the purpose of this chapter to acquaint the industrial and academic chemical scientists and engineers with the important international sources of compiled physico-chemical and thermodynamic data on properties of pure substances and mixtures that are indispensable in both basic and applied research activities.

4.2 EARLY HISTORY AND PERSPECTIVES

4.2.1 Early history

The earliest compilations of thermodynamic data did no more than extract reported property values from the open literature and present or compile them in tabular form for the users. With the publication of the International Critical Tables in the late twenties, the concept of critical evaluation of data and critical tables emerged. The first attempts at compiling numerical tables were by individual or small groups of workers operating privately or with the support of government or industrial organisations. Between 1930 and 1957, a limited number of excellent numerical data compilation centres were organised in the U.S. with coverage of restricted areas of science. On the international level, the *Landolt–Börnstein Tabellen* in Germany and the *Tables Annuelles de Constantes et Données Numeriques* in France continued to function; however, with time, these organisations found that it was more and more difficult to provide complete and effective coverage of their respective areas. After World War II, the information data situation became more critical.

In 1955, the U.S. National Academy of Sciences–National Research Council reviewed the entire technical and data situation in the U.S.A. and decided that a new approach was needed to provide science and technology in the U.S. with continuing up-to-date tables of critically evaluated numerical data. In 1957, the Office of Critical Tables was established by the National Academy of Sciences–National Research Council in the U.S. with the following responsibilities: to survey the needs of critical data, to stimulate and encourage existing data projects, and to assist in the establishment of new needed critical-data-compilation projects (Waddington[2]). From time to time numerous compilations have been produced, largely in response to an urgently-felt need of some part of the technical community, such as the petroleum and chemical industries, the military or the U.S. Federal Government in general. There was little or no coordination or standardisation of format or quality, and in some technical areas there were extensive duplications. Furthermore, the rate of appearance of new data in the world's literature in most fields was, and still is, far greater than the rate at which the data are extracted, evaluated, and compiled for general use. Recognition of these problems and of the financial costs to any one nation failing to cope with this problem are some of the reasons why the U.S. government decided to establish the National Standard Reference Data System in the last decade (National Bureau of Standards[3]).

4.2.2 The U.S. National Standard Reference Data System

The National Standard Reference Data System (NSRDS) was established in 1963 at the National Bureau of Standards (NBS) in Washington, D.C. by action of the Federal Council for Science and Technology and the President's

Office of Science and Technology (Brady and Wallenstein[4]). In 1968, this national policy was affirmed by the U.S. Congress which provided formal authority for funding and implementing the programme. The NSRDS is recognised as having responsibility for the total spectrum of data activities in the physical and engineering sciences within the U.S. Federal Government, as well as for all related private organisations that wish to participate. This office was initially directed by Dr. Edward L. Brady and is now under the direction of Dr. David R. Lide, Jr., with close liaison with the Office of Critical Tables, NRC, and more recently since 1969 by its counterpart, the Numerical Data Advisory Board of the U.S. National Research Council.

The main responsibilities of NSRDS (Brady[5]) are: to promote and co-ordinate critical data compilation work under the auspices of all government agencies, establish standards of quality for all products of the system, and establish standards of methodology and other functions to ensure the compatibility of all units of the NSRDS. Four types of activities are considered in the programme (Brady and Wallenstein[4]); data collection and evaluation, preparation of critical reviews, computation of useful functions, and experimental measurements. The data evaluation programme covers seven broad categories of properties: (a) thermodynamic and transport properties, (b) atomic and molecular data, (c) chemical kinetics, (d) solid-state data, (e) nuclear data, (f) colloid and surface properties, and (g) mechanical properties. As of June 1970 (Lide[6]) the NSRDS–NBS Series stands at 33 titles, of which nine were related to thermodynamic properties of chemical substances, 40–50 to other data compilations, bibliographies, and description of data handling techniques. Current figures show that about 70 000 copies of documents have been distributed. More information about NSRDS is available[6–9].

4.2.3 International co-operation

It became increasingly clear that the problem of data for science and technology is international. No single country has the resources of money and manpower to support the comprehensive effort needed in this area. To some extent, the evaluation and compilation of critical data on the properties of substances have been a joint activity of the world's scientists for many years[10]. The International Critical Tables, produced between 1926 and 1930, contained contributions from 408 scientists in 18 countries (Waddington[2]). The tables of Landolt–Börnstein, originally German in origin, now contain contributions by scientists from many countries. The establishment of NSRDS in the U.S. in 1963 has stimulated additional interest among scientists in other countries to the establishment of multilateral international programmes incorporating activities from all countries wishing to participate.

In January 1966, the general assembly of the International Council of Scientific Unions (ICSU) approved the appointment of a Committee on Data for Science and Technology (CODATA) to supply the necessary leadership in coordination of an international effort to improve compilations of critically selected numerical and other quantitative scientific data. The CODATA is the first world-wide organisation established officially for this special area. In its first 2 years of operation, 1966 to 1968, the CODATA office was temporarily located in Washington, D.C., U.S.A., under the direction of Dr. Guy Waddington. On July 1, 1968, the Central Office was moved to Frankfurt, Germany, with Dr. Christoph Schafer as the new Director. As of 1969, the CODATA Committee consisted of ten Union Members, seven National Members, one co-opted Member, and six Liaison Representatives from other organised international groups. In December 1970, the number of National Members increased to twelve[11]. CODATA issued a comprehensive compendium in 1969[12] which included a world-wide survey and analysis of those centres that evaluate, compile, and publish numerical data for science and technology. The CODATA constitution allows for the organisation of special purpose task forces. The Task Group on Key Values for Thermodynamics has completed and published[13] a first tentative set of key values which is available on request and free of charge from the CODATA Central Office.

4.3 SOURCES OF THERMODYNAMIC DATA

Many older sources of thermodynamic data are still considered useful today due to lack of new measurements for many substances. The first part of this section presents a brief review of these earlier but still important compilations. In turn, descriptions are given of recent continuing critical data compilations followed by brief notices to recent textbooks and monographs on thermodynamic properties of chemical compounds. Other miscellaneous publications are given at the end of this section.

4.3.1 Early compilations

4.3.1.1 Comprehensive compilations

(a) *Landolt–Bornstein tables*—First edition, 281 pp., 1883; fifth edition, 7457 pp., 1923–36; sixth edition, 8492 pp., 1950–59; new series 'Numerical Data and Functional Relationships in Science and Technology', 1961–.

(b) *Annual tables of constants and numerical data*—Ten volumes, 1910–30; 40 instalments, 1936–45; renamed as 'Tables of Selected Constants', 1947; 15 monographs on properties of organic and inorganic substances, 1969.

(c) *International critical tables*—Seven volumes, 1926–30; index, 1933.

(d) *Data on theoretical metallurgy*—U.S. Bureau of Mines Bulletins 542, 584, 592 and 601, 1930–64.

(e) *Selected values of chemical thermodynamic properties*—U.S. National Bureau of Standards Circular 500 (Rossini *et al.*[31]). Series I and II, 1268 pp., 1952.

4.3.1.2 Work on special properties

(a) *Thermochemische untersuchungen*—J. Thomsen, Vols. I–IV, 1882–86.

(b) *Thermochimie*—M. Berthelot, two volumes, 1897.

(c) *Thermochemia zwiazkow organicznych* – W. Swietoslawski, 1918.

(d) *Thermodynamics and the free energy of chemical substances* – G. N. Lewis and M. Randall, 1923; revised as *Thermodynamics* by K. S. Pitzer and L. Brewer, 1961.

(e) *Heats of combustion of organic compounds* – M. S. Kharasch, *J. Res. Natl. Bur. Std.* (1929). **2,** 359–429.

(f) *The free energies of some organic compounds* – Parks and Huffman, ACS Monograph No. 60, 1932.

(g) *Thermochemistry of the chemical substances* – F. R. Bichowsky and F. D. Rossini, 1936.

4.3.2 Recent continuing critical data compilations

4.3.2.1 Selected values of properties of hydrocarbons and related compounds, American Petroleum Institute Research Project 44

This project was established in 1942 at the U. S. National Bureau of Standards in Washington, D.C., with the sponsorship and financial support of the American Petroleum Institute as the American Petroleum Institute Research Project 44 (API RP44), and was directed by F. D. Rossini from its inception until 1960. In 1950, the project was moved to Carnegie Institute of Technology (now Carnegie–Mellon University), Pittsburgh, Pennsylvania; and in 1961, it was again re-located to Texas A&M University, College Station, Texas, with Professor Bruno J. Zwolinski as the new director. Since 1966, the project has been on a self-supporting basis with an API-appointed Advisory Committee providing scientific and technical guidance.

The API RP44 serves as the central agency of the API for the collection, evaluation, and publication of numerical data on the physical, thermodynamic, and spectral properties of all classes of hydrocarbons, together with a limited number of classes of sulphur and nitrogen derivatives of hydrocarbons present in petroleum. The critically selected numerical values are listed in tables which are available to the public as 'Selected Values of Properties of Hydrocarbons and Related Compounds[14]. These tables are issued as supplementary volumes twice yearly under dates of April 30 and October 31 of each year. The loose-leaf data sheet format provides a ready mechanism for revision and updating of the old data for publication of new and expanded data. In addition to the uninterrupted preparation and distribution of loose-leaf data sheets, two bound volumes were published in 1947 and 1953. One hard-cover volume, known as the 'green' volume, was issued at the National Bureau of Standards, Washington, D.C., in 1947 as NBS Circular 461 (Rossini *et al.*[15]). The 1953 revised hard-cover volume, known as the 'red' volume, was published at Carnegia Institute of Technology (Rossini *et al.*[16]). As of December 31, 1971, the total number of valid data sheets in the current publication is 2644, consisting of seven special API binder volumes. The Selected Spectral Catalogues (i.r., u.v., Raman, mass, and n.m.r.) consist of 8242 valid sheets in 22 volumes. For the benefit of individual users including students, some property tables will be collected and published as separate data handbooks with each handbook covering one or a few related properties. The first in the new series of API 44–TRC Publications in Science and Engineering, entitled 'Vapor pressures and heats of vaporisation of hydrocarbons and related compounds'[17], has now been published, and a second handbook entitled 'Chemical thermodynamic data of hydrocarbons and related compounds'[18] is in the press.

The major source of data is the open literature, together with contributions over the years from associated API Research Project 6, 35, 42, 45. 48, 52, 56, 58A, 58B, 60, and 62. Other sources of input are the unpublished data from contributing industrial, academic, and government research laboratories. Property values unavailable experimentally both inside and outside of the reported experimental and molecular structural ranges are calculated by tested precise correlation methods. Such values are, in many cases, even more reliable than the experimental ones. The numerical tables are arranged alphabetically according to property codes. Other details are described in the 'Comprehensive index of API 44–TRC selected data on thermodynamics and spectroscopy'[19] as also in the revised introduction to the API RP44 Tables.

The references used for evaluation of the best values for the compounds listed in the given property table are presented in the 'Specific references tables', which give the authors' last names followed by a superscript number which locates the place of that reference in the general list of references. In this general list, the complete literature sources for each specific reference are presented. The general list of references is arranged alphabetically by the last name of the first author. This is one of the few numerical data projects on organic compounds providing selected values of high quality data and operating continuously for 30 years since 1942.

4.3.2.2 Selected values of properties of chemical compounds

The Manufacturing Chemists' Association (MCA) established this critical data project in 1955 at Carnegie Institute of Technology. The aim of this project was to provide reliable physical and thermodynamic properties on all pure inorganic and organic substances outside of the hydrocarbon area. This project was also re-located with the API 44 Project in 1961 and incorporated into the Thermodynamics Research Center (TRC) at Texas A&M University under the direction of Professor Bruno J. Zwolinski. The project was renamed the Thermodynamics Research Center Data Project (TRCDP) on July 1, 1966, and it has been independently operated by the TRC at Texas A&M University on a self-supporting basis since July 1, 1964.

The physical and thermodynamic properties covered in the TRCDP tables, as also the method of evaluation and the format for presentation of the best values of the selected properties, are exactly the same as those described for the API 44 tables mentioned previously. As of December 31, 1970, there are [20] 1109 valid data sheets in four volumes on physical and thermodynamic properties and 2414 valid sheets on spectral data in five categories (i.r., u.v., Raman, mass and n.m.r.) in nine additional volumes. The inorganic compounds included in the TRCDP Tables are: the non-metallic elements and rare gases, the non-metallic oxides, halides, sulphides, nitrides, and phosphorus compounds; whereas the organic compounds are primarily oxygen, halogen, and nitrogen derivatives. A much wider variety of classes of organic (non-hydrocarbon) compounds are covered in the TRCDP Spectral Catalogs.

In order to facilitate retrieval of data for the physical, thermodynamic, and spectral data on chemical compounds contained in both API 44 and TRCDP critical tables, a special book[9] was issued in 1968. The main section of this comprehensive index is divided into two parts, namely, the 'compound name index' and the 'formula index', each of which lists in a line entry all the compounds and their physical, thermodynamic, and spectral properties included in the published valid tables as of April 30, 1968, for API 44 tables and June 30, 1968, for TRCDP tables. This index is updated every 3–4 years.

The supplementary volume dated December 31, 1970, contains a revised Introduction which includes a brief description of the 1969 atomic weights, 1968 International Practical Temperature Scale, SI Units, and the new 1969 fundamental constants. This work is one of the important sources of information on the physical and thermodynamic properties of non-hydrocarbon organic compounds and certain physical and thermodynamic properties of inorganic compounds which complements the NBS Circular 500 (see Section 4.3.1.1).

4.3.2.3 Selected values for the thermodynamic properties of metals and alloys

This compilation work was started in 1955 at the University of California, Berkeley, with Professor R. R. Hultgren as the project director. The published thermodynamic data on all metallic elements and binary alloy systems were collected and evaluated. Some values were obtained by extrapolation or calculation, and the results were issued in loose-leaf form. In 1963, the evaluated data were published in bound book form[21], which included 63 elements and 168 alloy systems. The fundamental constants listed by Pitzer and Brewer[22] were used for the calculations. The atomic weights were based on O = 16.000. Each element and alloy system was treated separately. Results were presented as tables and graphs. References of numerical data employed for evaluation were given. A systems index by formula and an author index for data references are provided. New and revised data sheets are issued at intervals as supplements to the previous book. The 1963 bound text is now being revised and updated for publication in the 1971–72 period by the NBS Office of Standard Reference Data.

The thermodynamic data for the elements include: heat capacity, enthalpy, and Gibbs energy which are tabulated at selected temperature intervals. The reported values are relative to the standard states. Also reported are the following properties: enthalpy and entropy of phase change; vapour pressure; Gibbs energy and enthalpy of vaporisation. For alloys, the properties covered are: enthalpy, Gibbs energy, entropy, excess Gibbs energy, and excess entropy of formation. Values for activities and activity coefficients and partial molar thermodynamic quantities over a range of composition are given. Phase diagrams are included for most of the alloy systems.

This project is sponsored by the U.S. Atomic Energy Commission, with assistance from the American Iron and Steel Institute and the International Copper Research Association. It is also associated with the National Standard Reference Data Program of the U.S. National Bureau of Standards. The thermodynamic properties of several listed metallic elements are also reported in NBS Circular 500 (Section 4.3.1.1), Thermodynamic Properties of Chemical Substances (Section 4.3.2.4), JANAF Thermochemical Tables (Section 4.3.2.7), and NBS Selected Values of Chemical Thermodynamic Properties (Section 4.3.2.11). This compilation is valuable especially for metallurgists and solid state scientists.

4.3.2.4 Thermodynamic properties of chemical substances

This project is sponsored by the U.S.S.R. Academy of Sciences. The compilation work has been continued for over 10 years in the Institute of Mineral Fuels of the U.S.S.R. Academy of Sciences and the U.S.S.R. State Institute of Applied Chemistry with V. P. Glushko as chief editor and L. V. Gurvich as director. The results were published in 1956 as a book entitled 'Thermodynamic properties of combustion product components'. The second edition appeared in 1962 in two volumes and was renamed as 'Thermodynamic properties of chemical substances', which contains the thermodynamic properties of elements and simple inorganic compounds in their standard reference states in different temperature ranges.

In Volume I, the methods of calculation and critical analysis of literature numerical data until 1960 are presented. The molecular constants, heats of formation and phase transition, specific heats, dissociation energies, limits of uncertainty of the evaluated values, and other constants related to the calculation of the tabulated thermodynamic properties listed in Volume II are also given. The tabulated data were mainly taken from the literature. Some unpublished data were provided by a number of research institutions within the Soviet Union. Estimates were made whenever experimental data were lacking.

The appendix contains: atomic weight, isotopic composition and nuclear spin of 32 chemical elements; fundamental constants and conversion factors employed in computation; 60 formulae for calculating the moments of inertia for 57 typical molecules having distinctive structures; formulae for calculation of vibrational frequencies for 10 different types of molecules; methods of calculating thermodynamic properties for real gases; and critical constants for simple inorganic compounds. All symbols and constants used in the text are explained and defined. A bibliography (4392 references), an author index with equivalent English names, and a formula index in Volumes I and II are included.

The thermodynamic properties for 335 gases in the ideal gas state, 44 liquids, and 45 crystals are presented in 380 tables and included in Volume II. These chemical compounds are formed from 31 chemical elements(O, H, He, Ne, Ar, Kr, Xe, F, Cl, Br, I, S, N, P, C, Si, Pb, Hg, Zr, B, Al, Be, Mg, Ca, Sr,

Ba, Li, Na, K, Rb, and Cs), isotopes of hydrogen (H, D, T), and electron gas (e^-). There are 55 tables which contain the thermodynamic properties for monatomic and ion gases; 110 tables for diatomic molecules, and 170 tables for polyatomic molecules. The reported thermodynamic properties for gaseous species in the ideal gas state at one atm are: Gibbs energy function $(G^O - H_0^O)/T$ or Φ_T^*, entropy, enthalpy, enthalpy of formation (ΔHf^O) or I_T^O, and equilibrium constant K_p and log K_p, where K_p is the constant of dissociation or ionisation. For compounds in the crystal and liquid states, values for heat capacity and vapour pressure are also given from 293.15 K to appropriate temperatures. The thermodynamic properties for the following 22 gases: e^-, H, H^+, H^-, H_2, O, O^+, O_2, O_2^+, OH, OH^+, H_2O, N, N^+, N_2, N_2^+, NO, NO^+, C, C^+, CO, and CO^+ are reported at temperatures (K): 293.15, 298.15, 400 to 6000 (at 100 K intervals), 6000 to 10 000 (at 200 K intervals), and 10 000 to 20 000 (at 500 K intervals). Those of the other 299 less stable gaseous species are given at the temperatures (K): 293.15, 298.15, 400 to 6000 (at 100 K intervals); and those of the 14 least stable gases are given at 293.15, 298.15, and 400 to 4000 K (at 100 K intervals).

This is an excellent source of information for critically evaluated thermodynamic properties of elements and simple inorganic compounds. There are enthalpies but no heat capacity data reported for the gaseous species. The first supplement to the main work was published in 1970. The original publication is in Russian; however, an un-edited rough draft English translation of Volume I is available from Clearinghouse, U.S.A., as AD 659660, AD 659659, and AD 659679 (March 1967). The thermodynamic properties of chemical elements and most of their compounds included in this compilation are also covered by NBS Circular 500 (Section 4.3.1.1), and JANAF Thermochemical Tables (Section 4.3.2.7). Those for the metallic elements and several non-metallic elements and their chemical compounds are also contained in Hultgren's Selected Values for the Thermodynamic Properties of Metals and Alloys (Section 4.3.2.3) and TRCDP Selected Values of Properties of Chemical Compounds (Section 4.3.2.2), respectively. Nevertheless, the text of this compilation has a detailed explanation of the calculation methods, critical review of molecular constants, etc. which are valuable for the users.

4.3.2.5 Phase diagrams for ceramists

The American Ceramic Society (ACS) and the U.S. National Bureau of Standards (NBS) served as co-sponsors of this compilation. The collection of literature data and critical evaluation were done mainly at NBS. The ACS provided editorial assistance, printing, and distribution services. The first compilation appeared in the *Journal of the American Ceramic Society* in 1933. The contents were updated and expanded at a later time. The 1964 edition (Levin *et al.*[23]) supersedes previous publications of the series and covers the literature up to 1962. A supplement with 2083 new diagrams, 1969 Part II edition, appeared in 1970.

The systems included are: metal–oxygen and metal oxide systems (including Si), systems containing halides, sulphides, cyanides, alone and with other substances, systems of water with metal oxides, and miscellaneous substances. Simple and complex phase diagrams for one-, two-, and multi-component systems are presented. The diagrams include isothermal and isoplethal sections of phase relations, temperature–composition projections, compatibility relations, etc. Melting points of metallic oxides are also listed. References are given below each diagram. An author index and a system index are also included. The introduction treats the following topics: theory, methods of calculation, silicate chemistry, phase rule, and experimental methods. It is a valuable source of information for ceramists.

4.3.2.6 Constitution of binary alloys

This work was compiled by Hansen and Anderko[24] at the Illinois Institute of Technology Research Institute (IITRI) under the sponsorship of the U.S. Air Force. It is a revised edition of Hansen's book entitled 'Der Aufbau der Zweistofflegierungen' published in 1936 and covers the literature through 1955 with limited coverage of those published in 1956 and 1957. The first supplement by Elliott[25] updates the 1958 edition through 1962 with a review of almost 2000 alloys which relate to the binary equilibrium for metal–metal, metal–metalloid, and metal–non-metal. The second supplement was compiled by Shunk[26] which contains reviews for data published from 1962 through 1964 on 1702 binary systems (313 of which have not been previously published). The reviews include not only data where one component has a significant vapour pressure but also high (applied) pressure data.

In addition to binary alloy systems, some borides, hydrides, oxides, chalcogenides, and phosphides are adopted for evaluation. In most cases, the evaluated results are incorporated in composite diagrams, with uncertainties indicated. Values for solubilities, eutectic points, melting points, crystal structure and cell dimensions are given. The numerical data employed for evaluation were obtained from the open literature. References were mainly from *Chemical Abstracts*, *Metallurgical Abstracts*, *Review of Metal Literature*, special abstract bulletins, reviews, and monographs. The table of contents to the Second Supplement is a cumulative listing of systems contained in the three volumes. The 1958 edition is a necessity for the use of the Supplements. The systems are arranged in alphabetical order by chemical symbols. Some physical properties and structural data of the elements are found in the 1958 edition and First Supplement. The Second Supplement has a table of crystal structures for the representative compounds.

In 1967, the work was re-located at the Binary Metal and Metalloid Constitution Data Center, Illinois Institute of Technology Research Institute, Chicago, with F. A. Shunk as director. The programme is now sponsored by the U.S. Atomic Energy Commission, Aerospace Research Laboratories, National Aeronautics and Space Administration, and the National Bureau of Standards (NBS). This programme is related to the National Standard Reference Data System of NBS. The compilation work is particularly useful to metallurgists.

4.3.2.7 JANAF thermochemical tables

The project was established at the Thermal Laboratory of the Dow Chemical Company, Midland, Michigan, in late 1959, with D. R. Stull as the project director. Technical assistance and guidance are provided by members of the JANAF Thermochemical Panel (1959–61) and Working Group (1961–70), which include representatives from government, industries, and universities. The project is supported by the U.S. Government and its main purpose is to supply critically evaluated thermodynamic data of rocket fuel combustion products for propellant performance calculations by the rocket industries. The results are presented in tables known as the Joint Army-Navy-Air Force (JANAF) Thermochemical Tables. The JANAF Thermochemical Tables were published after they were critically reviewed by the Panel and Working Group. Due to the urgent need by the rocket industry, most of the early tables were issued as interim tables printed on grey sheets and called 'grey' tables in which the thermodynamic data employed for calculation were not critically evaluated. The selected final values were printed on 'white' paper.

Two volumes containing over 400 tables were published in 1960. The later tables were issued as loose-leaf supplements which appeared quarterly. Originally only qualified recipients could receive these tables, but now they are available to the general public. In August 1965, all the JANAF tables (over 800 sheets) published through Supplement No. 17, dated March 31, 1965, were made available to the public as a U.S. Government Report (Stull *et al.*[27]). After that, three yearly addendums appeared (Stull *et al.*[28–30]). The entire work is to be published as a single book in 1971 or early 1972.

The JANAF tables include the thermodynamic properties in the solid, liquid and/or ideal gaseous states for 31 elements (Al, B, Ba, Be, Br_2, C, Ca, Cl_2, Co, Cs, Cu, F_2, Fe, H_2, Hg, I_2, K, Li, Mg, Mo, N_2, Na, O_2, P, Pb, S, Si, Sr, Ti, W, and Zr), one electron gas (e^-), and their oxides, halides, hydrides, nitrides, silicates, etc. As of December 31, 1970, there were 1117 tables issued of which 31 are reference state tables, 293 crystal tables, 157 liquid tables, and 636 ideal gas tables. Included in the ideal gas species are 65 ions (41 positive and 24 negative ions). The reference state tables contain the thermodynamic properties for elements. The values listed for each phase in these tables are real properties. However, in the other single-phase tables, only those for the named phase were real, and the property values presented for the other phases in the table were obtained by extrapolation. The thermodynamic properties for the different phases in the reference state table and those for the different crystalline phases in the crystal table are separated by solid lines at appropriate temperatures. The values of thermodynamic properties for the real phase and the adjacent phase in the single-phase table are separated by dotted lines at the appropriate temperature of phase change for that given compound. The issued tables are subject to revision when new data appear which are significantly different from the old table values. The dates of each revision are indicated at the bottom of each table.

The thermodynamic properties given are: heat capacity, entropy, Gibbs energy function, enthalpy, enthalpy of formation, Gibbs energy of formation, and logarithm of equilibrium constant of formation. These properties are listed at 100 degree intervals from 0 to 6000 K for ideal gases, including those at 298.15 K. For crystals and liquids, the property values were terminated above the melting and boiling points by several hundred degrees, respectively. The molecular and spectroscopic constants, and pertinent thermal data employed in the calculations are listed and briefly discussed on the back of each table. These tables are arranged alphabetically according to the molecular formula of the compounds. Only seven simple hydrocarbon species (CH, CH_2, CH_3, CH_4, C_2H, C_2H_2, and C_2H_4) are included. Most of the inorganic compounds are also covered by the U.S.S.R. 'Thermodynamic Properties of Chemical Substances' (see Section 4.3.2.4 for details), and some of them are included in Texas A&M Thermodynamics Research Center Data Project (see Section 4.3.2.2). It is an important source of information on thermodynamic properties of chemical compounds of interest to the rocket industry and other related chemical industries.

4.3.2.8 *Chemical thermodynamics in non-ferrous metallurgy*

A series of handbooks in eight volumes initiated in the U.S.S.R. in 1960 covers the thermodynamic and physico-chemical properties of industrially important non-ferrous and rare metals. Substances included are: Zn, Cu, Pb, Sn, Ag, W, Mo, Ti, Zr, Nb, Ta, Al, Sb, Mg, Ni, Bi, and Cd, and their compounds. The properties reported are: atomic volume, density, fusion and boiling points, vapour pressure, heat capacity, enthalpy and entropy of formation and phase changes, Gibbs energy of formation, and enthalpy and Gibbs energy functions. This series was compiled by J. I. Gerassimov, A. N. Krestovnikov, and A. S. Shakhov and was published by Metallurgical Publishing House, Moscow. Volume I, which appeared in 1960, included the following contents: theoretical introduction, thermodynamic properties of important gases, thermodynamics of zinc and its important compounds. Volume II, published in 1961, included: thermodynamics of copper, lead, tin, silver, and their important compounds. Thermodynamics of tungsten, molybdenum, titanium, zirconium, niobium, tantalum, and their important compounds were the contents of volume III (1963) of which an English translation is available from the Clearinghouse for Federal Scientific and Technical Information, Springfield, Va. 22151. In 1966, Volume IV was published which is the latest volume available as of 1970. It included: thermodynamics of aluminium, antimony, magnesium, nickel, bismuth, cadmium, and their important compounds. This series is valuable for industrial application of chemical thermodynamics. Part of the data was also covered by Sections 4.3.2.1–4.3.2.4 and 4.3.2.7.

4.3.2.9 *Thermochemistry for steelmaking*

This programme was established about 9 years ago under the direction of J. F. Elliott, Department of Metallurgy, Massachusetts Institute of Technology, Cambridge, Massachusetts 02139, U.S.A. Volumes I and II were published in 1960 and 1963, respectively. Selected elements and compounds (carbides, nitrides, oxides, phosphides, silicides, and sulphides) involved in the chemistry and technology of steelmaking were treated in Volume I. In addition, Volume II also treats binary and ternary iron alloys and solutions, complex oxide systems and slags. The following thermodynamic properties were included: temperature and enthalpy of phase changes; heat capacity; enthalpy, entropy, and Gibbs energy function at 100 K intervals; and enthalpy, Gibbs energy, and log K_p of formation. Phase diagrams, which take up about 30% of Volume II, include binary diagrams for elements, oxides, and iron solutions and show the effect of the addition of a second metallic element upon the solubility of graphite in iron. This compilation is unique in that it includes useful thermodynamic properties of slag and metal solutions, interaction coefficient, and diffusion coefficients in melts. It is an important source of data on chemical thermodynamics for the steel industry. Part of the covered data was also reported in publications described in Sections 4.3.2.3, 4.3.2.4, 4.3.2.7 and 4.3.2.8.

4.3.2.10 Thermodynamic functions of gases

This well-known British compilation covers the *PVT* and related thermodynamic properties of the following industrially important gases: NH_3, CO_2, CO, air, C_2H_2, C_2H_4, C_3H_8, Ar, CH_4, N_2, and C_2H_6. The properties included were as follows: temperature–entropy diagrams, entropy, enthalpy, and volume for single phases and/or two phases in equilibrium, specific heats, and Joule–Thomson coefficients. The following physical properties were also included: density, molecular volume, boiling point, triple point temperature and pressure, and critical constants. The first two volumes treated eight gases and were published in 1956 and reprinted in 1962; Volume 3, having data on the other three gases, appeared in 1965. This series was started in 1948 under the supervision of the Thermodynamics Committee of the Mechanical Engineering Research Board of the Ministry of Technology of Great Britain, with Dr. F. Din as general editor. This project was discontinued and replaced by a more comprehensive and international programme, namely, the IUPAC Chemical Thermodynamic Tables under Dr. Selby Angus in London (see Section 4.3.2.14), to continue this work. The contents of this series on hydrocarbons to a certain extent overlap with the j-tables of the API Research Project 44 (see Section 4.3.2.1).

4.3.2.11 Selected values of chemical thermodynamic properties

The U.S. National Bureau of Standards Circular 500 was published[31] in 1952 under the direction of Dr. F. D. Rossini (see Section 4.3.1). Revision of this comprehensive compilation was started in the early 1960s by the Chemical Thermodynamic Data Group, Thermochemistry Section, Physical Chemistry Division of the Institute for Basic Standards of NBS, Washington, D.C. The updated version of Series I was issued as the NBS Technical Note 270 series. As of 1970, four issues have been published[32]. The first two issues 270–1 and 270–2, contained tables for the first 23 elements and the elements 24 through 32 in the Standard Order of Arrangement and appeared in 1965 and 1966, respectively. These tables were later superseded by those listed in the next issue, 270–3, which covered the thermodynamic properties for the first 34 elements in the Standard Order of Arrangement and appeared in 1968. The fourth issue, 270–4, was published in 1969 and included the tables for elements 35 through 53. These Technical Notes are available from the Superintendent of Documents, U.S. Government Printing Office, Washington, D.C. 20402.

In the revision, the 1963 fundamental constants and the IUPAC Table of Relative Atomic Weights 1961, based on the atomic mass of $^{12}C = 12$ (exactly), were used. Usages for nomenclature, symbols, basic physical constants, and conversion factors recommended by international committees and bodies such as the IUPAC are followed. The introduction covers the following topics: symbols and abbreviations, conventions regarding pure substances and solutions, unit of energy and fundamental constants, internal consistency of the tables, uncertainties, and arrangement of the tables. The Standard Order of Arrangement of the Elements is shown in Figure 4.1.

The NBS compilation is one of the most comprehensive and authoritative works on inorganic substances which provides self-consistent tables of

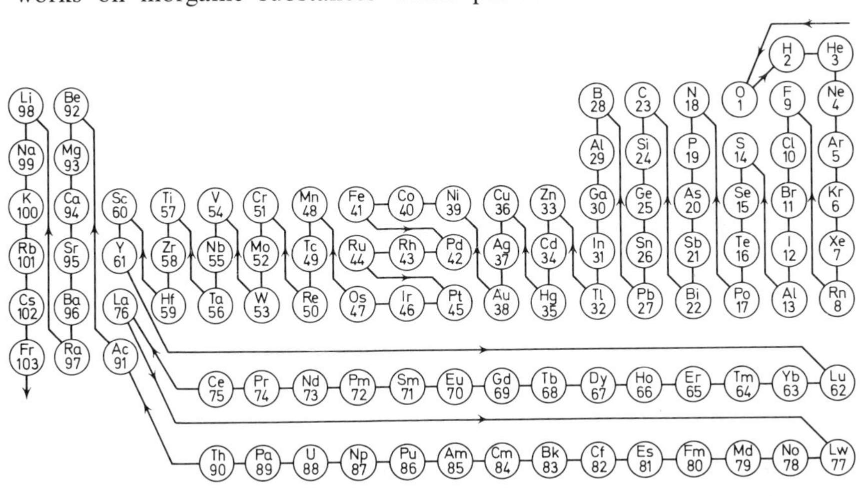

Figure 4.1 U.S.A. standard order of arrangement of the elements. (Compounds are located in tables under the element (in the compound) occurring latest in the list

selected values for the enthalpy of formation at 0 and 298.15 K; Gibbs energy of formation, entropy and heat capacity at 298.15 K; and phase transition data for all inorganic substances, organic molecules containing only one and two carbon atoms, and organo-metallic compounds and complexes. Part of the contents overlaps those cited in Sections 4.3.2.1–4.3.2.4, 4.3.2.7 and 4.3.2.8. Since each compilation has its own merit and purpose to serve, such partial overlapping of efforts should not be considered as serious. However, it is important to mention that, if possible, all compilers should try to use the same set of basic data for all calculations. For the benefit of the users, all the compiled tables should in turn be consistent with each other.

4.3.2.12 Thermodynamic constants of substances

In addition to the publication of 'Thermodynamic Properties of Chemical Substances' in 1962 (see Section 4.3.2.4), the U.S.S.R. Academy of Sciences sponsored this compilation project which has been directed by V. P. Glushko, chief editor, and V. A. Medvedev. The contents of this handbook were planned to be published in 10 parts. As of 1970, the first four parts have been completed and issued. Part I appeared in 1965 containing the thermodynamic properties for O, H, D, T, F, Cl, Br, I, At, ^{3}He, He, Ne, Ar, Kr, Xe, Rn, and their compounds. Similar properties for S, Se, Te, Po and N, P, As, Sb, Bi, and their compounds were published in 1966 and 1968 as Parts II and III, respectively. Part IV has two volumes. The first volume was issued in 1970, including the thermodynamic properties of the elements C, Si, Ge, Sn, Pb, and their compounds. The references and appendix will be in the second volume issued in 1971.

The thermodynamic properties listed are as follows: enthalpy of formation at 0 and 298.15 K; Gibbs energy of formation at 298.15 K; dissociation energy; enthalpy, entropy, heat capacity at 298.15 K; vapour pressure and temperature at phase transitions; and enthalpy and entropy of polymorphic transition, fusion, vaporisation, and sublimation. Symmetry and structure type of crystalline substances, ionisation potentials, and critical temperature,

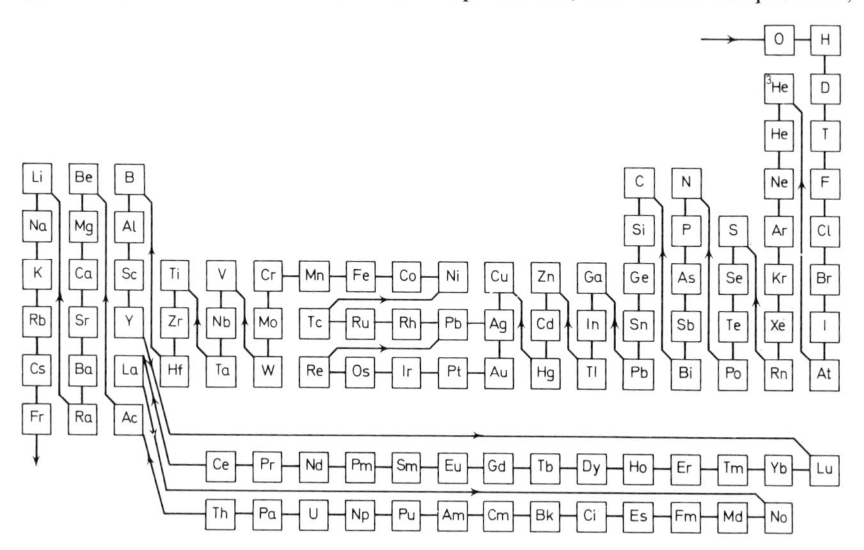

Figure 4.2 Standard order of arrangement of the elements adopted by the U.S.S.R. Acadeny of Sciences

pressure, density, and volume, and a formula index for the compounds contained in each part are given in the appendix. Thermal functions for the ideal gaseous state and the condensed phases as a function of temperature are not given.

The thermodynamic properties included in this U.S.S.R. compilation are very similar to those listed in the NBS 'Selected Values of Chemical Thermodynamic Properties' (see Section 4.3.1 and 4.3.2.11), except that the NBS compilation contains carbon compounds having no more than two carbon atoms and this compilation has no such limitation. The U.S.S.R. order of compilation of thermodynamic properties for the elements and their compounds is very similar to the U.S.A. Standard Order of Arrangement of the Elements and Compounds used by NBS, API Research Project 44, and the TRC Data Project. The two minor differences between these two orders of arrangement are (a) that the U.S.A. system puts the elements He and other inert gases directly after element H, and the U.S.S.R. order presented in Figure 4.2 puts the halogens before the inert gases; (b) the U.S.S.R. order includes the manganese sub-group together with the Group VIII metals and numbers these twelve metals from bottom to top instead of in reverse order.

This compilation covers more carbon compounds (although still limited to one or two carbon atoms) than those listed in 'Selected Values of Chemical Thermodynamic Properties' by NBS and also lists specific references related to the selected values for each property of the given compounds which the NBS work will publish after the completion of the revision. It should be noted that since the listed property values were evaluated based on the data available at that time, it is possible that the selected values given in the two compilations for the same property may be different. It is to the user's advantage to have two excellent sources of information for choice rather than only one. Some of the listed properties were also covered by other less comprehensive compilations as presented previously. However, where self-consistencies are concerned, the property values given in the comprehensive compilations, like this one, are preferred to those from other more limited compilation sources.

4.3.2.13 High-temperature behaviour of inorganic salts

The purpose of this programme is the evaluation and publication in concise form of the thermodynamic and kinetic information on the high temperature behaviour of important classes of inorganic salts. For each salt, the following three kinds of information are under investigation: the reactions by which salts decompose in various temperature ranges; Gibbs energy changes, equilibrium constants, and partial pressures for these reactions; and rate constants, activation energies, and mechanisms of these reactions. The programme was initiated in 1963 in the Electrochemistry Section of the Institute for Basic Standards of the U.S. National Bureau of Standards. In 1968, this work was moved to the Electrochemistry Branch, Naval Research Laboratory, Washington, D.C., which is supported by the Office of Standard Reference Data of NBS and directed by Kurt H. Stern. A series of publications has already been planned. The first volume, entitled 'High Temperature Properties and Decomposition of Inorganic Salts, Part I. Sulphates' was issued in 1966, containing 60 cations. The properties covered were: phase transition temperatures above 298.15 K; enthalpy of formation and entropy of reactants and products at 298.15 K; densities at 298.15 K; and kinetics data on thermal decomposition. In Part 1, the experimental methods and theory were briefly outlined, and the quality and treatment of the data were reviewed. Future volumes will include carbonates and nitrates, among other classes.

4.3.2.14 The Thermodynamic Tables Project of the International Union of Pure and Applied Chemistry

This project was established by the IUPAC Division of Physical Chemistry in 1964. Its aim is the publication of tables on the thermodynamic and transport properties of industrial gases and liquids of interest to science and industry. International working panels have been organised – one panel for each fluid or groups of fluids. The following countries are represented on panels by one or more members; France, Germany, India, Japan, the Netherlands, the U.K., the U.S.A., and the U.S.S.R. Individual panels have been established for the following fluids: carbon dioxide (CO_2); atmospheric gases (O_2, N_2, and air); quantum fluids (H_2, D_2 and He); inert fluids (Ne, Ar, Kr, Xe); aliphatic hydrocarbons (CH_4, C_2H_6, C_2H_4, etc.); halogenated hydrocarbons; multicomponent systems (to be selected); and ammonia (NH_3). The thermodynamic data such as pressure, temperature, enthalpy, and entropy over a wide range of the variables will be included in the published tables. Experts in the specific fields will evaluate the data, and the presentation of the best values will be done in a consistent way. The first issues of this comprehensive publication will be the properties of argon to be published in 1972 (Butterworths, London).

4.3.3 Selected recent books on thermodynamic properties of chemical substances

Many books have been published recently which contain thermodynamic data. Some of the published data were critically evaluated and presented as tables. In most cases, however, the listed values were taken directly from the original sources, without critical review and were reported without comments. Obviously, critical evaluation of thermodynamic data is a specialised field which needs experienced scientists, adequate financial support, and appropriate facilities. Therefore, published data on reliable critical values for thermodynamic properties of chemical substances are still rather limited.

During the last 2 years, several excellent books containing thermodynamic properties of organic and organometallic compounds appeared. Brief discussions on a few of these books are presented as follows.

Thermochemistry of Organic and Organometallic Compounds by Cox and Pilcher appeared in 1970 and gives values for the enthalpy of formation and vaporisation at 25 °C for about 3000 organic and organometallic compounds. This is the first modern comprehensive review in English. It includes a critical compilation of thermodynamic data published on these compounds since 1930 when accurate measurement in thermochemistry began. Values of low precision or those obtained by suspect methods were excluded. The data are listed with estimates of error and details of the methods of measurements. Methods for the estimation of thermochemical data by the use of modern bond energy schemes are also described. Selected values are offered in the case of two or more independent determinations. The compilation is therefore both critical and comprehensive.

Handbook of Metal Ligand Heats and Related Thermodynamic Quantities by Christensen and Izatt is another excellent source of thermodynamic information on this special field in which significant progress has been made recently.

The book, *The Chemical Thermodynamics of Organic Compounds* by Stull, Westrum and Sinke, deals mainly with thermodynamic data on hydrocarbons, organic oxygen, nitrogen, halogen, and sulphur compounds. Selected values were given for the enthalpy of formation and entropy at 298.15 K for approximately 4400 species. The thermodynamic functions from 298 to 1000 K for 741 species were also tabulated. Sources of all the values presented were listed in the bibliography of 1656 references. A very useful compilation, indeed.

A selected list of recent books which includes the thermodynamic properties of chemical species is presented below. These books are arranged according to the following categories: general, elements, inorganic, organic and organometallic. Within each category, the books are arranged chronologically and in turn alphabetically by title. Usually, but not always, more recent data implies better quality of the information.

4.3.3.1 General

Thermodynamic and Thermochemical Constants by K. V. Astakhov (ed.), Science Publishing House, Moscow, 1970, 480 pp.

Thermophysical Properties of Matter: Volume 4. Specific Heat, Metallic Elements and Alloys; Volume 5. Specific Heat, Non-metallic Solids; Volume 6. Specific Heat, Non-metallic Liquids and Gases by Y. S. Touloukian and E. H. Buyco (Volumes 4 and 5); Y. S. Touloukian and T. Makita (Volumes 6). Plenum, 1970. 830 pp.; 1738 pp.; and 384 pp.

Thermophysical Properties of Substances and Materials, 2nd edition, by V. A. Rabinovich (ed.), State Service for Standard and Reference Data (GSSSD), Moscow, 1970, 299 pp.

Reference Book on Thermophysical Properties of Gases and Liquids by N. B. Vargaftik, Science Publishing House, Moscow, 1969.

The Virial Coefficients of Gases by J. H. Dymond and E. B. Smith, Clarendon Press, Oxford, England, 1969, 231 pp.

Thermophysical Properties of Gases, Liquids, and Plasmas at High Temperatures by I. I. Novikov and A. I. Gordon, GSSSD, Moscow, 1969, 407 pp.

Thermophysical Properties of Solid Substances at High Temperatures by I. I. Novikov and A. I. Gordon, GSSSD, Moscow, 1969, 496 pp.

4.3.3.2 Elements

Handbook of the Rare Elements—Vol. I. Trace Elements and Light Elements, Vol. II Refractory Elements, Vol. III. Radioactive Elements and Rare Earth Elements by M. A. Filiand and E. I. Semenova. M. E. Alferieff (Trans. and Ed.). Boston Technical Publishers, Boston, Mass., U.S.A., 1968, 1970, 1970. 265 pp. 417 pp, 303 pp.

Provisional Thermodynamic Functions for Helium–4 for Temperatures from 2 to 1500 K with Pressures to 100 MN/m^2 by R. D. McCarty. NBS Report 9762. U.S. Government Printing Office, Washington, D.C., 1970.

The Properties of Helium: Density, Specific Heats, Viscosity, and Thermal Conductivity at Pressures from 1 to 100 Bar and from Room Temperature to About 1800 K by H. Petersen. Danish Atomic Energy Commission, Research Establishment Riso Report No. 224, 1970.

The Thermodynamic Properties of Nitrogen by T. C. Coleman and R. B. Stewart. Research Report No. 11, University of Idaho, 1970.

Handbook of the Physicochemical Properties of the Elements by G. V. Sansonov. Plenum, New York, 1968, 941 pp.

Properties of Liquid and Solid Hydrogen by B. N. Eselson, V. G. Manzhely *et al.* Publishing House for State Standards, Moscow, 1968.

Thermodynamic and Thermophysical Properties of Helium by N. V. Tsederberg, V. N. Popov and N. A. Morozova. A. F. Alyab'ev (ed.), Atomizdat, Moscow, 1969.

Cesium Data Sheets, ORNL-4186 by S. J. Rimshaw and E. E. Ketchen. Clearinghouse, Springfield, Va., U.S.A., 1967, 27 pp.

Curium Data Sheets, ORNL-4187 by S. J. Rimshaw and E. E. Ketchen. Clearinghouse, Springfield, Va., U.S.A., 1967, 52 pp.

Metallic and Chemical Properties of Elements of the Periodic System by I. I. Kornilov, *et al.* Science Publishing House, Moscow, 1967.

Metallurgical Thermochemistry, 4th edition, by O. Kubaschewski, E. El. Evans, and C. B. Alcock, Pergamon Press, New York, 1967, 504 pp.

Metals Reference Book, Vols. I, II, and III. 4th edition, by C. J. Smithells. Plenum, New York, 1967, 370 pp, 683 pp, 1147 pp.

Numerical Values and Functions of Physics, Chemistry, Astronomy, Geophysics, and Technology, 6th edition, Vol. IV, Part 4a, *Heat Technology,* by H. R. Landolt–Börnstein, Springer Verlag, Berlin, 1967, 944 pp.

Strontium Data Sheets, ORNL-4188 by S. J. Rimshaw and E. E. Ketchen. Clearinghouse, Springfield, Va., U.S.A., 1967, 45 pp.

Zero Pressure Thermodynamic Properties of Nitrogen Gas, IC8319 by R. E. Barieau and P. C. Tally. U. S. Department of the Interior, 1967, 48 pp.

4.3.3.3 Inorganic substances

Steam Tables (Industrial) in SI Units by Ministry of Technology, U.K., Edward Arnold Ltd., London, 1970, 160 pp.

Phase Diagrams for Ceramists. 1969 Supplement, by E. M. Levin, C. R. Robbins, and H. F. McMurdie, American Ceramic Society, Inc., Columbus, Ohio, U.S.A., 1969, 625 pp.

Properties of Water and Steam in SI-Units, 0–800 °C and 0–1000 Bar by E. Schmidt, Springer Verlag, New York, 1969, 205 pp.

Steam Tables: Thermodynamic Properties of Water, Including Vapour, Liquid, and Solid Phases (English Units) by J. H. Keenen, F. G. Keyes, P. G. Hill, and J. G. Moore, Wiley, New York, 1969, 162 pp.

Physicochemical Properties of Oxides by G. V. Samsonov, Publishing House for State Standards, Moscow, 1968.

Thermodynamic Functions for Ideal Gases at Temperatures Up to 6000 K by H. D. Baehr, H. Hartmann, and H. Schomacker, Springer Verlag, Berlin, 1968.

Thermodynamic Properties of Ammonia as an Ideal Gas by L. Haar, U.S. Government Printing Office, Washington, D.C., 1968, 10 pp.

Thermodynamic Tables in SI Units by R. W. Haywood, Cambridge University Press, New York, 1968, 41 pp.

Thermophysical Properties of Alkali Metals by V. E. Kirillin (ed.), Publishing House for State Standards, Moscow, 1968.

Tables of the Thermodynamic Properties of Carbon Dioxide by W. A. Stein, Braunschweig University, 1967.

Thermophysical Properties of Metals and Alloys by R. E. Kryzhizhanovsky, Metallurgia Publishing House, Moscow, 1967.

Plutonium: Physicochemical Properties of Its Compounds and Alloys by O. Kubaschewski (ed.). International Atomic Energy Agency, Vienna, 1966, 112 pp.

Thermodynamic Properties of Metals and Alloys by P. D. Anderson, UCRL-11821, Clearinghouse, Springfield, Va., U.S.A., 1966, 136 pp.

Thermodynamics of Certain Refractory Compounds. Vol. II. Thermodynamic Tables, Bibliography and Property File by H. L. Schick, Academic Press, Inc., New York, 1966, 792 pp.

4.3.3.4 Organic substances

Handbook on Vapor Pressures and Heats of Vaporization of Hydrocarbons and Related Compounds by R. C. Wilhoit and B. J. Zwolinski, Thermodynamics Research Center, Texas A&M University, College Station, Texas, U.S.A., 1971, 337 pp.

Handbook on Chemical Thermodynamic Properties of Hydrocarbons and Related Sulfur Compounds by B. J. Zwolinski, R. C. Wilhoit and Jing Chao, Thermodynamics Research Center, Texas A&M University, College Station, Texas, U.S.A., 1971, 518 pp.

Handbook of Biochemistry, 2nd edition, by H. A. Sober (ed)., The Chemical Rubber Co., Cleveland, Ohio, U.S.A., 1970, 1700 pp.

Heat Capacities of Linear High Polymers by B. Wunderlich and H. Baur, Springer Verlag, New York, 1971, 218 pp.

Technical Data Book—Petroleum Refining, 2nd edition, American Petroleum Institute, Washington, D.C., 1970, 824 pp.

Thermochemistry of Organic and Organometallic Compounds by J. D. Cox and G. Pilcher, Academic Press, New York, 1970, 643 pp.

Thermophysical Properties of Freon-22 by A. V. Kletskii, Publishing House for State Standards, Moscow, 1970.

Thermophysical Properties of Organic Coolants by M. P. Vukalovich *et al.,* Atomizdat Publishing House, Moscow, 1970, 288 pp.

The Chemical Thermodynamics of Organic Compounds by D. R. Stull, E. F. Westrum, Jr., and G. C. Sinke, Wiley, New York, 1969, 895 pp.

Thermodynamic Properties of Ethylene. Parts 1 and 2 by W. Featherstone and M. R. Gibson, ICI (Agricultural) Ltd., London, 1968 and 1969.

Thermophysical Properties of Gaseous and Liquid Methane by V. A. Zagoruchenko and A. M. Zhuravlev, Publishing House for State Standards, Moscow, 1969.

Abridged Thermodynamic and Thermochemical Tables with Charts in British Units by F. D. Hamblin, Pergamon Press, New York, 1968, 73 pp.
Physical Constants of Linear Homopolymers by O. G. Lewis, Springer Verlag, New York, 1968, 173 pp.
Polymer Handbook by J. Brandrup and E. H. Immergut (ed.), Interscience Publishers, New York, 1966.

4.3.3.5 *Organometallic compounds*

Handbook of Metal Ligand Heats and Related Thermodynamic Quantities by J. J. Christensen and R. M. Izatt, Marcel Dekker, New York, 1970, 324 pp.
Organometallic Compounds of Arsenic, Antimony, and Bismuth by G. O. Doak and L. D. Freedman, Wiley, New York, 1970, 509 pp.
Thermochemistry of Transition Metal Complexes by S. J. Ashcroft and C. T. Mortimer, Academic Press, New York, 1970, 478 pp.
The Thermal Properties of Transition–Metal Amine Complexes by W. W. Wendlandt and J. P. Smith, American Elsevier Publishing Co., New York, 1967, 235 pp.

4.3.4 Other pertinent sources of information

Pertinent information on thermodynamic properties of chemical substances has also been published as special reports (e.g. NSRDS–NBS Series) and series of journal papers on special topics (e.g. publications by National Physical Laboratory, Teddington, England; Bartlesville Energy Research Center, Bartlesville, Oklahoma; and NBS Cryogenic Data Center, Boulder, Colorado). Many important thermodynamic data have appeared in the well-known journals, such as *Journal of Chemical and Engineering Data, Journal of Chemical Thermodynamics, Journal of Chemical Society, Transactions of the Faraday Society, Journal of Chemical Physics, Journal of Physical Chemistry, Journal of the American Chemical Society*, to name a few.

The annual publications of the *Bulletin of Thermodynamics and Thermochemistry*, prepared under the auspices of the IUPAC, are another excellent source of information. The Bulletin contains two parts. The first part furnishes the information, from private communications, about research works which are in progress, or completed but still unpublished by scientists in thermodynamics and thermochemistry from all countries of the world. The second part gives the Substance–Property Index to abstracts and bibliography of 25 physical and thermodynamic properties of organic and inorganic substances and their mixtures. In the Appendix, recent books, monographs, reviews and other related materials of interest to thermodynamicists are listed for reference. It is most valuable to those who do not have enough time and/or appropriate library facilities for searching recent literature.

4.4 CRITICAL EVALUATION OF THERMOCHEMICAL DATA

4.4.1 Thermochemical input data from an ideal experimental programme

The kinds of literature thermochemical information needed for preparation of complete and internally consistent tables of critical data will be determined by the kinds of materials under investigation; namely, inorganic, organic pure substances, or their mixtures. Thus, it is necessary to discuss the nature and kinds of input data for critical table preparation in terms of a specific group of materials such as the pure organic substances (Zwolinski and Wilhoit[33]).

An ideal experimental programme of measurements for desirable input data should include: (a) synthesis, purification, and characterisation of samples of certified purity; (b) measurements of simple physical properties with highest precision and accuracy; (c) spectroscopic and molecular studies, e.g. infrared, Raman, microwave and electron diffraction; (d) thermochemistry, including combustion, solution and other related reaction calorimetry; (e) thermodynamics: (i) low- and high-temperature calorimetry, 5 to 1000 K, (ii) vapour-flow calorimetry, (iii) P–V–T measurements, and (iv) equilibrium measurements; and (f) theoretical studies on intramolecular and intermolecular energetics in order to extend data on key compounds and create complete tables of entire classes of closely related compounds. In such an ideal thermodynamic data programme, *all* measurements should be made on the *same* sample with the *same* equipment by the *same* experienced investigators. Currently, only two laboratories approach these requirements, i.e. the Thermodynamics Research Laboratory, Bartlesville Energy Research Center of the U.S.A., and the Chemistry Division, National Physical Laboratory of England.

4.4.2 Imperfections of thermochemical input data from the open literature

Seldom do data extracted from the literature meet the requirements described in Section 4.4.1, because *different* investigators provide measurements on *different* samples and in *different* kinds of equipment and seldom of necessary ranges of thermodynamic variables. The reported values in the literature are usually very incomplete, inconsistent, and at times inaccurate. Very often, in the preparation of critical data tables on closely related substances, it is necessary to estimate missing or incomplete input data through theoretical or semi-empirical molecular structural correlations. Furthermore, correlation procedures for extrapolating and interpolating structure-wise along the molecular framework of a closely related group of substances are of inestimable value for judging the reliability of experimental data in many instances.

4.4.3 Evaluation and selection of best values

There is no simple recipe that one can follow to produce tables of selected or critically evaluated data except to note that such studies require intellectual capacities on a high level of scientific competence. If a physicist is to evaluate nuclear data, he should be intimately familiar with all physical and theoretical aspects of nuclear physics. Similarly for chemical thermodynamic data, the compiler or evaluator should be thoroughly grounded and experienced in all aspects of experimental chemistry and of theory, such as thermodynamics and quantum chemistry. Thus, a worker in this area has to be an expert in both experiment and theory; and in our experience, seldom does a first-rate physical chemist become an effective compiler in less than 5 years of continuing experience and work in the area of chemical thermodynamic data.

The first and most cardinal rule is for the evaluator or appraiser to be extremely critical of every physical and chemical step in an experimental study. By physical features, it is meant whether the best of instrumentation was used with required calibration and recognition of all systematic and random errors and complete disclosure of all fundamental constants (Taylor *et al.*[34]), conversion units, temperature ('The Comité International des Poids et Mesures'[35], Rossini[36], Bedford *et al.*[37, 38], Douglas[39], Hust[40], and mass scales (IUPAC Commission of Atomic Weights[41]). By chemical features, it is meant that the substance(s) has been unequivocally defined with respect to composition and purity. In addition, scientists are human beings and commit errors in observation and in calculation in addition to the usual number of typographical and printing errors. Once the original literature data have been reduced to current SI units, and defined standard conditions with ascertained magnitudes of uncertainty – *a numerical result whose uncertainty is completely unknown is completely valueless* – the process of evaluation begins.

The second cardinal rule in the evaluation of data is to work with closely related groups or chemical families of substances and never, if possible, with data for only a single substance. From here on, the evaluation process reduces to all the usual tools of the analyst, consisting of graphical, analytical methods, coupled with computer techniques, and of all necessary physical and chemical theories or equations to test the reliability and internal consistency of the data. Once the best or selected critical numerical data are at hand, the range of the uncertainties should be stated clearly. A useful general rule for magnitudes of uncertainty is that the uncertainty may be 3 to 30 units in the *last* significant figure, except for thermochemical data, such as enthalpies of formation, where the uncertainty is larger by roughly a factor of seven to may be 20 to 200 units in the *last* significant figure. Naturally it is to be expected that critical tables prepared by any one compilation effort will be fully documented and internally consistent. Of even greater importance to the world-wide users of these tables is that the tables in the *same area* of science also be consistent with each other. The recently issued CODATA 'key substances' represent a giant and historical step in this direction.

Two final comments are in order with regard to revision and distribution. Every compilation group should make provisions for ease of continual revision of the selected best values as dictated by availability of more precise, more complete, and more accurate literature data. Finally, the data should be published in whatever format would permit ease of distribution and use by every scientific worker in every region of the globe. The techniques of storage and retrieval of critical data should be dictated by the maximum availability and maximum usefulness of the critical data information to the largest segments of the scientific society. Critical data in most instances should not be compiled for archival purposes alone.

4.5 USE OF ELECTRONIC COMPUTER FOR HANDLING THERMODYNAMIC DATA

The information explosion due to the incredibly growing availability of scientific data must be controlled and directed. Information handling by electronic means is the only feasible way to resolve this serious problem. Computers and the computer-based techniques are now available to assist the data centre in the generation, evaluation, storage, retrieval, and dissemination of numerical and graphic data. In spite of the high cost of installation, operation, and maintenance, they offer considerable advantages where speed, accuracy, and overall control are important.

As mentioned previously, the calculations of thermodynamic functions from input data for chemical species in the crystal, liquid, and gaseous states are all performed by use of an appropriate electronic computer. Today, the digital computers are used as freely as desk mechanical calculators and slide rules were used in the past. In scientific and industrial communities, increasing interests are developing in having the critical data, produced by various thermodynamic property compilation groups, and in many instances the correlation equations themselves, stored on magnetic tapes for use in various calculations. The JANAF Thermochemical Data has been stored on magnetic tapes and offered for sale at a reasonable price for many years. The Office of Standard Reference Data, U.S. National Bureau of Standards, is exploring the feasibility of developing a magnetic memory bank from its associated data centres in order to facilitate the accessibility of important numerical data for the users[8].

The lack of comprehensive bibliographic control of the literature of science and technology was a major obstacle in the compilation of data for evaluation and publication. The best state of the art in bibliography and documentation is not widely used at present. Librarians and documentalists certainly bear a share of the responsibility for failing to communicate their techniques to the scientists and engineers.

The U.S. Government favours the establishment of the Specialized Information Center[4] as a major key to the nationalisation of the information system in the U.S. Ultimately, the Center will become the accepted retailer of information, switching, interpreting, and otherwise processing information for the large wholesale depositories and archival journals to the individual user. The Center for Computer Sciences and Technology, NBS, U.S.A., conducts research and provides technical services designed to aid U.S. Government agencies in the selection, acquisition, and effective use of automatic data processing equipment and serves as the principal focus for the development of federal standards for automatic data processing equipment, techniques, and computer languages (Stevens[43]). A series of general-purpose programmes for text editing, data retrieval, file management, and computer-assisted typesetting has been scheduled for publication by NBS. One of this series appeared in 1970 (Messina and Hilsenrath[44]). The Cryogenics Division, NBS, Boulder, Colorado, U.S.A., has established a cryogenic data bank in which all the pertinent information published in the open literature is stored on magnetic tapes and available to the public users with a reasonable service charge for the computer searching and preparation of a report.

Comprehensive computer programmes have been compiled for general and estimation of thermodynamic and thermophysical property values for gases and liquids by the American Institute of Chemical Engineers in the U.S. and the U.K. National Engineering Laboratory in Scotland, which can provide useful engineering data for substances of importance in industrial plant and process design and operation. A simple but versatile computer programme for generating data for hydrocarbon process computations was compiled by Turnbull[44]. It makes good use of modern developments in the theory of the thermodynamic behaviour of hydrocarbon materials.

In the Federal Republic of Germany, the feasibility of a national materials data bank has been studied in detail. The Task Group on Computer Use, CODATA, currently completed a survey in several countries of computer applications to numerical data generation, storage, and retrieval.

The above brief review illustrates the increasing use of computerised techniques in the handling of numerical data in many countries. The successful realisation of these ventures will require efficient organisation and adequate support of the projects by stimulating ideas and financial means, as well as national and international co-operation. For detailed information on the new developments in this field, the reader is referred to the Amer. ˋem. Soc. *Journal of Chemical Documentation, The Information Scientist,* d other related reviews, books and reports.

ˌcknowledgement

The partial support of the American Petroleum Institute, the Thermodynamics Research Center Data Project and the NBS Office of Standard Reference Data is duly acknowledged.

References

1. Rossini, F. D. (1967). *J. Chem. Doc.*, **7,** 2
2. Waddington, G. (1969). *J. Chem. Doc.*, **9,** 174
3. National Bureau of Standards (1963). *National Standard Reference Data Program, Background Information,* NBS Tech. Note 194. (Washington, D.C.: U.S. Government Printing Office)
4. Brady, E. L. and Wallenstein, M. B. (1964). *National Standard Reference Data System, Plan of Operation,* NSRDS-NBS 1. (Washington, D.C.: U.S. Government Printing Office)
5. Brady, E. L. (1967). *J. Chem. Doc.*, **7,** 6
6. Lide, D. R. (1970). *Critical Evaluation of Data in the Physical Science,* NBS Tech. Note 553. (Washington, D.C.: U.S. Government Printing Office)
7. Brady, E. L. (1966). *Status Report, National Standard Reference Data System,* NBS Tech. Note 289. (Washington, D.C.: U.S. Government Printing Office)
8. Alt, F. L. (1966). *Information Handling in the National Standard Reference Data System,* NBS Tech. Note 290. (Washington, D.C.: U.S. Government Printing Office)
9. Brady, E. L. (1968). *Status Report, National Standard Reference Data System,* NBS Tech. Note 448. (Washington, D.C.: U.S. Government Printing Office)
10. Brady, E. L. and Wallenstein, M. B. (1967). *Science,* **156,** 754
11. The Committee on Data for Science and Technology (December 1970). *CODATA Newsletter,* **5,** 2, 15, 19
12. The Committee on Data for Science and Technology (1969). *International Compendium of Numerical Data Projects.* (New York: Springer-Verlag)
13. The Committee on Data for Science and Technology (November 1970). *CODATA Bull.*, **2**
14. American Petroleum Institute Research Project 44 (1971). *Selected Values of Properties of Hydrocarbons and Related Compounds.* (College Station, Texas, U.S.A.: Texas A&M University)
15. Rossini, F. D., Pitzer, K. S., Taylor, W. J., Ebert, J. P., Kilpatrick, J. E., Beckett, C. W., Williams, M. G. and Werner, H. G. (1947). *Selected Values of Properties of Hydrocarbons,* NBS Circ. 461. (Washington, D.C.: U.S. Government Printing Office)

16. Rossini, F. D., Pitzer, K. S., Arnett, R. L., Braun, R. M. and Pimentel, G. C. (1953). *Selected Values of Physical and Thermodynamic Properties of Hydrocarbons and Related Compounds*. (Pittsburgh: Carnegie Press)
17. Zwolinski, B. J. and Wilhoit, R. C. (1971). *Handbook of Vapor Pressures and Heats of Vaporization of Hydrocarbons and Related Compounds*. (College Station, Texas, U.S.A.: Texas A&M University)
18. Zwolinski, B. J., Wilhoit, R. C. and Chao, J. (1971). *Handbook of Chemical Thermodynamic Data of Hydrocarbons and Related Compounds*. (College Station, Texas, U.S.A.: Texas A&M University)
19. Thermodynamics Research Center (1968). *Comprehensive Index of AP144-TRC Selected Data on Thermodynamics and Spectroscopy*, Publication 100. (College Station, Texas, U.S.A.: Texas A&M University)
20. Thermodynamics Research Center (1971). *Selected Values of Properties of Chemical Compounds*. (College Station, Texas, U.S.A.: Texas A&M University)
21. Hultgren, R. R. (1963). *Selected Values of the Thermodynamic Properties of Metals and Alloys*. (New York: McGraw-Hill Book Co.)
22. Pitzer, K. S. and Brewer, L. (1961). *Thermodynamics*, 723. (New York: McGraw-Hill Book Company)
23. Levin, E. M., Robbins, C. R. and McMurdie, H. F. (1964). *Phase Diagram for Ceramists*. (Columbus, Ohio: American Ceramic Society)
24. Hansen, M. and Ankerko, K. (1958). *Constitution of Binary Alloys*. (New York: McGraw-Hill Book Co.)
25. Elliott, R. P. (1965). *Constitution of Binary Alloys*, First Supplement. (New York: McGraw-Hill Book Co.)
26. Shunk, F. A. (1969). *Constitution of Binary Alloys*, Second Supplement. (New York: McGraw-Hill Book Co.)
27. Stull, D. R. and staff (1965). *JANAF Thermochemical Tables*, PB 168370. (Springfield, Virginia, U.S.A.: Clearinghouse)
28. Stull, D. R., Chao, J., Hu, A. T., Karris, G. C., Phillips, E. W., Prophet, H., Sinke, G. C., Syverud, A. N. and Wollert, S. K. (1966). *JANAF Thermochemical Tables*, First Addendum, PB 168370-1. (Springfield, Virginia, U.S.A.: Clearinghouse)
29. Stull, D. R., Chao, J., Hu, A. T., Karris, G. C., Prophet, H. and Syverud, A. N. (1967). *JANAF Thermochemical Tables*, Second Addendum, PB 168370-2. (Springfield, Virginia, U.S.A.: Clearinghouse)
30. Stull, D. R., Chao, J., Hu, A. T., Karris, G. C., Prophet, H., Syverud, A. N. and Webb, D. U. (1968). *JANAF Thermochemical Tables*, Third Addendum, PB 168370-3. (Springfield, Virginia, U.S.A.: Clearinghouse)
31. Rossini, F. D., Wagman, D. D., Evans, W. H., Levine, S. and Jaffe, I. (1952). *Selected Values of Chemical Thermodynamic Properties*, NBS Circ. 500. (Washington, D.C.: U.S. Government Printing Office)
32. Wagman, D. D., Evans, W. H., Halow, I., Parker, V. B., Bailey, S. M. and Schumm, R. H. (1965–1971). *Selected Values of Chemical Thermodynamic Properties*, NBS Tech. Notes 270-1 to 270-5. (Washington, D.C.: U.S. Government Printing Office)
33. Zwolinski, B. J. and Wilhoit, R. C. (1966). *Proc. Amer. Petrol. Inst.*, **46**, 125
34. Taylor, B. N., Parker, W. H. and Langenberg, D. N. (1969). *Rev. Mod. Phys.*, **41**, 375
35. The Comité International des Poids et Mesures. (1969). *Metrologia*, **5**, 35
36. Rossini, F. D. (1970). *J. Chem. Thermodynamics*, **2**, 447
37. Bedford, R. E., Durieux, M., Muijlwijk, R. and Barber, C. R. (1969). *Metrologia*, **5**, 47
38. Bedford, R. E., Preston-Thomas, H., Durieux, M. and Muijlwijk, R. (1969). *Metrologia*, **5**, 45
39. Douglas, T. B. (1969). *J. Res. Nat. Bur. Std.*, **73A**, 451
40. Hust, J. G. (1969). *Cryogenics*, **10**, 443
41. The Commission on Atomic Weights of IUPAC (1970). *Pure Appl. Chem.*, **21**, 91
42. Kent, A. (1965). *Specialized Information Centers*, 290. (Washington, D.C.: Spartan Books)
43. Stevens, M. E. (1970). *Research and Development in the Computer and Information Sciences*, Vol. 1 NBS Monograph 113, 165. (Washington, D.C.: U.S. Government Printing Office)
44. Messina, C. G. and Hilsenrath, J. (1970). *Edit-Insertion Programs for Automatic Typesetting for Computer Printout*, NBS Tech. Note 500, 47. (Washington, D.C.: U.S. Government Printing Office)

45. Turnbull, J. N. (1969). *The Organization of Data for Process Computation in the Petroleum Industry*, private communication, 1970

5
Thermodynamics of Organic Mixtures

H. V. KEHIAIAN
Université de Provence, Laboratoire de Chimie Générale, Marseille

---

5.1 THEORETICAL WORK 123
- 5.1.1 *Regular solution theory* 125
- 5.1.2 *Quasi-lattice and related theories* 126
- 5.1.3 *Flory theory* 127
- 5.1.4 *Theory of associated mixtures* 127

5.2 EXPERIMENTAL WORK 131
- 5.2.1 *Hydrocarbon mixtures* 132
  - 5.2.1.1 *n-Alkanes* 132
  - 5.2.1.2 *Branched alkanes + n-alkanes* 132
  - 5.2.1.3 *Cycloalkanes + n-alkanes* 133
  - 5.2.1.4 *Cycloalkanes* 133
  - 5.2.1.5 *Alkenes + alkanes* 134
  - 5.2.1.6 *Alkynes + alkanes* 134
  - 5.2.1.7 *Aromatic hydrocarbons + n-alkanes* 134
  - 5.2.1.8 *Aromatic hydrocarbons + branched alkanes* 135
  - 5.2.1.9 *Aromatic hydrocarbons + cycloalkanes* 135
  - 5.2.1.10 *Aromatic hydrocarbons + alkenes* 136
  - 5.2.1.11 *Aromatic hydrocarbons* 136
- 5.2.2 *Perhalogenoalkanes + alkanes* 136
  - 5.2.2.1 *Tetrachloromethane + n-alkanes* 136
  - 5.2.2.2 *Tetrachloromethane + branched alkanes* 136
  - 5.2.2.3 *Tetrachloromethane + cycloalkanes* 137
  - 5.2.2.4 *Perfluoro-n-alkanes + n-alkanes* 137
- 5.2.3 *Aliphatic* X-*compounds without* X—H *bonds + alkanes* 137
  - 5.2.3.1 *Tertiary amines* 137
  - 5.2.3.2 *Ethers* 137
  - 5.2.3.3 *Chloroalkanes* 138
  - 5.2.3.4 *Bromoalkanes* 138
- 5.2.4 *Aliphatic* X-*compounds without* X—H *bonds + aromatic hydrocarbons* 138

5.2.4.1 *Tertiary amines* 139
5.2.4.2 *Ethers* 139
5.2.4.3 *Chloroalkanes* 140
5.2.5 *Perhalogenoalkanes + aromatic hydrocarbons* 140
5.2.5.1 *Perfluoroalkanes* 140
5.2.5.2 *Tetrachloromethane* 140
5.2.6 *Aromatic* X-*compounds without* X—H *bonds + alkanes* 141
5.2.6.1 *Tertiary amines* 141
5.2.6.2 *Ethers* 142
5.2.6.3 *Chlorocarbons* 142
5.2.6.4 *Fluorocarbons* 142
5.2.7 *Aromatic* X-*compounds without* X—H *bonds + aromatic hydrocarbons* 142
5.2.7.1 *Tertiary amines* 143
5.2.7.2 *Ethers* 143
5.2.7.3 *Chlorocarbons* 143
5.2.7.4 *Fluorocarbons* 143
5.2.8 *Aliphatic* X-*compounds without* X—H *bonds + perhalogenoalkanes* 144
5.2.8.1 *Tertiary amines + tetrachloromethane* 144
5.2.8.2 *Ethers + tetrachloromethane* 144
5.2.8.3 *Thioethers + tetrachloromethane* 144
5.2.9 *Aromatic* X-*compounds without* X—H *bonds + perhalogenoalkanes* 144
5.2.10 *Miscellaneous mixtures of aliphatic and aromatic* X-*compounds without* X—H *bonds* 145
5.2.10.1 *Trichloromethane + tertiary amines* 145
5.2.10.2 *Trichloromethane + ethers* 145
5.2.11 *Aliphatic* X-*compounds containing* X—H *bonds* 145
5.2.11.1 *Aliphatic amines + alkanes* 146
5.2.11.2 *Alcohols + alkanes* 146
5.2.11.3 *Amines + aromatic hydrocarbons* 146
5.2.11.4 *Alcohols + aromatic hydrocarbons* 147
5.2.11.5 *Amines + tetrachloromethane* 147
5.2.11.6 *Alcohols + tetrachloromethane* 147
5.2.12 *Aromatic* X-*compounds containing* X—H *bonds* 147
5.2.12.1 *Aromatic amines* 147
5.2.12.2 *Phenols* 148
5.2.13 *Miscellaneous mixtures of aliphatic* X-*compounds containing* X—H *bonds* 148
5.2.13.1 *Amines* 148
5.2.13.2 *Alcohols* 148
5.2.13.3 *Amines + ethers* 148
5.2.13.4 *Alcohols + ethers* 149
5.2.13.5 *Amines + alcohols* 149
5.2.14 *Miscellaneous mixtures of aromatic and aliphatic* X-*compounds containing* X—H *bonds* 149
5.2.15 *Organic compounds containing complex functional groups* 149

5.2.15.1 *Alkanones* 149
5.2.15.2 *Alkanoates* 150
5.2.15.3 *Alkanoic acids* 150
5.2.15.4 *Sulphoxides and sulphones* 150
5.2.15.5 *Nitroalkanes* 150
5.2.15.6 *Amides* 151

ACKNOWLEDGEMENTS 151

The annual output of papers relevant to the thermodynamics of non-electrolyte solutions amounts at present to about 1100 publications dealing with the properties of more than 2000 different systems[1,2]. To avoid excessive length, this Review restricts itself to studies made on two-component mixtures of liquid organic substances, with both components well below their critical temperatures. Preference has been given to key substances and key mixtures.

This review is based essentially upon papers published since 1968 and referred to in *Chemical Abstracts* over the period from January 1969 to November 1971. Papers published in 1968 and discussed elsewhere[3–8] have been in part excluded.

Section 5.1 deals with theories which permit the interpretation, prediction and correlation of the thermodynamic properties of systems of the kind under review.

Section 5.2 describes, for each class of mixtures separately, the experimental work performed, the methods of correlation used, and the interpretation given to the observed data, in terms of molecular interactions.

5.1 THEORETICAL WORK

Two classes of mixtures are open to more fundamental treatment by statistical mechanical methods, namely, mixtures of simple spherical molecules (i.e. molecules with spherically symmetrical conformal pair interaction potentials) and mixtures of simple chain molecules (i.e. polymers built up of quasi-identical segments). The theories applicable to these two classes of mixtures have been reviewed and discussed by Rowlinson[3,6,9], Flory[7], Scott and Fenby[5], Hijmans and Holleman[10], Cruickshank and Hicks[11], Williamson and Scott[12], and Patterson *et al.*[13,14]. Only a few of the substances under review (CH_4, CF_4, CCl_4, neopentane and *n*-alkanes) classify as one or other of the two types of simple molecules. All the other organic substances consist of molecules that are more or less complex both with regard to their geometry and to their force field. Indeed, for most organic molecules, the intermolecular potential contains (besides the repulsive term and the attractive dispersion term) additional terms arising from the highly unsymmetrical permanent charge distribution within the molecules. New contributions to the statistical theory of dipolar interactions in liquid mixtures have been made by Barriol *et al.*[15]. The complication in the case of organic molecules is that they are big compared with the average intermolecular distances in the liquid state, and can therefore hardly be treated as point dipoles. This and other circumstances render the statistical treatment of organic mixtures, in terms of the intermolecular potential, extremely difficult.

A theory of solutions is regarded as successful when it correctly predicts the properties of mixtures from those of the pure components. Unfortunately, even for simple mixtures, such predictions lack precision because of the uncertainties with regard to the combining rules. For simple molecules the correction factor l_{AB} in Berthelot's geometric mean rule of the energy parameters ε_{ij}

$$\varepsilon_{AB} = l_{AB}(\varepsilon_{AA}\varepsilon_{BB})^{\frac{1}{2}} \tag{5.1}$$

is slightly less than unity[16]. A quantitative estimate of l_{AB} from molecular data being, at present, not feasible, it is gratifying when one finds that l_{AB}, adjusted to fit one property, e.g. the excess Gibbs free energy G^E, predicts correctly the other properties, e.g. the excess volume V^E and the excess enthalpy H^E. For other than simple molecules the l_{AB} factors are as complex quantities as are the interchange parameters X_{AB} (cf. below).

In default of a better statistical treatment, theories of organic mixtures make use, not directly of the ε_{ij} parameters, but of the (less informative) interchange parameters X_{AB}, defined formally by

$$X_{AB} = (\varepsilon_{AA}+\varepsilon_{BB})/2-\varepsilon_{AB} \tag{5.2}$$

To obtain the X_{AB} parameters one needs to have both experimental data on mixtures and a theoretical model relating the experimental data to X_{AB}.

The real meaning of the calculated X_{AB} depends on how effectively the particular theoretical model accounts for the liquid state properties (equation-of-state contributions[7]) and for the non-randomness of mixing, induced by differences in sizes[17] and/or in interaction energies (orientation effects) between the molecules. The better the theory, the closer is the calculated X_{AB} to the original definition, equation (5.2).

Like the earlier cell theories and corresponding-state theories, the new theory of Flory (cf. Ref. 7) accounts, in a simplified manner, for the equation-of-state contributions, but is not applicable to systems in which orientation effects are important. On the contrary, quasi-lattice and related theories (QLT) may account approximately for energy-induced non-randomness effects, but neglect completely equation-of-state contributions. It is widely, but not unanimously[7], assumed that the latter contributions could be eliminated by using experimental data, in particular energies of mixing, E^E_V, corrected for a mixing process at constant volume, along with QLT formulae. The question remains open.

As a matter of fact, no known theory of organic mixtures leads to X_{AB} parameters whose significance is perfectly clear. The numerical value of X_{AB} depends often on the kind of experimental data used in the calculations. Consequently, one is obliged to consider interchange parameters separately for G^E, H^E, etc. This dictates that care be used when interpreting X_{AB} values in terms of molecular interactions, but it does not discredit the approach as a whole, since it enables one to correlate satisfactorily the properties of a series of related systems by means of a small number of parameters and to obtain valuable information on the relative strengths of the intermolecular forces by comparing properly selected systems, in which energy effects other than interchange are likely to be the same (e.g. homomorphic systems).

As a rule, in systems with only weak dispersive interactions (small X_{AB}, slight deviations from ideality, no orientation effects) the equation-of-state contributions should be explicitly taken into account. On the other hand, for systems with large differences between the interaction energies (large X_{AB}, orientation effects, usually large deviations from ideality) the equation-of-state contributions play a secondary role (in part they are included, discretely, in the interchange parameters) which grants some success to the more rudimentary models of solutions. Keeping this in mind, one should not expect to obtain agreement with experiment by QLT and related theories, to better than tens of joules per mole for the molar excess energies, and it is not helpful to improve the agreement by increasingly arbitrarily the number of adjustable parameters.

When appropriate, the interchange parameters are defined not for the molecules as a whole, but separately for a number of relative orientations. This aims to distinguish between different segments, contact points, surface regions, or groups on each type of molecule.

5.1.1 Regular solution theory

Hildebrand, Prausnitz and Scott[4] (cf. also reference 18) have given a full and up-dated account of this theory and its applications.

Recent studies[19] prove that the Scatchard–Hildebrand equation

$$E_V^E = (x_A V_A + x_B V_B)\, \phi_A\, \phi_B X_{AB} \tag{5.3}$$

in which x_i, V_i and ϕ_i refer to the mole fraction, molar volume and volume fraction, predicts fairly well the symmetry of the E_V^E curve with its maximum situated near the smaller component. The characteristic assumption of the Scatchard–Hildebrand equation is expressed by the relation

$$X_{AB} = (\delta_A - \delta_B)^2 \tag{5.4}$$

between the interchange parameter X_{AB} and the solubility parameters δ_i.

In general, for non-polar systems, the predictions made by equation (5.4) are in qualitative agreement with experiment. Striking disagreement with the predicted values indicates strong like, or strong unlike, interactions. The latter are easier to detect, since they lead to X_{AB} much inferior to $(\delta_A - \delta_B)^2$ and, in extreme cases, to $X_{AB} < 0$. Strong like interactions between polar molecules in a non-polar solvent are best reflected by the maximum of the H^E curve being shifted from the position given by equation (5.3), towards the non-polar component.

The simplicity of the solubility parameter concept has continued to tempt authors to extend this concept to polar and H-bonded systems[20–22], and to seek relationships between δ_i and the chemical structure of substances (splitting of δ_i into different terms representing group contributions)[23–25].

More important are the refinements of the regular solution theory which permit one to analyse V^E within the framework of this originally constant-volume approach, and to allow for deviations from the geometric-mean combining rule[4, 18, 26–28]. Bagley *et al.*[22] have emphasised the usefulness of experimental measurements of the internal pressure of liquids, which is shown to describe a volume-dependent δ_i (cf. also reference 18). Undoubtedly, by such successive refinements and extensions, the solubility parameter theory will finally merge into other theories.

5.1.2 Quasi-lattice and related theories

Cruickshank *et al.*[29–34] have shown the possibility of correlating infinite dilution activity coefficients, γ_A^∞, of various hydrocarbon solutes (A) in *n*-alkane solvents (B) by means of QLT. For this purpose, γ_A^∞ is separated into a configurational part given, in the Flory approximation, by

$$\ln \gamma_{A(conf.)}^\infty = \ln (V_A/V_B) + [1-(V_A/V_B)] \tag{5.5}$$

and an interaction part given by

$$\ln \gamma_{A(int.)}^\infty = X_{AB}/RT \tag{5.6}$$

The overall free-energy interchange-parameter X_{AB} is decomposed into a sum of terms, each term corresponding to the process of interchanging two different types (s,t) of segments or contact points.

$$X_{AB} = -\tfrac{1}{2}\sum_s \sum_t (\alpha_{As}-\alpha_{Bs})(\alpha_{At}-\alpha_{Bt})X_{st} \tag{5.7}$$

This expression is equivalent to that proposed by Redlich *et al.*[35] in a group- or surface-interaction version of the theory. The parameter α_{As} represents the 'surface fraction' of type s surface on the molecule. Kehiaian *et al.*[36, 37] have correlated H^E data by means of equation (5.7), regarding the X_{st} terms as 'enthalpy of interchange' parameters.

A still more empirical version of QLT is one dealing in terms of a 'solution of groups'. The possibilities of this approach for correlating and predicting activity coefficients of structurally related substances have been reviewed by Deal and Derr[38]. Ratcliff *et al.*[39] have presented a group solution model in which the principle of congruence is used to account for the differences between the molecular sizes.

Equation (5.7) assumes random mixing of the intermolecular contact surfaces. The common way of accounting for orientation effects, within the framework of the various versions of QLT, is essentially Guggenheim's quasi-chemical approach. Barker's generalised QLT (contact-point version) has continued to be widely applied to many systems, e.g. cyclohexane + *n*-alkanes[40], aromatic hydrocarbon mixtures[41–43], ethers + CCl_4 or $CHCl_3$[44–46], hydrocarbons + CCl_4[47], or C_6H_5F[48], C_6F_5H + 1,4-dioxane[49], etc. The theory is usually compared with experimental data corrected for mixing at constant total volume. This correction does not eliminate, however, the need for two different sets of interchange parameters, one for G^E and another for E_V^E. There are more important factors which affect the agreement.

Kershaw and Malcolm[45] have emphasised that for the correlation to be successful, the groups or atoms, characterised by a given set of interchange parameters, must be in the same intramolecular environment in all the substances under consideration. The need for a more systematic determination of the interchange parameters has also been emphasised[36, 37].

Huggins' theory of solutions[50] is essentially the surface version of Barker's QLT. In a recent extension of his theory, Huggins[51] attempts to interpret V^E by an equation similar to that derived for H^E, in terms of an additional set of adjustable parameters for the volumes of interchange.

The group interaction theory of Redlich *et al.*[35] has been developed by Chao *et al.*[52], to correlate H^E in polar systems, e.g. *n*-alcohols + *n*-alkanes.

5.1.3 Flory theory

The essential features of this theory have been outlined recently by Flory[7]. Scott and Fenby[5] and Patterson and Delmas[14] have compared this theory with other corresponding-states theories, and Cruickshank and Hicks[11] have made the comparison with the simple QLT.

In constructing the partition function of the system the basic assumption is made that the intermolecular energy $\mathscr{U}$ is inversely proportional to the volume. Otherwise, $\mathscr{U}$ is assumed to depend on the contact surface areas of the molecules, as in the surface formulation of the random-mixing QLT. The superiority of the Flory theory is its ability to predict H^E and V^E with the same interaction parameter X_{AB}. The properties of structurally related substances may be correlated by splitting X_{AB} into a sum of individual contributions, cf. equation (5.7). Unfortunately, in order to obtain the molar excess entropy, S^E, and thus also G^E, one must introduce an additional set of parameters to represent the interaction entropy of interchange[7, 53].

Recent applications of the Flory theory include the following systems: *n*-alkanes[11], cycloalkanes[54, 55], aromatic hydrocarbons[41–43], aromatic hydrocarbons + cycloalkanes[56] or CCl_4 [57], etc.[58, 59].

5.1.4 Theory of associated mixtures

When writing the configurational partition function of a mixture A + B in terms of the intermolecular potential energies, one assumes separability of the internal degrees of freedom from the external ones. If the molecular interactions are strong enough to affect profoundly the internal partition functions Z_A^* and/or Z_B^* (i.e. through the occurrence of 'chemical interactions'), it is justifiable to treat the mixture as consisting not solely of 'monomer' species A and B, but additionally of as many other kinds of 'true molecular species', A_iB_j, as are necessary to grant separability to the degrees of freedom in the enlarged system ('associated mixtures'). All purely physical interactions between A, B and A_iB_j appear then in the true configurational partition function Z_{conf}, whereas the chemical interactions are included in the internal partition functions, $Z_{A_iB_j}^*$.

However clear it might appear, the above definition of associated mixtures is not of much practical help in establishing whether a given real mixture is associated or not. The mixing of two liquids is probably always accompanied by a more or less important change of the internal degrees of freedom. Depending on both system and property, this change may prove sometimes to be relevant, and other times insignificant. Spectroscopic techniques provide a valuable tool for detecting perturbations at the molecular or atomic levels, but, usually, they do not supply sufficient information as to the stoichiometry of the complexes. Moreover, there seems to be a poor parallelism between the spectral evidence for complex formation and the thermodynamic behaviour of the system.

At present we cannot write a general intermolecular potential function for purely physical interactions and even if we could, the calculation of the configurational integral would require a number of drastic approximations. Consequently, for lack of a well-defined purely physical reference system, we cannot yet trace a precise borderline between associated and non-associated mixtures.

Fortunately, the history of solution thermodynamics shows that in the battle between 'chemical' and 'physical' theories, the former lose more and more ground in favour of the latter. This evolution is to the benefit of both theories since it leads to an increasingly sharper distinction between chemical and physical forces.

It is important to realise that the theory of associated mixtures must not be developed ignoring the theory of simple non-associated mixtures, but should follow as closely as possible the progress achieved in the latter. It is true that the form of Z_{conf} of a given associated mixture cannot be established directly, because we cannot inhibit the chemical reactions occurring in the system. A plausible Z_{conf} can be written, however, on the basis of a careful examination of the properties of systems of similar non-associating molecules and of classes of structurally related substances.

The different models of associated mixtures are named in accordance with the assumed form of Z_{conf} (ideal associated, athermal associated[60], etc.).

Many papers have been published in recent years on associated systems. Only a few, dealing primarily with mixtures (not with diluted solutions), will be referred to here.

The model for the ideal associated mixture has been used to interpret the following systems (the assumed types of complexes are given in parentheses): ethers + $CHCl_3$ [61,62] (AB); $CH_3OH + C_2H_5COOH$ [63] (AB_2); tetrahydofuran + $CHCl_3$ [64] ($AB + AB_2$); pyridine + $CHCl_3$ [64] and aromatic hydrocarbons + CCl_4 [65] ($A_2 + AB$); i-C_3H_7OH + i-$C_3H_7NH_2$ [61] ($A_2 + B_2 + AB$); $CH_3OH + n\text{-}C_6H_{14}$ [66] and $C_2H_5OH + n\text{-}C_7H_{16}$ [67] ($A_2 + \ldots A_r$). In general, the theoretical equations may be fitted to the experimental data, which is not surprising in view of the relatively large number of disposable parameters. But the approach as a whole should not be regarded as satisfactory. A comparison with the properties of other simpler systems, e.g. $CHCl_3 +$ n-alkanes or i-C_3H_7OH + n-alkanes, shows that the interchange energies between $CHCl_3$ and the hydrocarbon segments in ethers are not negligible, that i-C_3H_7OH association cannot be accounted for in terms of dimerisation, and so on. It is particularly erroneous to neglect size effects in treating associated mixtures, e.g. mixtures of alcohols or amines with n-alkanes. One finds always that H^E depends on the size of the n-alkane[68].

The Flory theory of athermal associated mixtures[69] takes account of differences in sizes in terms of Flory's combinatorial factor for chain molecules. The theory of Saroléa[70] uses Guggenheim's combinatorial factor. It reduces to Flory's theory when making the lattice coordination number $z \to \infty$.

As is well known, in Flory's statistics concentration must be expressed in volume fractions[69] (whereas surface fractions are used in Guggenheim's statistics) in order to obtain concentration-independent classical association constants

$$K^{\phi}_{i,i+1} = \phi_{A_{i+1}}/\phi_{A_i}\phi_A \tag{5.8}$$

The important point is that the combinatorial factor, once fixed in Z_{conf}, determines the expression for the combinatorial entropy of mixing the true components, the proper concentration scale, and also the combinatorial part in the entropy of association. The latter leads to a certain relationship between the consecutive thermodynamic association constants $K_{i,i+1}$ ($i = 1, 2, \ldots$) of a given monomer. In Flory's statistics[69]

$$K_{i,i+1} = K\frac{i+1}{i}(i \geqslant 1) \tag{5.9}$$

where K is independent of i, whereas for ideal systems

$$K_{i,i+1} = K(i \geqslant 1)\ \text{(Mecke–Kempter)} \tag{5.10}$$

Other factors, pertaining to the internal partition functions $Z^*_{A_i}$, may induce further differentiation between the $K_{i,i+1}$ values. These are usually accounted for empirically in two-parametric models with equation (5.9) valid for $i > 1$ and an independent $K_{1,2}$. No justification exists, however, for increasing the number of different association constants within the model of ideal associated mixtures, in order to repair its inherent deficiencies.

Kehiaian and Treszczanowicz[71–73] have studied in detail the properties G^E, H^E, S^E, and the molar excess heat capacity C^E_p, for various models of athermal associated mixtures using Flory's statistics. For the continuous association model their basic equations are identical with those of Kretschmer and Wiebe[74].

In Wiehe and Bagley's approach[75] (cf. also references 73, 76) the Flory entropy of mixing of the true components is used along with the Mecke–Kempter correlation, equation (5.10). This makes the approach inconsistent from the point of view of statistical thermodynamics.

The simplest procedure to account for physical contributions in associated mixtures theory (cf. reference 5) is to supplement the chemical term in the equations describing the excess functions, with a physical term of the Scatchard–Hildebrand form, equation (5.3)[74, 77]. A larger variety of behaviour may be predicted if we consider, instead of one overall interchange parameter X_{AB}, as many interchange parameters $X_{A_{i_1}B_{j_1}, A_{i_2}B_{j_2}}$ as there are pairs of true components A_iB_j ('regular associated mixtures')[78–80]. This model involves a considerable number of parameters. In applications some simplifying correlations between the parameters have been postulated, e.g. Renon and Prausnitz[81] interpreted several alcohol + hydrocarbon mixtures with the Flory–Kretschmer–Wiebe equations, assuming $X_{A_i,B}$ and $X_{A_{i_1},A_{i_2}}$ to be independent of the i values. This led to a physical term of the form given by equation (5.3). Other investigations along similar lines have been made on the systems: $C_6H_5OH + CCl_4$ [80], C_2H_2 + organic solvents[82], $C_2H_5OH + CCl_4$ [83], C_6F_5H + 1,4-dioxane[49], etc.

The statistical theories of associated mixtures developed so far have been based on QLT. The number of published papers is very limited. Among recent ones we note a review article by Smirnova[84] and two papers by Pouchlý *et al.*[85], the latter dealing more specially with polymer solvation. McGlashan *et al.*[86] have used a simple non-athermal A + B + AB model (cf. reference 87) to interpret the properties of aromatic hydrocarbon + CCl_4 mixtures.

The surface formulation of QLT can be easily developed to include chemical interactions[88]. We assume that besides the intermolecular physical interactions occurring between any two regions of the molecular contact surfaces, certain regions interact chemically through the formation of complexes A_iB_j. The physical interactions are treated as random. In the configurational partition function of the enlarged system

$$Z_{conf} = g \exp\{-U/RT\} \tag{5.11}$$

g is Guggenheim's combinatorial factor and U the intermolecular energy expressed in terms of surface interactions. The overall interchange parameter X_{AB} of the pair of monomers A and B is then given by equation (5.7).

Let us assume that the contact surface of the complex A_iB_j is composed of the joint total surfaces of i molecules A and j molecules B. This is only an approximation, since the chemically-reacted surface regions in A_iB_j do not come into physical contact with the surroundings. In this 'surface additivity assumption' the overall interchange parameter $X_{A_{i_1}B_{j_1},A_{i_2}B_{j_2}}$ is given by

$$X_{A_{i_1}B_{j_1},A_{i_2}B_{j_2}} = \left[\frac{q_A q_B (i_1 j_2 - i_2 j_1)}{(i_1 q_A + j_1 q_B)(i_2 q_A + j_2 q_B)}\right]^2 X_{AB} \tag{5.12}$$

where q_A and q_B are the surface areas of the monomers. Note that for $i_1 = i$, $j_1 = i_2 = 0$, and $j_2 = 1$, $X_{A_i,B} = X_{AB}$, whereas for $j_1 = j_2 = 0$, $X_{A_{i_1},A_{i_2}} = 0$, in agreement with the assumptions of Renon and Prausnitz[81] (see above).

Equation (5.12) leads to excess functions consisting of a chemical term and a physical term of the Scatchard–Hildebrand form, equation (5.3) (with surface fractions instead of volume fractions). Classical association constants have to be expressed in surface fractions in order to be independent of concentration. The novel result of this theory is an equation that correlates the association constants of substances forming homologous series, such as K_{AB} constants of $CHCl_3$ + ethers or $K_{i,i+1}$ constants of n-alcohols, etc. For example, one obtains

$$K_{i,i+1} = \frac{i+1}{iq_A} K' \tag{5.13}$$

where K' is but slightly dependent on i. For a given monomer, q_A is constant and equation (5.13) reduces to equation (5.9). For a homologous series of substances, equation (5.13) predicts association constants inversely proportional to their total contact surface q_A.

The physical interaction parameters X_{AB} can also be correlated for members of homologous series. For n-alcohols + n-alkane systems, cf. equation (5.7), X_{AB} should be roughly inversely proportional to q_A^2.

Assuming that the contact surfaces of the molecules are proportional to their volumes (this is, in fact, the physical meaning of Flory's approximation $z \rightarrow \infty$), one obtains all the equations of the Flory–Kretschmer–Wiebe theory.

The surface-additivity assumption leads to solvent- and concentration-independent classical association constants in disagreement with experimental evidence. In order to interpret solvent effects within the framework of this theory, it is necessary to account for the fact that different solvents interact differently with the chemically active regions of the solute molecules. This is not yet sufficient to give a physical interpretation of the solvent effect, and for this we must specify the surface structure of the complexes more realistically than in the surface-additivity assumption.

The solvent effect raises arguments (see, e.g. references 89, 90) as to the choice of the 'best' solvent for determining association constants, etc. The problem presents difficulties, since in liquid media association constants represent a complicated average free energy of interchange of chemical contacts in A_iB_j for physical contacts with the surrounding medium. For association constants referred to the pure liquid, A and B, this medium is the (hypothetical) physical mixture $iA + jB$, and for infinite dilution association constants it is the solvent. Only a complete statistical treatment of the liquid mixture can give the true significance of these constants and the solvent-dependency of the latter.

5.2 EXPERIMENTAL WORK

In Rowlinson's monograph[3], systems are classified in several broad classes (alkanes, aromatic hydrocarbons, fluorocarbons, etc.) according to their chemical structure. Hildebrand and Scott[18] have emphasised that in dealing with molecular interactions one should recognise distinct types of molecules depending on their electronic structure. As far as we are concerned these types are: (1) molecules with none but bonding electrons; (2) molecules with π-electrons; (3) molecules with lone-pair non-bonding electrons; (4) molecules capable of forming H bonds and (5) molecules containing strong permanent dipoles.

The classification adopted in this Review follows the above pattern. We consider hydrocarbon compounds and hetero-organic compounds (briefly, X-compounds). In the latter, the heteroatom X is N, O, F, Cl, and occasionally Br or S.

We distinguish two classes of X-compounds, (i) those without X–H bonds and (ii) those with X–H bonds. In general the former are not capable of forming H-bonds ($CHCl_3$ is an exception), whereas the latter, e.g. alcohols, can do so.

Moreover, we distinguish X-compounds in which (a) X is linked to C by formal σ-bonds, C–X, e.g. ethers; (b) X is contained in a heterocyclic ring, C$\overset{\cdots}{-}$X, e.g. pyridine; (c) X is linked to C (or to other heteroatoms) by multiple or coordinate bonds, e.g. ketones, sulphones, nitroalkanes, etc.

This review deals mainly with hydrocarbons and with X-compounds of types (a) and (b). X-compounds of type (c) belong to the category of molecules with strong permanent dipoles and are only briefly treated in Section 5.2.15.

5.2.1 Hydrocarbon mixtures

5.2.1.1 n-Alkanes

An account of the general features of *n*-alkane mixtures, and a compilation of accurate measurements prior to 1968 has been given by Rowlinson[3]. Recent studies include liquid–vapour (L–V) equilibria for: $CH_4+C_2H_6$ [91], C_3H_8 [91, 92], *n*-C_5H_{12} [93, 94]; $C_2H_6+C_3H_8$ [91, 95, 96], *n*-C_6H_{14} [97], *n*-C_8H_{18} [98], *n*-$C_{12}H_{26}$ [99]; C_3H_8+n-C_4H_{10} [91, 96, 100], *n*-C_5H_{12} [100], *n*-C_6H_{14} [101], *n*-C_7H_{16} [101]; and *n*-$C_4H_{10}+n$-C_6H_{14} [102]. Densities and V^E have been reported for: $CH_4+C_3H_8$ [92]; C_2H_6+n-$C_{12}H_{26}$ [99]; and other mixtures of lower (C_1–C_4) [103] or higher (C_6–C_{16}) [104, 105] homologues.

The relatively small deviations from ideality of *n*-alkane mixtures are mainly due to equation-of-state contributions. The difference in force fields surrounding CH_3 and CH_2 groups is likely to be very small[13]. QLT attributes the entire non-ideality, apart from size effects, to a non-zero X_{st} between CH_3 and CH_2 segments. The theory successfully correlates γ_A^∞ values (cf. equations (5.5) and (5.6)), as obtained by gas–liquid partition chromatography (GLPC)[29–33, 106]. Maffiolo *et al.*[107] have reviewed the subject.

Williamson and Scott[12], Cruickshank and Hicks[11] and Patterson and Bardin[13] have discussed the different theoretical treatments of *n*-alkane mixtures. Stoeckli and Staveley[92] have obtained a reasonably good estimate of G^E, H^E and V^E of $CH_4+C_3H_8$ with the very recent McGlashan one-fluid theory[108] of simple mixtures. The unusually small value derived for l_{AB}, equation (5.1), is probably due to the non-spherical shape of C_3H_8.

The principle of congruence has been tested on V^E data[103, 104] and shown to be inadequate for the lower homologues (C_1–C_4)[103], but to hold for the higher[104].

5.2.1.2 Branched alkanes + n-alkanes

Of interest to the theory of simple mixtures are the recent L–V equilibrium studies on neo-$C_5H_{12}+CH_4$ by Rogers and Prausnitz[109] and on neo-$C_5H_{12}+$ *n*-C_5H_{12} by Hoepfner *et al.*[102]. Other recent L–V equilibrium data, for more technical purposes, include the systems: i-$C_4H_{10}+CH_4$ [110], C_2H_6 [91], C_3H_8 [91, 96], *n*-C_4H_{10} [96]; i-$C_5H_{12}+CH_4$ [93]; and neo-$C_5H_{12}+CH_4$ [93].

As is well known[3], branched alkane + *n*-alkane mixtures exhibit more positive deviations from ideality than the corresponding *n*-alkanes. Tewari *et al.*[106] have measured, by GLPC, γ_A^∞ of branched isomeric hexanes, heptanes, and octanes in *n*-alkanes. The QLT leads to inconsistent results[29, 106], which is not surprising in view of the large differences in structure between the isomers.

5.2.1.3 Cycloalkanes + n-alkanes

Cyclohexane is the only cyclic alkane studied in mixtures with *n*-alkanes. Mostly reported were H^E data with n-C_6H_{14} [19, 111–116], n-C_7H_{16} [19], n-C_8H_{18} [40] and n-$C_{10}H_{22}$ [40]. The spate of measurements with n-C_6H_{14} followed McGlashan's proposal to use cyclohexane + n-C_6H_{14} as the standard system for testing heat-of-mixing calorimeters. This proposal has now been accepted internationally by the IUPAC Commission I.2. The experimental results may be represented by the polynomial

$$H^E = x_B(1-x_B)\left[H_0+H_1(1-2x_B)+H_2(1-2x_B)^2+H_3(1-2x_B)^3\right] \quad (5.14)$$

where x_B is the mole fraction of n-C_6H_{14}. The coefficients H_i(298.15 K) J mol^{-1} have been determined by Ewing *et al.*[115] with high precision using three different designs of apparatus. The three series of experiments lead to results agreeing within 0.3 J mol^{-1}. One set of coefficients is $H_0 = 864.55$, $H_1 = 247.70$, $H_2 = 99.92$, $H_3 = 35.46$. The maximum value of H^E (298.15 K) is 221.0 J mol^{-1} at $x_B = 0.421$, and C_p^E at the maximum is -1.39 J K^{-1} mol^{-1}.

Other properties of this class of mixtures have received little attention recently. G^E has been reported for cyclohexane + n-C_8H_{18} [59] and V^E for cyclohexane + n-C_5H_{12} [117].

The excess functions of cyclohexane + *n*-alkane mixtures are considerably more positive than those of *n*-alkane mixtures of comparable sizes, e.g. H^E of cyclohexane + n-C_7H_{16} is ~240 J mol^{-1}, whereas H^E of n-C_6H_{14} + n-C_7H_{16}, ~15 J mol^{-1}*. It is noteworthy[36] that methylcyclohexane behaves similarly to *n*-alkanes, and its H^E with n-C_7H_{16} does not exceed 35 J mol^{-1}. This suggests that particular structural factors in liquid cyclohexane are responsible for the observed non-ideality.

Jain and Yadav[59] have interpreted G^E, H^E and V^E for cyclohexane + *n*-alkanes in terms of the Flory theory. Vesely and Pick[40] have correlated H^E data by use of Barker's QLT.

The solid–liquid phase diagram of cyclopentane + CH_4 [118] reveals partial miscibility of the components below 166 K.

Further studies in this area can be expected to include the following compounds: the series of cycloalkanes $(CH_2)_n$ ($n = 3$–10), their monomethyl derivatives, the *cis*- and *trans*-decalins, and globular bicyclo-[2,2,2]-octane.

5.2.1.4 Cycloalkanes

The assumed globular shape and the regular structure (CH_2 segments only) of cycloalkanes have interested investigators in cycloalkane systems, and

---

*All excess functions given in this review relate equimolar mixtures at ~25 °C.

careful studies have been made on cyclopentane+cyclo-octane[119, 120], cyclopentane+cyclohexane[121, 122], *cis*- and *trans*-decalins+cyclohexane[19, 55] and *cis*-+*trans*-decalins[14, 54, 123]. The studies have revealed the complex behaviour of the mixtures and it appears that current statistical theories do not interpret the data satisfactorily[54, 55, 120]. Studies of $(CH_2)_n$+*n*-alkane mixtures (cf. Section 5.2.1.3) might show that there are important structural and energetic differences between the cycloalkane homologues.

5.2.1.5 *Alkenes + alkanes*

Recent investigations have dealt mainly with L–V equilibria in: propene+ CH_4 [124], C_2H_6 [96], C_3H_8 [125–127]; 1- and 2-butenes+*n*-C_4H_{10} [128]; propadiene+ C_3H_8 [96, 127] and 1,3-butadiene+*n*-alkanes[129]. Though few, and of only moderate accuracy, these data indicate that 1-alkanes, despite their π-electrons, do not show larger deviations from ideality in mixtures with *n*-alkanes than do *n*-alkanes or branched alkanes. This peculiarity is confirmed by the systematic GLPC measurements on γ_A^{∞} in 1-alkene solutes in *n*-alkanes[33, 106]. The interchange parameters of the CH_3 and $CH_2{=}CH$ groups with CH_2 are similar.

A successful correlation of the γ_A^{∞} data was obtained[106] by use of the first-order perturbation treatment developed by Luckhurst and Martire[130], using Kreglewski's method[163] of calculating the depth of the intermolecular potential-well of the solute.

For a better knowledge of this interesting class of mixtures, systematic G^E, H^E and V^E data are needed. Since 2-butene+*n*-C_4H_{10} is reported[128] to exhibit positive deviations, we may expect 2-alkenes to behave differently from 1-alkenes. Further studies are needed to determine the effect of the relative position of the double bonds on the properties of alkadienes+ *n*-alkanes.

5.2.1.6 *Alkynes + alkanes*

A few disparate L–V equilibrium studies have been performed for technical purposes[128, 131, 132]. As expected, hexyne and heptyne+*n*-C_7H_{16} exhibit positive deviations[133]. In view of the H-donor ability of 1-alkynes, which could manifest itself through a sort of 'self-association', a comparative study of 1-, 2-, and 3-alkyne+*n*-alkane systems would be of interest. This might better permit the separation of the contribution of like from unlike interactions in mixtures of alkynes+acetone[134] or other proton acceptors.

5.2.1.7 *Aromatic hydrocarbons + n-alkanes*

Much work has been carried out on this important class of mixtures. L–V equilibria (mostly isobaric) have been measured for C_6H_6+*n*-C_6H_{14} [135] and *n*-C_7H_{16} [136]; toluene+*n*-C_6H_{14} [135] and *n*-C_7H_{16} [137]; xylenes+*n*-C_6H_{14} and *n*-C_7H_{16} [135]. G^E has been calculated from isothermal L–V data for

$C_6H_6 + n\text{-}C_6H_{14}$ [138–140] and $n\text{-}C_7H_{16}$ [139, 141]. A series of papers report H^E of alkylbenzenes + n-alkanes[113, 142, 143] and V^E of C_6H_6 and alkylbenzenes + n-alkanes[117, 144–146].

As is well known[3], aromatic hydrocarbon + alkane mixtures have much larger values for G^E and H^E than do mixtures of alkanes. This is probably due to the relatively strong interactions between aromatic π-electrons (π–π interactions[147]) compared with the aliphatic–aliphatic and aromatic–aliphatic interactions. Alkylation of the benzene ring, reducing the aromatic surface fraction on the molecules, produces a decrease of H^E and of the interactional G^E. In these terms one may correlate the γ_A^∞ [30, 34, 106] and H^E [36] data on C_6H_6 or n-alkylbenzene + n-alkane systems. For polyalkylbenzenes this approach becomes more ambiguous since one has the choice between adjusting geometric factors[143] or interaction parameters[106].

The Flory theory has been shown to predict H^E and V^E fairly well by means of a single interchange parameter[117, 144, 145].

5.2.1.8 Aromatic hydrocarbons + branched alkanes

Funk and Prausnitz[26] have made a very useful contribution in this area. Using L–V equilibrium and H^E data, they have calculated G^E of 23 mixtures of C_6H_6 or toluene with various alkanes, corrected to 25, 50 and 75 °C, and reduced with the Redlich–Kister, NRTL and Wilson equations. The G^E and H^E data have been interpreted with the refined regular solution theory[18]. A linear relation was found between l_{AB}, equation (5.1), and the so-called branching parameter, i.e. the ratio of number of CH_3 groups to total number of C atoms in the alkane.

The effect of branching on γ_A^∞ of alkanes in n-alkylbenzenes has also been studied by Gainey and Pecsok[34].

5.2.1.9 Aromatic hydrocarbons + cycloalkanes

As in the past[3], C_6H_6 + cyclohexane has been frequently investigated, and the values of H^E [111, 115, 148] and V^E [149, 150] are known at present with an accuracy of, respectively, 1.1 J mol^{-1} and 0.0005 cm^3 mol^{-1}. G^E has been re-determined from isothermal L–V equilibria [141, 151].

Other recent investigations include L–V equilibria[152] and H^E [113] of toluene + cyclohexane, H^E of isomeric xylenes + cyclohexane[153], and H^E and V^E of toluene + methylcyclohexane[154].

The existing data indicate that the interaction of the aromatic ring is stronger with the cyclohexane ring than with a non-cyclic alkane. According to the correlation given by Funk and Prausnitz[26] (cf. Section 5.2.1.8), for cycloalkanes the branching parameter being equal to zero, ε_{AB} equals $(\varepsilon_{AA}\varepsilon_{BB})^{\frac{1}{2}}$.

Benson and Singh[56] have shown that published H^E and V^E data on cyclohexane + C_6H_6 or toluene can be fairly well interpreted by Flory's theory, whereas other approaches proved less satisfactory.

5.2.1.10 Aromatic hydrocarbons + alkenes

The first systematic studies in this field are due to Vera and Prausnitz[155]. L–V equilibrium and Redlich–Kister and Wilson parameters for G^E have been reported. A comparison of G^E, H^E and V^E [146] of C_6H_6+cyclohexane with the corresponding functions of C_6H_6+cyclohexane show that the deviations from ideality are smaller in mixtures with alkenes than with alkanes. This is probably due to the compensating effect of unlike π–π interactions between benzene and alkene.

5.2.1.11 Aromatic hydrocarbons

Extensive H^E and V^E measurements have been carried out by Benson *et* al.[41–43], Rastogi *et al.*[156], and other authors[19, 157–161]. Accurate G^E has been obtained for C_6H_6+*m*-xylene only[162] (cf. also reference 135).

The deviations from ideality are relatively small because of the compensation of like and unlike π–π interactions. H^E is larger, the larger the difference between the aromatic surface fractions of the molecules. Polyalkylbenzenes are not sufficiently closely related structurally to permit an accurate correlation of their properties, unless one introduces an excessively large number of parameters, cf. for example, reference 43. For a given system, Flory's theory is shown[41–43] to relate V^E reasonably well to H^E.

5.2.2 Perhalogenoalkanes+alkanes

This represents one of the simplest classes of organic mixtures. We shall confine ourselves to CCl_4 and perfluoro-*n*-alkane systems.

5.2.2.1 Tetrachloromethane + n-alkanes

It is only very recently that G^E [138, 164], H^E[165, 166] and V^E [145, 165] have been measured systematically for these mixtures. Both QLT [37] and Flory theory[145, 166] are likely to provide a reasonable correlation of the data.

5.2.2.2 Tetrachloromethane + branched alkanes

CCl_4+neo-C_5H_{12} is an excellent system for testing theories of simple mixtures of spherical molecules of nearly equal sizes[3]. A recent paper by Chang and Westrum[167] reports the heat capacities of this system.

Rodger *et al.*[47, 168] have measured L–V equilibria for 23 CCl_4+hydrocarbon systems and interpreted these in terms of group contributions.

5.2.2.3 *Tetrachloromethane + cycloalkanes*

CCl_4 + cyclohexane is one of the most thoroughly investigated non-electrolyte mixtures. Accurate H^E measurements have been performed recently by Ewing and Marsh[121]. Ocon *et al.*[169] have reported V^E and L–V equilibrium data.

CCl_4 + cyclopentane has been extensively studied by Boublik *et al.*[170]. Equations derived from a first-order perturbation treatment of hard spheres provided a satisfactory estimate of the excess functions[171].

5.2.2.4 *Perfluoro-n-alkanes + n-alkanes*

In a notable paper, Young[172] has reported the upper critical solution temperatures of C_4F_{10} and C_7F_{16} + n-alkanes (C_5–C_{15}). A three-segment-type (CH_3, CH_2, and 'average' fluorocarbon) QLT has been shown to provide an excellent correlation of the data.

5.2.3 Aliphatic X-compounds without X–H bonds + alkanes

Due to the more-or-less polar character of the C—X bond, like interactions in the X-compounds are strengthened by an electrostatic term. On the other hand, the net dispersion forces between N, O and F are probably weaker than between the respective isoelectronic groups CH, CH_2 and CH_3 [173]. The excess functions of this class of mixtures are thus determined largely by the balance of these two counteracting effects in a manner which merits detailed investigation.

The first studies in this field are of very recent date. To our knowledge no measurements have yet been reported on mixtures of alkanes with alkyl fluorides, thioethers or trialkyl phosphines.

5.2.3.1 *Tertiary amines*

Mixtures of homomorphs are nearly ideal [68, 174], e.g. $(C_2H_5)_3N + (C_2H_5)_3CH$ has $H^E \sim 40$ J mol^{-1}. Other investigated systems, viz., $(C_2H_5)_3N + n\text{-}C_7H_{16}$ [175], $(CH_3)_3N$ [177] or $(C_2H_5)_3N$ [178] with $n\text{-}C_6H_{14}$ or $n\text{-}C_7H_{16}$, N-methylpiperidine + methylcyclohexane[176] also exhibit relatively small departures from ideality.

5.2.3.2 *Ethers*

Two classes of systems present a particular interest, viz., n-ethers + n-alkanes and cyclic ethers + cycloalkanes. The first H^E measurements on n-ethers + n-alkanes have been reported by Sosnkowska-Kehiaian *et al.*[36, 173]. A systematic research programme on cyclic ethers + hydrocarbons has been initiated by Andrews and Morcom[179]. In the last few years, H^E, V^E and a few G^E data have been reported on cyclohexane + trimethylene oxide[179], tetrahydropyran[122, 180, 181], and 1,3- and 1,4-dioxanes[122, 179].

Structural requirements being fully met for the chain-like *n*-ethers + *n*-alkanes, a satisfactory correlation was obtained[36] by use of the surface-version QLT. It will be interesting to test a similar treatment on the cyclic ethers.

5.2.3.3 *Chloroalkanes*

A systematic investigation of chloroalkanes should deal primarily with the following series: 1-chloro-, α, ω-dichloro-, 1,1-dichloro- and 1,1,1-trichloroalkanes.

The first studies to be made on 1-chloroalkanes + *n*-alkanes have been reported recently, and give measurements of γ_A^∞ [182], V^E [183] and H^E [184].

1,2-Dihalogenoethanes have long aroused interest owing to their conformational isomerism. The 1,2-$C_2H_4Cl_2$ + cyclohexane mixture has been most thoroughly investigated, mainly by Kohler *et al.*[185–187]. Bissel *et al.*[165] have reported values for H^E and V^E for CH_2Cl_2 + *n*-alkanes.

Trichloromethane has been investigated very frequently in mixtures with complex substances, but not, until very recently[37, 165, 181], in mixtures with *n*-alkanes. A comparison[34] of H^E in the series:

	Cl—CCl_3 + *n*-C_7H_{16}	CH_3—CCl_3 + *n*-C_7H_{16}	H—CCl_3 + *n*-C_7H_{16}
H^E/J mol^{-1}	340	660	780

suggests that the like interactions in $CHCl_3$ are of the dipole–dipole type rather than due to C—H···Cl H-bonding.

H^E [184] and γ_A^∞ [182] values for 1-chloroalkane + *n*-alkane systems can be fairly-well correlated by use of QLT. The properties of α, ω-dichloroalkane systems may be predicted with the same interchange parameters, provided that the Cl atoms are separated by at least three to four CH_2 groups[184]. The force field surrounding a Cl atom in 1-chloroalkanes is, however, not the same as, for example, in 1,1-dichloroalkanes, which explains why the γ_A^∞ coefficients of both groups of substances fail to correlate with a single set of parameters[182].

5.2.3.4 *Bromoalkanes*

Noteworthy papers have reported G^E for 1-bromopentane + *n*-C_5H_{12} [188], γ_A^∞ in *n*-alkanes[182] and a complete account of the 1,2-$C_2H_4Br_2$ + cyclohexane system[185, 187].

5.2.4 Aliphatic X-compounds without X–H bonds + aromatic hydrocarbons

Many authors (e.g. Kohler[80]) have reported on an unusual, and relatively strong, interaction between various organic polar compounds and aromatic hydrocarbons. Different mechanisms (electrostatic, charge-transfer, H-bonding) have been, and still are, proposed to account for this.

A comparative study of homomorphic systems[174] (cf. Section 5.2.4.1) showed that a similar interaction occurs between tertiary aliphatic amines and aromatic hydrocarbons, despite the very weak polarity of the C—N bond. It seemed likely that the interaction mainly involves the dipole of the lone-pair electrons of the N atom (rather than the C—N bond dipole) and the easily polarisable π-electrons of the aromatic hydrocarbon ('n–π interaction')[68, 174]. We would suggest that n–π interaction takes place between any X-compound containing lone-pair electrons and aromatic hydrocarbons.

In the case of perhalogenated carbons, e.g. CCl_4, where the C—X bond dipoles seem less exposed for exerting electrostatic interactions than are the dipoles of the lone-pair electrons of the peripheral X atoms, the observed interaction with aromatic hydrocarbons may be predominantly a n–π interaction.

For molecules in which the bond moments are larger and better exposed, e.g. in 1-chloroalkanes, and especially in the case of molecules containing strong polar groups (ketones, sulphones, etc., cf. Section 5.2.15) bond–π interactions undoubtedly contribute to the observed attraction for aromatic hydrocarbons.

The concept of n–π interaction continues[203, 208, 209] to stimulate experimental work, and provides a qualitative explanation of many of the observed excess functions.

Other π-electron-containing compounds can be expected to behave like the aromatic hydrocarbons. Alkenes and, especially dialkenes (with conjugated and with separated double bonds), should be investigated.

5.2.4.1 *Tertiary amines*

H^E for $(C_2H_5)_3CH + C_6H_6$ is larger by $\sim 600\ J\ mol^{-1}$ than H^E for $(C_2H_5)_3N + C_6H_6$. This indicates that the effect of like π–π interactions (which make the former system strongly endothermic) is partially compensated in the latter system by an exothermic effect, attributed to unlike n–π interactions[139, 174].

Bittrich *et al.*, Letcher and Bayles and Wóycicki and Sadowska have determined H^E and V^E for aromatic hydrocarbons with $(C_2H_5)_3N$[116, 175, 189], or with *N*-methylpiperidine[176]. The observed excess functions have been discussed qualitatively[116, 176, 189] in terms of π–π and n–π interactions. A quantitative approach, similar to that made for ether + hydrocarbon mixtures (cf. Section 5.2.4.2) has not been yet attempted.

5.2.4.2 *Ethers*

Kehiaian *et al.*[36] have measured H^E for a series of *n*-ethers with C_6H_6 and have correlated their results using the surface-version QLT. Andrews and Morcom[179] have measured H^E and V^E for cyclic ethers with C_6H_6 (cf. also reference 192). The unlike interactions in these systems are fairly strong (for certain ethers $H^E < 0$) and are probably due to n–π interactions[174]. Other authors[190] have suggested H-bonding between $(C_2H_5)_2O$ and C_6H_6. It is also noteworthy that Flory's theory requires a negative interchange parameter for ether + C_6H_6 systems[160, 191].

5.2.4.3 *Chloroalkanes*

For a systematic study, the four series of derivatives mentioned above (cf. Section 5.2.3.3) should be taken. Data reported so far give only H^E for 1-chlorobutane + C_6H_6 and *n*-alkylbenzenes[184], H^E for $CH_2Cl_2 + C_6H_6$ [111] and H^E and V^E for $CHCl_3 + C_6H_6$ and alkylbenzenes[193]. Interchange parameters derived from Barker's QLT suggest that alkylation of the C_6H_6 ring strengthens the chlorine–aromatic interaction[184]. An interesting calorimetric study by Abello *et al.*[194] reinforces the view that the overall interaction of $CHCl_3$ with alkylbenzenes follows the same pattern. The problem arises[193] as to whether the interaction of $CHCl_3$ with C_6H_6 is concerned only with the chlorinated part of $CHCl_3$ (cf. Section 5.2.5.1), or also involves the H atom. Here again, comparison of H^E in the series:

	$Cl—CCl_3 + C_6H_6$	$CH_3—CCl_3 + C_6H_6$ [195]	$H—CCl_3 + C_6H_6$
$H^E/J\ mol^{-1}$	115	50	−420

(cf. Section 5.2.3.3) is enlightening. The relatively small decrease in H^E, on passing from CCl_4 to CH_3CCl_3, despite the lesser attraction of C_6H_6 for CH_3 than for Cl, may be attributed to the molecular dipole in CH_3CCl_3 (bond–π, superposed on n–π interactions). $CHCl_3$ has the same number of Cl atoms but a smaller dipole moment than CH_3CCl_3. The strong decrease in H^E on passing from CH_3CCl_3 to $HCCl_3$, indicates specific C—H···π bonding, in agreement also with spectral evidence.

The interaction between 1,2-$C_2H_4Cl_2$ and C_6H_6, which remains controversial (cf., for example, references 80, 196), may be qualitatively explained in terms of n–π and bond–π interactions. A comparative study with higher α, ω-dihalogenoalkane homologues has been initiated[184].

5.2.5 Perhalogenoalkanes + aromatic hydrocarbons

5.2.5.1 *Perfluoroalkanes*

No measurements seem to have been reported recently on perfluoroalkane + aromatic hydrocarbon mixtures. G^E for these mixtures are of the same order of magnitude as for those with aliphatic hydrocarbons[3]. This may well be due to compensation of the endothermic effect of like π–π interactions by the exothermic effect from the unlike n–π interactions.

5.2.5.2 *Tetrachloromethane*

Mixtures with CCl_4 have been extensively investigated during recent years. Very accurate H^E [111, 148, 197] and V^E [198] data have been reported for CCl_4 +

C_6H_6, and H^E and V^E for CCl_4+alkylbenzenes have been measured by McGlashan *et al.*[86], Bittrich *et al.*[116, 199], Rastogi *et al.*[156] and Ocon *et al.*[159]. The G^E functions of most of these systems are not known with the same accuracy. L–V equilibria have been measured for a few systems[168, 200].

The particular theoretical interest in these systems resulted from the assumption of charge-transfer complex formation in the mixture. McGlashan *et al.*[86] interpreted the data quantitatively with a model for non-athermal associated mixtures (cf. Section 5.1.4), while other authors have attempted merely to separate the 'physical' from the 'chemical' contributions[116, 156]. Howell *et al.*[57] showed that Flory's theory gives a reasonably good estimate of V^E. The shape of the H^E curves[86] leaves no doubt as to the presence of orientational forces between the components. However, as with other systems (cf., for example, reference 201), the contribution of charge transfer effects may be negligible as compared with electrostatic (presumably n–π) interactions.

5.2.6 Aromatic X-compounds without X–H bonds+alkanes

Typical of this class of mixtures are the very large positive deviations from ideality, e.g. H^E for *N,N*-dimethylaniline+cyclohexane[202] is $\sim$1140 J mol^{-1}, which is $\sim$750 J mol^{-1} more than found for the homomorphic system isopropylbenzene+cyclohexane[174]. This extra endothermic contribution to H^E may be attributed to the presence of like n–π interactions between the molecules of the aromatic X-compound.

The basic series of substances in need of study derive from the formula $C_6H_5(CH_2)_xX(CH_3)_y$, where X = N ($y = 2$), O ($y = 1$), F ($y = 0$), etc., and $x = 3, 2, 1, 0$. When X and C_6H_5 are separated by a sufficient number of CH_2 groups, the force fields surrounding them should be nearly the same as in $C_6H_5(CH_2)_xCH_3$ and $CH_3(CH_2)_xX(CH_3)_y$, respectively. The properties of the aromatic X-compounds are then predictable from information gleaned by the study of aromatic hydrocarbons and aliphatic X-compounds[36, 184]. The study of substances in which the distance between C_6H_5 and X is shortened gives information on the reciprocal influence of these two groups with regard to their interactions with a series of solvents (alkanes, C_6H_6, CCl_4, etc.).

5.2.6.1 Tertiary amines

Recent studies have concentrated mainly on pyridine bases. Extensive H^E and V^E data have been reported by Wóycicki and Sadowska[154] and by Murakami *et al.*[203]. The n–π interaction has been suggested[174] as the plausible mechanism for the 'self-association' of pyridine bases. Indeed, one finds that H^E for pyridine+cyclohexane (1450 J mol^{-1}) is larger by 650 J mol^{-1} than for the homomorphic system, C_6H_6+cyclohexane. The values found for the partial molar enthalpies at infinite dilution[203, 204] of alkylpyridines in cyclohexane demonstrate the reality of the steric effect exerted by the *o*-CH_3 group on n–π interaction.

It is quite interesting to note that *N*-methylpyrrole + cyclohexane[122, 205] has a comparatively small H^E value[205] (only $\sim 300\ J\ mol^{-1}$). This can be readily explained because in pyrrole the lone-pair electrons of the nitrogen are involved in the aromatic π-electron sextet and are not available for intermolecular n–π interactions (cf. also Section 5.2.6.2).

5.2.6.2 *Ethers*

Kehiaian *et al.*[36] have determined H^E for compounds of the series $C_6H_5(CH_2)_xOCH_3$. As expected, π–π and n–π interactions predominate, e.g. $C_6H_5OCH_3 + n\text{-}C_7H_{14}$ has $H^E \sim 1300\ J\ mol^{-1}$ whereas in the homomorphic system, $C_6H_5CH_2CH_3 + n\text{-}C_7H_{16}$, $H^E \sim 545\ J\ mol^{-1}$.

In furan, the π-electron sextet involves only *one* of the lone-pair electrons of the oxygen atom, the second lone-pair being available for n–π interactions. This explains the relatively large H^E [181] ($1160\ J\ mol^{-1}$) of furan + cyclohexane (cf. Section 5.2.6.1, pyrrole).

5.2.6.3 *Chlorocarbons*

Studies of $C_6H_5(CH_2)_xCl + n$-alkane systems[184] have been made similar to those on ethers (cf. Section 5.2.6.2). Use of Cl–aliphatic and Cl–aromatic interchange enthalpies, derived from independent studies, permits close prediction of the measured H^E ($1470\ J\ mol^{-1}$) in $C_6H_5(CH_2)_2Cl + n\text{-}C_7H_{16}$. Near approach of the Cl atom to the C_6H_5 group is shown to weaken considerably the like molecular interactions, e.g. the experimental H^E for $C_6H_5Cl + n\text{-}C_7H_{16}$ [175] is $648\ J\ mol^{-1}$ whereas the calculated value is $2200\ J\ mol^{-1}$. V^E of $C_6H_5Cl + n\text{-}C_5H_{12}$ [117] and $n\text{-}C_7H_{16}$ [175] have been also reported.

5.2.6.4 *Fluorocarbons*

C_6F_6 is the most thoroughly investigated compound of this group. Recent papers report V^E [206] and H^E [207] for C_6F_6 + cycloalkanes, and there is a study of C_6H_5F + methylcyclohexane[48]. Unpublished H^E data exist for C_6F_5H + cyclohexane[2].

As already known, C_6F_6 + alkane mixtures exhibit a relatively smaller deviation from ideality (hence, better mutual solubility) than do perfluoroalkane + alkane mixtures. This may well be due to the π–π interactions in C_6F_6 which improve somewhat the characteristic weakness of the like interactions between aliphatic perfluorocarbons and render aromatic fluorocarbons more like aliphatic hydrocarbons.

5.2.7 Aromatic X-compounds without X–H bonds + aromatic hydrocarbons

The behaviour of this class of mixtures is determined by the balance between like and unlike π–π and n–π interactions[174]. This leads to deviations from ideality much less positive than in mixtures with aliphatic hydrocarbons (cf. Section 5.2.6), which may even become negative. Due to the orientational character of these interactions, the excess functions are frequently represented by S-shaped curves.

5.2.7.1 *Tertiary amines*

Pyridine + C_6H_6 is the most carefully investigated system of this class. New H^E data have been reported by Garrett *et al.*[209] and compared with previous measurements[203, 208]. L–V equilibrium data[210] and V^E [208, 211] are also available. Alkylpyridines have been studied by Wóycicki and Sadowska[208] and Murakami *et al.*[203]. A qualitative discussion in terms of π–π and n–π interactions has been given[203, 208].

5.2.7.2 *Ethers*

The observed relatively small H^E values for $C_6H_5(CH_2)_2OCH_3 + C_6H_5C_2H_5$ mixtures have been correctly predicted with the interchange enthalpies determined from simpler systems[36]. Furan + C_6H_6 also show a small H^E value[179].

5.2.7.3 *Chlorocarbons*

G^E [139], H^E [184] and V^E values[146, 175, 212] for $C_6H_5Cl + C_6H_6$ and H^E values for C_6H_5Cl with *n*-alkylbenzenes[184] have been determined, and conform to the general behaviour of aromatic X-compounds + aromatic hydrocarbon mixtures.

5.2.7.4 *Fluorocarbons*

Thermodynamic studies on fluorocarbon solutions performed before 1970 have been reviewed[3, 4, 213]. Recent systematic investigations including H^E [207], V^E [206] and L–V critical temperatures[214] of C_6F_6 + hydrocarbon mixtures have thrown new light on the whole subject. Contrary to previous views[215] in which the formation of charge-transfer complexes in $C_6F_6 + C_6H_6$ mixtures was advocated, the data are interpreted by Powell *et al.*[214] in terms of purely physical angle-dependent forces. The successful application of Rowlinson–Sutton's theory to these mixtures is noteworthy. As to the nature of the forces, Powell *et al.*[214] suggest they may result from a combination of simple shape factors and additional electrostatic dipole–quadrupole interactions involving the C—F bond dipole and the π-electron quadrupole[206]. Thus, briefly, the authors[214] suggest a bond–π mechanism. To be consistent with our own views (cf., for example, Section 5.2.6.4), we suggest that unlike π–π and n–π interactions also contribute to the observed attraction between

C_6F_6 and C_6H_6, which would allow an understanding of the negative deviations from ideality in this mixture.

5.2.8 Aliphatic X-compounds without X–H bonds+ perhalogenoalkanes

Among the numerous classes of systems under this heading, only ethers + CCl_4 have been studied in detail (cf. Section 5.2.8.2). Particular interest attaches to the study of *n*-ether + perfluoro-*n*-alkane mixtures, for which liquid–liquid phase equilibrium curves have already been determined by Malesińska[2]. Structurally these compounds are well suited to application of QLT and the interesting point lies in the determination of the O—F interchange parameter and its comparison with the O—Cl interchange parameter. As is well known, a charge-transfer interaction is postulated between O and Cl.

5.2.8.1 Tertiary amines + tetrachloromethane

Only a few data are available, namely, G^E of $(CH_3)_3N + CCl_4$ [177] and H^E of $(C_2H_5)_3N + CCl_4$ [116]. Both functions are negative, indicating a specific N—Cl interaction.

5.2.8.2 Ethers + tetrachloromethane

Mixtures of mono- and poly-ethers with CCl_4 have been investigated carefully by Malcolm *et al.*[44, 45, 216] and by Williamson *et al.*[46, 217, 218]. Barker's QLT has been applied to correlate the E_V^E and G^E data. Kehiaian[36] has re-analysed the H^E data by use of the random-mixing surface-version QLT. Other studies include H^E of CCl_4 + tetrahydrofuran[219] and 1,4-dioxane[111].

5.2.8.3 Thioethers + tetrachloromethane

Gray *et al.*[217] have investigated the $(CH_3)_2S + CCl_4$ mixture to find a negative H^E (-427 J mol^{-1}), indicative of strong unlike interactions. A quantitative estimate of the S—Cl interchange parameter requires a preliminary study of thioether + *n*-alkane mixtures.

5.2.9 Aromatic X-compounds without X–H bonds+ perhalogenoalkanes

For the few X-compounds investigated, the X atom is either involved in the aromatic ring or is directly linked to it. Recent papers report V^E data for $CCl_4 + C_6H_5X$ (X = F, Cl, Br)[199] and H^E [217] and G^E [207] for CCl_4 + pyridine.

5.2.10 Miscellaneous mixtures of aliphatic and aromatic X-compounds without X–H bonds

This is a broad group which we have not divided into smaller classes for lack of sufficient information. The simplest class would include mixtures of homologues belonging to the same series, e.g. *n*-ethers[36] or 1-chloroalkanes[221], or to different series, e.g. *n*-ethers + 1-chloroalkanes. The literature usually gives data on much more complex systems.

Roveillo and Gomel[222] have measured V^E of heterocyclic amines + ethers. Interesting, though difficult to interpret, are the measurements on C_6F_6 + 1,4-dioxane[223] and various aliphatic, aromatic and heterocyclic amines[224].

$CHCl_3$, a proton donor, has been frequently investigated in mixtures with proton acceptors (cf. Sections 5.2.10.1 and 5.2.10.2). C_6F_5H is equally a proton donor, and Watson *et al.*[49] have determined G^E, H^E and V^E for C_6F_5H + 1,4-dioxane and interpreted the data by use of Barker's QLT and the theory of associated mixtures. The QLT treatment needs revision, by using the correct aliphatic–O and O—F parameters.

5.2.10.1 *Trichloromethane + tertiary amines*

The H^E [116] (-3818 J mol^{-1}) and V^E [225] (~ -1.9 cm^3 mol^{-1}) values for $(C_2H_5)_3N + CHCl_3$ leave no doubt as to the presence of strong unlike interactions between the components. Since H^E of $(C_2H_5)_3N + CCl_4$ is only -611 J mol^{-1} [116], the amine–$CHCl_3$ interaction must be predominantly C—H···N bonding. Findlay and Kenyon[226] have measured H^E, G^E and V^E, and Becker *et al.*[64, 227] H^E for $CHCl_3$ + pyridine. Although the interactions in this system are more complex, the role of H-bonding is evident from the data.

5.2.10.2 *Trichloromethane + ethers*

This class of mixtures has been investigated recently by Malcolm *et al.*[44, 45, 216], Williamson *et al.*[46, 217, 218], Becker and Kiefer[228] and Chevalier and Barès[62]. The observed H^E and G^E data have been interpreted either by use of Barker's QLT[44, 46] or by association models[62, 228]. H^E data for $CHCl_3$ + furan, 2-methylfuran and tetrahydrofuran (cf. also reference 219) have been reported by Murakami *et al.*[181] and discussed in terms of C—H···O, C—H···π, and π–π interactions.

5.2.11 Aliphatic X-compounds containing X–H bonds

The complexity in behaviour of these H-bonded systems depends largely on the nature of the solvent. Alkanes are the most 'inert' solvents. The observed excess functions are mainly determined by the like X—H···X interactions and to a lesser, although not negligible extent, by the like dipole–dipole interactions and by the differences in dispersion forces (cf. Section 5.2.3). Aromatic hydrocarbons contribute to this through like π–π and unlike X—H···π and n–π interactions. CCl_4 contributes (apart from the relatively small differences in dispersion forces) mainly through the specific X—Cl attraction.

5.2.11.1 Aliphatic amines + alkanes

The observed positive deviations from ideality are primarily due to N—H···N bonding. We have seen (cf. Section 5.2.3.1) that tertiary aliphatic amines + alkanes are nearly ideal. These systems conform closely to the model for athermal associated mixtures[68]. However, for the lower amines (e.g. CH_3NH_2, which are strongly associated) and when the ratio of N to C atoms is relatively large, the contribution of other than N—H···N interactions is quite important and may even lead to liquid–liquid (L–L) phase separation at low temperatures[229].

Letcher and Bayles[175] have made extensive H^E and V^V measurements on mixtures containing aliphatic amines. Recent G^E data have been reported by Siedler and Bittrich[178], for n-$C_4H_9NH_2$ + n–C_5H_{12} and $(C_2H_5)_2NH$ + n–C_6H_{14}, and by Wolff and Würtz[230], for $(CH_3)_2NH$ + n–C_6H_{14}. L–V equilibria have been measured for cyclohexylamine and N-methylcyclohexylamine + cyclohexane by Kern *et al.*[231]. H^E [232] and V^E [122, 232] data have been reported for mixtures of heterocyclic amines + cycloalkanes.

5.2.11.2 Alcohols + alkanes

n-Alcohols + n-alkanes provide the simplest systems of this class and have been studied extensively[233, 234]. Many authors have also investigated the properties of isomeric alcohols in various alkanes. An outstanding paper, reporting correlated G^E and H^E data for mixtures of normal and branched alcohols with n-C_6H_{14}, has been published by Brown *et al.*[235]. Smirnova and Kurtynina[236] have tabulated the thermodynamic functions of alcohol + hydrocarbon mixtures. Vesely and Pick[237] have reported H^E for the series of n-alcohols + cyclohexane.

Other recent papers include: L–V equilibria and calculated G^E data for CH_3OH + n-C_6H_{14} [66, 238], n-$C_{16}H_{34}$ [239], cyclohexane[240, 241], and methycyclohexane[238]; C_2H_5OH + n-C_6H_{14} [140, 151], n-C_7H_{16} [242, 243] and n-C_8H_{18} [244]; n-C_3H_8OH + n-C_6H_{14} [243], n-C_7H_{16} [243], cyclohexane[241] and methylcyclohexane[245]; n-C_4H_9OH + cyclohexane[242, 246] and methylcyclohexane[245, 247]; H^E data for CH_3OH + cyclohexane[240, 248]; C_2H_5OH + n-$C_{14}H_{30}$ [249]; n-C_3H_8OH + cyclohexane[250]; n-C_4H_9OH + cyclohexane[250]; i-C_4H_9OH + cyclohexane[250]; tert-C_4H_9OH + n-alkanes[251]; isomeric octanols + cyclohexane[252]; n-$C_{10}H_{21}OH$ + cyclohexane[250]; C_p^E data[253] for n-C_3H_7OH + n-C_6H_{14}, n-$C_{16}H_{34}$ and cyclohexane[254].

5.2.11.3 Amines + aromatic hydrocarbons

The systems studied are $C_2H_5NH_2$ [175] and $(C_2H_5)_2NH$[175, 189] with C_6H_6, and piperidine and 2-methylpiperidine with C_6H_6 or toluene[255]. We note also a study[256] of the $NH_2CH_2CH_2NH_2$ + C_6H_6 system.

5.2.11.4 *Alcohols + aromatic hydrocarbons*

The above-mentioned paper by Brown *et al.*[235] (cf. Section 5.2.11.2) gives a detailed account on H^E and G^E for alcohols + C_6H_6 mixtures and a qualitative discussion of the observed data in terms of O—H···O, O—H···π, and π–π interactions. To complete the picture and to give a plausible explanation of the negative H^E observed in C_2H_5OH + toluene mixtures at low temperatures[77], n–π interactions between C_6H_6 and alcohol associates should also be taken into account.

Other recent contributions include L–V equilibria and G^E data for $CH_3OH + C_6H_6$ [257, 258] or toluene[259]; $C_2H_5OH + C_6H_6$ [129]; n-C_4H_9OH + toluene[247]; 3-methylbutan-1-ol + ethylbenzene[260]; V^E for $CH_3OH + C_6H_6$ or toluene[261] and C_p^E for n-$C_3H_8OH + C_6H_6$ or mesitylene[254].

5.2.11.5 *Amines + tetrachloromethane*

The strong N—Cl interaction[262], noted in tertiary amines + CCl_4 mixtures (Section 5.2.8.1), manifests itself also in primary and secondary amine + CCl_4 mixtures. It results in a considerable lowering of the self-association constants of the amines in CCl_4 [263, 264], and in less positive (or even negative) excess functions for amine + CCl_4 mixtures (as compared to solutions in alkanes), e.g., Bittrich *et al.*[116] report $H^E = -485$ J mol^{-1} for $(C_2H_5)_2NH$ + CCl_4.

5.2.11.6 *Alcohols + tetrachloromethane*

Volumetric properties of mixtures of this type have been investigated systematically by Brusset *et al.*[265]. Ocon *et al.*[261] have reported V^E for CCl_4 + CH_3OH. L–V equilibrium studies include those on systems containing CCl_4 and CH_3OH [268], i-propanol[267] and t-butanol[268].

5.2.12 Aromatic X-compounds containing X–H bonds

For these compounds the like interactions are quite complex since they include not only X–H···X bonds, but also π–π, X–H···π, and n–π attractive forces. In mixtures with alkanes the deviation from ideality is therefore large enough to produce partial immiscibility of the components. In mixtures with aromatic hydrocarbons or with CCl_4 the observed non-ideality is smaller due to specific unlike interactions which cancel the effect of the like interactions in the aromatic X-compound.

5.2.12.1 *Aromatic amines*

We note a L–V equilibrium study [269] on N-methylaniline + cyclohexane and H^E [205] and V^E [122] data have been reported for pyrrole + cyclohexane. Campbell and Kartzmark[248] have determined H^E for aniline + n-C_6H_{14}.

5.2.12.2 *Phenols*

Recent studies include isothermal L–V equilibria[270] in mixtures of C_6H_5OH and *o*-cresol with *n*-$C_{10}H_{22}$ and other hydrocarbons, and V^E of C_6H_5OH and isomeric cresols with cyclohexane[271]. Papers of note on the self-association of C_6H_5OH in non-polar solvents have been published by Murakami and Fujishiro[91], Kohler[80] and by Whetsel and Lady[272] (spectroscopic measurements).

5.2.13 Miscellaneous mixtures of aliphatic X-compounds containing X—H bonds

Despite the complexity of the molecular interactions in these mixtures, a reasonably satisfactory theoretical approach seems possible. The *sine qua non* condition is to base the approach on a theory which is valid for each of the components separately in mixtures with alkanes. Most, if not all, interchange parameters are then determined from the study of alkane-containing mixtures, and there remains only the evaluation of the cross-interaction parameters. The application of this approach to mixtures of different representatives of the same homologous series (e.g. *n*-alcohol mixtures), would be a valuable test of the theory of, e.g., *n*-alcohol + *n*-alkane mixtures.

5.2.13.1 *Amines*

1-Aminoalkane mixtures have not been yet investigated. Bittrich *et al.*[274] have continued their study of the $(C_2H_5)_2NH + (C_2H_5)_3N$ system. Letcher and Bayles[175] have measured H^E and V^E in *n*-$C_4H_9NH_2$, $(C_2H_5)_2NH$, $(C_2H_5)_3N$ binaries. Wóycicki and Sadowska[176] have reported H^E and V^E data for heterocyclic amines.

5.2.13.2 *Alcohols*

Benson *et al.*[275,276], Diaz Pena *et al.*[277] and Ramalho and Delmas[233] are the leading investigators of alcohol mixtures. The available data have not yet been properly interpreted, and at most have been used to test the applicability of the principle of congruence. L–V equilibrium data have been reported for mixtures of isomeric butanols[278] and pentanols[279].

5.2.13.3 *Amines + ethers*

Roveillo and Gomel[280] have measured V^E for cyclic amines + cyclic ethers. A detailed study of 1-aminoalkanes + *n*-ethers would add to understanding of the properties of the more complex alcohol + amine mixtures.

5.2.13.4 *Alcohols + ethers*

Unfortunately, no systematic study of *n*-alcohol + *n*-ether systems has been made so far. It is essential for the interpretation of the properties of the technically important cellosolves. Papers of note report V^E for CH_3OH and C_2H_5OH + isopropyl ether[281] and H^E for C_2H_5OH + 1,4-dioxane[282]. Martire and Riedl[283] have derived equilibrium constants, enthalpies and entropies of H-bonding of alcohols to ethers by GLPC.

5.2.13.5 *Amines + alcohols*

Very interesting studies on these mixtures are due Nakanishi *et al.*[281, 284, 285]. The excess functions show that the O—H···N bonding is relatively stronger than are O–H···O and N–H····N bondings, e.g., for $CH_3OH + (C_2H_5)_2NH$, H^E is as low as -4500 J mol^{-1}. A similar result was found for i-C_3H_7OH + i-$C_3H_7NH_2$ [61] ($H^E \sim -2290$ J mol^{-1}). Krichevtsov *et al.*[286] have determined L–V equilibra in several amine + alcohol systems.

5.2.14 Miscellaneous mixtures of aromatic and aliphatic X-compounds containing X—H bonds

This is a very complex class of organic mixtures. We content ourselves by mention of a paper by Findlay and Copp[287] on G^E, H^E and V^E values for pyridine mixtures with *n*-alcohols, a L–V equilibrium study on CH_3OH + pyridine[288] and a paper reporting H^E and V^E data for mixtures of heterocyclic amines[289].

5.2.15 Organic compounds containing complex functional groups

Under this broad title we assemble all X-compounds containing functional groups with multiple and/or coordinate bonds. Their common feature is the strong dipole moment of the interatomic bonds, and usually of the whole molecule. This leads to strong angle-dependent interactions, the molecules behaving in a manner which recalls H-bonded species, even if they do not possess X—H bonds. However, an approach in terms of chemical association equilibria is probably not justified. The theoretical treatment of this type of mixture presents specific difficulties.

In mixtures with alkanes (which should be investigated first), the like interactions are so strong as to lead to partial immiscibility of the components. In mixtures with aromatic hydrocarbons, unlike dipole–induced dipole interactions favour the reciprocal miscibility of the components. Strong unlike interactions are manifest also in mixtures with CCl_4. We shall refer below to only a few representative classes of substances.

5.2.15.1 *Alkanones*

The L–L equilibrium in acetone + alkane mixtures has been investigated by Riccardi *et al.*[290] and by Edwards *et al.*[291]. A thermodynamic study of the acetone + n-$C_{10}H_{22}$ system has been initiated by Edwards and Rodriguez[292]. Campbell and Chatterjee[293] have investigated the L–V equilibrium of acetone + C_6H_6 up to the L–V critical state. Nagata *et al.*[294] and Paz Andrade and Casas[295] have measured the H^E value of methyl ethyl ketone + n-alcohols.

5.2.15.2 *Alkanoates*

The L–L equilibrium curves of $HCOOCH_3$ + n-alkanes (C_5–C_9), determined by Franzosini *et al.*[291,296], an isothermal L–V equilibrium study of $C_3H_7COOC_2H_5$ + n-C_7H_{16} by Scheller *et al.*[297] and the H^E curves of several alkanoate + alkane systems measured by Grolier[298], have provided useful information on the strength of the like interactions between alkanoate molecules. Several studies have been published on alkanoate + alcohol mixtures[257,267,281,294,298,299].

5.2.15.3 *Alkanoic acids*

We note first recent papers reporting L–V equilibrium data for CH_3COOH + n-C_7H_{16}[137], toluene[137] and CCl_4[300,301] and V^E data for CH_3COOH + cyclohexane, C_6H_6 and CCl_4[302].

Miksch *et al.*[300] have calculated G^E from L–V equilibrium data, taking into account the strong non-ideality of the vapour phase. The interpretation of the properties of alkanoic acid-containing mixtures raises serious difficulties, despite the fact that the predominant species in the mixture are well known to be the monomer and the dimer. According to Affsprung *et al.*[302], in addition to the monomer–dimer equilibrium, there remains a strong interaction between the dipole of the monomer and the easily polarisable dimer.

5.2.15.4 *Sulphoxides and sulphones*

Nissema[303] has published an extensive study on $(CH_3)_2SO$ + cyclohexane or C_6H_6, including L–L and L–V equilibria and V^E, G^E and H^E data. Lau *et al.*[304] have measured V^E for $(CH_3)_2SO + C_6H_6$, CCl_4 and $CHCl_3$, whilst Quitzsch *et al.*[305] have determined H^E and V^E data for $(CH_3)_2SO + CCl_4$. In a valuable recent paper, Chareyron and Clechet[306] report V^E data for $(CH_3)_2SO$ with the series of chloro- and bromo-methanes. For $(CH_3)_2SO + CCl_4$, $V^E = -0.61\ cm^3\ mol^{-1}$, a result interpreted[306] as proving complex formation.

Benoit and Charbonneau[307] have determined G^E, H^E and V^E values of sulpholane + C_6H_6 or CH_2Cl_2 mixtures. From a volumetric study of sulpholane + CH_3OH mixtures Jannelli *et al.*[308] infer that sulpholane is a solvent similar to CH_3CN or CH_3NO_2, with rather weak H-bonding ability.

5.2.15.5 *Nitroalkanes*

Franzosini *et al.*[290,309] and Malesińska[2] have investigated the L–L equilibrium curves of nitroalkane + hydrocarbon mixtures.

5.2.15.6 Amides

We merely note a paper by Quitzsch *et al.*[310] on L–L and L–V equilibria, V^E and H^E data of *N,N*-dimethyl- and *N,N*-diethylformamides + *n*-C_7H_{16} and a paper by Ababi and Balba[311] on L–V equilibria in mixtures of *N,N*-dimethylformamide + aromatic hydrocarbons.

Acknowledgements

This review is based essentially on experience accumulated in Warsaw during a decade of activity at the Institute of Physical Chemistry of the Polish Academy of Sciences. It has been written in part in Paris, during the tenure of a fellowship at the Centre National de la Recherche Scientifique, Laboratoire de Cinétique Chimique, and completed in Marseille, at the Université de Provence, Laboratoire de Chimie Générale.

It is a pleasure to express here my gratitude to Prof. M. Smialowski (Warsaw), Prof. G. Pannetier (Paris) and Prof. M. Laffitte (Marseille) for the excellent conditions of work offered during the respective periods and to all my colleagues who have contributed so much with their scientific publications and personal discussions to the preparation of this review.

References

1. IUPAC. (1970). *Bull. Thermodynamics and Thermochemistry,* **13,** (Univ. Michigan, Ann Arbor, Michigan, USA)
2. IUPAC. (1971). *Bull. Thermodynamics and Thermochemistry,* **14,** (Univ. Michigan: Ann Arbor, Michigan, USA)
3. Rowlinson, J. S. (1969). *Liquids and Liquid Mixtures,* 2nd Edition, (London: Butterworths)
4. Hildebrand, J. H., Prausnitz, J. M. and Scott, R. L. (1970). *Regular and Related Solutions* (New York: Van Nostrand Reinhold Company)
5. Scott, R. L. and Fenby, D. V. (1969). *Ann. Rev. Phys. Chem.,* **20,** 111
6. Rowlinson, J. S. (1970). *Discuss Faraday Soc.,* **49,** 30
7. Flory, P. J. (1970). *Discuss Faraday Soc.,* **49,** 7
8. Battino, R. (1971). *Chem. Rev.,* **71,** 5
9. Rowlinson, J. S. and Watson, I. D. (1969). *Chem. Eng. Sci.,* **24,** 1565
10. Hijmans, J. and Holleman, Th. (1969). *Advan. Chem. Phys.,* **16,** 223
11. Cruickshank, A. J. B. and Hicks, C. P. (1970). *Discuss Faraday Soc.,* **49,** 106
12. Williamson, A. G. and Scott, R. L. (1970). *Trans. Faraday Soc.,* **66,** 335
13. Patterson, D. and Bardin, J. M. (1970). *Trans. Faraday Soc.,* **66,** 321
14. Patterson, D. and Delmas, G. (1970). *Discuss Faraday Soc.,* **49,** 98
15. Barriol, J. and Boule, P. (1969). *J. Chim. Phys., Physicochim. Biol.,* **66,** 1385; Boule, P. (1970). *Ph.D. Thesis* (University of Nancy, France); Greffe, J.-L. (1971). *Ph.D. Thesis* (University of Nancy, France)
16. Leland, T. W., Rowlinson, J. S., Sather, G. A. and Watson, I. D. (1969). *Trans. Faraday Soc.,* **65,** 2034
17. Leland, T. W., Rowlinson, J. S. and Sather, G. A. (1968). *Trans. Faraday Soc.,* **64,** 1447
18. Hildebrand, J. H. and Scott, R. L. (1962). *Regular Solutions* (Englewood Cliffs, N. J.: Prentice Hall, Inc.)
19. Sturtevant, J. M. and Lyons, P. A. (1969). *J. Chem. Thermodynamics,* **1,** 201
20. Hansen, C. M. (1969). *Ind. Eng. Chem., Prod. Res. Develop.,* **8,** 2
21. Helpinstill, J. G. and Van Winkle, M. (1968). *Ind. Eng. Chem., Process Design Develop.,* **7,** 213

22. Bagley, E. B., Nelson, T. P. and Scigliano, J. M. (1971). *J. Paint Technol.*, **43**, 35
23. Rheineck, A. E. and Lin, K. F. (1968). *J. Paint Technol.*, **40**, 611
24. Watanabe, A. and Sugiyama, S. (1969). *Kogyo Kagaku Zasshi*, **72**, 1195
25. Konstam, A. H. and Feairheller, W. R., Jr. (1970). *AIChE J.*, **16**, 837
26. Funk, E. W. and Prausnitz, J. M. (1970). *Ind. Eng. Chem.*, **62**, 8; Funk, E. W. (1970). *Ph.D. Thesis* (Univ. California, Berkeley)
27. Bazua, E. R. and Prausnitz, J. M. (1971). *Cryogenics*, **11**, 114
28. Vitoria, M. and Walkley, J. (1969). *Trans. Faraday Soc.*, **65**, 57; 62
29. Cruickshank, A. J. B., Gainey, B. W. and Young, C. L. (1968). *Trans. Faraday Soc.*, **64**, 337
30. Gainey, B. W. and Young, C. L. (1968). *Trans. Faraday Soc.* **64**, 349
31. Young, C. L. (1968). *Trans. Faraday Soc.*, **64**, 1537
32. Hicks, C. P. and Young, C. L. (1968). *Trans. Faraday Soc.*, **64**, 2675
33. Cruickshank, A. J. B., Gainey, B. W., Hicks, C. P., Letcher, T. M. and Young, C. L. (1969). *Trans. Faraday Soc.*, **65**, 2356
34. Gainey, B. W. and Pecsok, R. L. (1970). *J. Phys. Chem.*, **74**, 2548
35. Redlich, O., Derr, E. L. and Pierotti, G. J. (1959). *J. Amer. Chem. Soc.*, **81**, 2283
36. Kehiaian, H. V., Sosnkowska-Kehiaian, K. and Hryniewicz, R. (1971). *J. Chim. Phys. Physicochim. Biol.*, **68**, 922
37. Kehiaian, H. V. (1971). *J. Chim. Phys. Physicochim. Biol.*, **68**, 935
38. Deal, C. H. Jr. and Derr, E. L. (1968). *Ind. Eng. Chem.*, **60**, 28; Derr, E. L. and Deal, C. H. Jr. (1969). *Proc. Int. Symp. Distill.*, **3**, 40
39. Ratcliff, C. A. and Chao, K. C. (1969). *Can. J. Chem. Eng.*, **47**, 148; Nguyen, T. H. and Ratcliff, A. G. (1971). Ibid., **49**, 120; Maripuri, V. C. and Ratcliff, G. A. (1971). Ibid., **49**, 375, 506
40. Vesely, F. and Pick, J. (1969). *Collect. Czech. Chem. Commun.*, **34**, 1792
41. Singh, J., Pflug, H. D. and Benson, G. C. (1968). *J. Phys. Chem.*, **72**, 1939
42. Murakami, S., Lam, V. T. and Benson, G. C. (1969). *J. Chem. Thermodynamics*, **1**, 397
43. Lam, V. T., Murakami, S. and Benson, G. C. (1970). *J. Chem. Thermodynamics*, **2**, 17
44. Malcolm, G. N., Baird, C. E., Bruce, G. R., Cheyne, K. G., Kershaw, R. W. and Pratt, M. C. (1969). *J. Polymer. Sci.*, **7**, 1495
45. Kershaw, R. W. and Malcolm, G. N. (1968). *Trans. Faraday Soc.*, **64**, 323
46. Beath, L. A., Watson, I. D. and Williamson, A. G. (1971). *Proc. 1st Intern. Conf. Calorimetry and Thermodynamics*, 773 (Warsaw: Polish Scientific Publishers-PWN)
47. Rodger, A. J., Hsu, C. C. and Furter, W. F. (1969). *Can. J. Chem. Eng.*, **47**, 490
48. Bhattacharyya, S. N., Mitra, R. C. and Mukherjee, A. (1968). *J. Phys. Chem.*, **72**, 56, 63
49. Watson, I. D., Knight, R. J., McKinnon, I. R. and Williamson, A. G. (1968). *Trans. Faraday Soc.*, **64**, 1763
50. Huggins, M. L. (1970). *J. Phys. Chem.*, **74**, 371; (1971). Ibid., **75**, 1255; (1971). *Polymer* (London), **12**, 389; (1971). *Macromolecules*, **4**, 274; (1969). *J. Paint Technol.*, **41**, 509
51. Huggins, M. L. (1971). *Macromolecules*, **4**, 274
52. Chao, K.–C., Robinson, R. L., Jr., Smith, M. L. and Kuo, C. M. (1967). *Chem. Eng. Progress Symp. Ser.*, **63**, 121; Kuo, C. M., Robinson, R. L., Jr. and Chao, K.–C. (1970). *Ind. Eng. Chem., Fundam.*, **9**, 564
53. Eichinger, B. E. and Flory, P. J. (1968). *Trans. Faraday Soc.*, **64**, 2035
54. Lal, M. and Swinton, F. L. (1969). *J. Phys. Chem.*, **73**, 2883
55. Benson, G. C., Murakami, S., Lam, V. T. and Singh, J. (1970). *Can, J. Chem.*, **48**, 211
56. Benson, G. C. and Singh, J. (1968). *J. Phys. Chem.*, **72**, 1345
57. Howell, P. J., Skillerne de Bristowe, B. J. and Stubley, D. (1971). *J. Chem. Soc. A*, 397
58. Nigam, R. K. and Singh, P. P. (1969). *Indian J. Chem.*, **7**, 156
59. Jain, D. V. S. and Yadav, O. P. (1971). *Indian J. Chem.*, **9**, 342
60. Kehiaian, H. V. and Treszczanowicz, A. (1966). *Bull. Acad. Polon. Sci., Ser. Sci. Chim.*, **14**, 891
61. Becker, F., Pflug, H. D. and Kiefer, M. (1968). *Z. Naturforsch. A*, **23**, 1805
62. Chevalier, J.–L. and Barès, D. (1969). *J. Chim. Phys. Physiochim. Biol.*, **66**, 1448; Chevalier, J.–L. (1969). Ibid., **66**, 1457
63. Apelblat, A. (1970). *J. Phys. Chem.*, **74**, 2214
64. Becker, F., Fries, E. W., Kiefer, M. and Pflug, H. D. (1970). *Z. Naturforsch., A*, **25**, 677
65. Kind, R. and Bittrich, H.–J. (1970). *J. Prakt. Chem.*, **312**, 641
66. Wolff, H. and Hoeppel, H.–E. (1968). *Ber. Bunsenges. Phys. Chem.*, **72**, 710; 722

67. Haskell, R. W., Hollinger, H. B. and Van Ness, H. C. (1968). *J. Phys. Chem.*, **72,** 4534
68. Kehiaian, H. V. (1966). *Bull. Acad. Polon. Sci., Ser. Sci. Chim.*, **14,** 703
69. Flory, P. J. (1944). *J. Chem. Phys.*, **12,** 425
70. Saroléa-Mathot, L. (1953). *Trans. Faraday Soc.*, **49,** 8
71. Kehiaian, H. V. and Treszczanowicz, A. (1969). *Bull. Soc. Chim. Fr.*, 1561; (1970). *Bull. Acad. Polon. Sci., Ser. Sci. Chim.*, **18,** 693; Treszczanowicz, A. and Kehiaian, H. V. (1968). Ibid., **16,** 171; (1970). Ibid., **18,** 155; 723; 729
72. Treszczanowicz, A. (1968). *Bull. Acad. Polon. Sci., Ser. Sci. Chim.*, **16,** 71; 439; (1970). Ibid., **18,** 159
73. Kehiaian, H. V. and Treszczanowicz, A. (1968). *Bull. Acad. Polon. Sci., Ser. Sci. Chim.*, **16,** 445
74. Kretschmer, C. B. and Wiebe, R. (1954). *J. Chem. Phys.*, **22,** 1697
75. Wiehe, I. A. and Bagley, E. B. (1967). *Ind. Eng. Chem., Fundam.*, **6,** 209
76. Biais, J., Dos Santos, J. and Lemanceau, B. (1970). *J. Chim. Phys. Physicochim. Biol.*, **67,** 806; Dos Santos, J., Biais, J. and Pineau, P. (1970). Ibid., **67,** 814
77. Hwa, S. C. P. and Ziegler, W. T. (1966). *J. Phys. Chem.*, **70,** 2571
78. Lacmann, R. (1962). *Z. Phys. Chem. (Frankfurt am Main)*, **35,** 86
79. Kehiaian, H. V. (1963). *Bull. Acad. Polon. Sci., Ser. Sci. Chim.*, **11,** 153
80. Kohler, F. (1969). *Monatsh. Chem.*, **100,** 1151
81. Renon, H. and Prausnitz, J. M. (1967). *Chem. Eng. Sci.*, **22,** 299, 1891
82. Harris, H. G., Jr. and Prausnitz, J. M. (1969). *Ind. Eng. Chem., Fundam.*, **8,** 180
83. Lempe, D. and Bittrich, H.-J. (1968). *Z. Phys. Chem. (Leipzig)*, **238,** 403
84. Smirnova, N. A. (1968). *Khim. Termodin. Rastvorov*, **2,** 3
85. Pouchly, J. (1969). *Collect. Czech. Chem. Commun.*, **34,** 1236; Pouchly, J., Solc, K. and Zivny, A. (1969). Ibid., **34,** 2716
86. McGlashan, M. L., Stubley, D. and Watts, H. (1969). *J. Chem. Soc. A*, 673; Howell, P. J. and Stubley, D. (1969), Ibid., 2488
87. Bellemans, A. (1963). *Bull. Soc. Chim. Belg.*, **72,** 465
88. Kehiaian, H. V. (1969). *Conf. on Mixts. of Non-Electrolytes and Intermol. Forces* (Rostock, Germany); cf. (1970) *Mitt. Chem. Ges. DDR*, **1,** 9
89. Christian, S. D. and Tucker, E. E. (1970). *J. Phys. Chem.*, **74,** 214
90. Fletcher, A. N. (1969). *J. Phys. Chem.*, **73,** 2217; (1970). Ibid., **74,** 216
91. Skripka, V. G., Nikitina, I. E., Zhdanovich, L. A., Sirotin, A. G. and Benyaminovich, O. A. (1970). *Gazov. Prom.*, **15,** 35
92. Stoeckli, H. F. and Staveley, L. A. K. (1970). *Helv. Chim. Acta*, **53,** 1961
93. Williams, B. and Prodany, N. W. (1971). *J. Chem. Eng. Data*, **16,** 1
94. Berry, V. M. and Sage, B. H. (1970). *Nat Stand. Ref. Data Ser., Nat. Bur. Stand.* 1970, *NSRDS-NBS*, **32,** 66p.
95. Djordjevich, L. and Budenholzer, R. A. (1970). *J. Chem. Eng. Data*, **15,** 10
96. Hirata, M., Suda, S., Hakuta, T. and Nagahama, K. (1969). *Mem. Fac. Technol., Tokyo Metrop. Univ.*, **19,** 103
97. Zais, E. J. and Silberberg, I. H. (1970). *J. Chem. Eng. Data*, **15,** 253
98. Rodrigues, A. B. J., McCaffray, D. S., Jr. and Kohn, J. P. (1968). *J. Chem. Eng. Data*, **13,** 164
99. Lee, K. H. and Kohn, J. P. (1969). *J. Chem. Eng. Data*, **14,** 292
100. Kay, W. B. (1970). *J. Chem. Eng. Data*, **15,** 46
101. Kay, W. B. (1971). *J. Chem. Eng. Data*, **16,** 137
102. Hoepfner, A., Kreibich, U. T. and Schaefer, K. (1970). *Ber. Bunsenges. Phys. Chem.*, **74,** 1016
103. Shana'a, M. Y. and Canfield, F. B. (1968). *Trans. Faraday Soc.*, **64,** 2281
104. Sims, M. J. and Winnick, J. (1969). *J. Chem. Eng. Data*, **14,** 164
105. Benkovskii, V. G., Nauruzov, M. Kh., Bogoslovskaya, T. M. and Serikov, Zh. (1970). *Tr. Inst. Khim. Nefti Prir. Solei, Akad. Nauk Kaz. SSR*, **1,** 16
106. Tewari, Y. B., Martire, D. E. and Sheridan, J. P. (1970). *J. Phys. Chem.*, **74,** 2345
107. Maffiolo, G., Lenoir, J. Y. and Renon, H. (1970). *Chem. Eng. Sci.*, **25,** 1847
108. McGlashan, M. L. (1970). *Trans. Faraday Soc.*, **66,** 18
109. Rogers, B. L. and Prausnitz, J. M. (1971). *J. Chem. Thermodynamics*, **3,** 211
110. Barsuk, S. D., Skripka, V. G. and Benyaminovich, O. A. (1970). *Gazov. Prom.*, **15,** 38
111. Murakami, S. and Benson, G. C. (1969). *J. Chem. Thermodynamics*, **1,** 559
112. McGlashan, M. L. and Stoeckli, H. F. (1969). *J. Chem. Thermodynamics*, **1,** 589

113. Paz Andrade, M. I., Baluja, M. C. and Nunez, L. (1970). *An. Quim.*, **66**, 949
114. Marsh, K. N. and Stokes, R. H. (1969). *J. Chem. Thermodynamics*, 1, 223; Ewing, M. B. and Marsh, K. N. (1970). Ibid., **2**, 295
115. Ewing, M. B., Marsh, K. N., Stokes, R. H. and Tuxford, C. W. (1970). *J. Chem. Thermodynamics*, **2**, 751
116. Bittrich, H.-J., Kupsch, C., Gotter, R. and Bock, G. (1971). *Proc. 1st Intern. Conf. Calorimetry and Thermodynamics*, 783 (Warsaw: Polish Scientific Publishers-PWN)
117. Mahl, B. S., Nigam, R. K., Chopra, S. L. and Singh, P. P. (1971). *J. Chem. Thermodynamics*, **3**, 363
118. Preston, G. T., Funk, E. W. and Prausnitz, J. M. (1971). *J. Phys. Chem.*, **75**, 2345
119. Stokes, R. H., Marsh, K. N. and Tomlins, R. P. (1969). *J. Chem. Thermodynamics*, **1**, 377; Ewing, M. B., Levien, B. J., Marsh, K. N. and Stokes, R. H. (1970). Ibid., **2**, 689
120. Marsh, K. N. (1971). *J. Chem. Thermodynamics*, **3**, 355
121. Ewing, M. B. and Marsh, K. N. (1970). *J. Chem. Thermodynamics*, **2**, 351
122. Roveillo, J. and Gomel, M. (1968). *C. R. Acad. Sci. Paris*, **266**, 845
123. Lagrange, G., Laurence, G. and Darmois, R. (1971). *Chim. Anal. (Paris)*, **53**, 107
124. Blagoi, Yu. P. and Orobinskii, N. A. (1968). *Sb. Nauch. Tr. Fiz.-Tekh. Inst. Nizk. Temp., Akad. Nauk Ukr. SSR*, **1**, 3
125, Funk, E. W. and Prausnitz, J. M. (1971), *AIChE J.*, **17**, 254
126. Hirata, M. and Hakuta, T. (1968). *Mem. Fac. Technol., Tokyo Metrop. Univ.*, **18**, 1595
127. Hakuta, T., Nagahama, K. and Hirata, M. (1969). *Bull. Jap. Petrol. Inst.*, **11**, 10
128. Chirikov, V. V., Galata, L. A., Chirikova, Z. P. and Kofman, L. S. (1969). *Teor. Osn. Khim. Tekhnol.*, **3**, 766
129. Rozhnov, M. S. and Efremova, G. D. (1970). *Khim. Prom. Ukr.*, 33
130. Luckhurst, G. R. and Martire, D. E. (1969). *Trans. Faraday Soc.*, **65**, 1248
131. Korotkova, V. N., Pavlov, S. Yu., Karpacheva, L. L., Serafimov, L. A. and Kofman, L. S. (1969). *Prom. Sin. Kauch., Nauch.-Tekh. Sb.*, **6**, 6
132. Mislavskaya, V. S. and Khodeeva, S. M. (1969). *Zh. Fiz. Khim.*, **43**, 1013; 1887
133. Kudryavtseva, L. S., Viit, H. and Eisen, O. (1968). *Eesti. NSV Tead. Akad. Toim., Keem, Geol.*, **17**, 242
134. Hirth, L. J., Harris, H. G. and Prausnitz, J. M. (1968). *AIChE J.*, **14**, 812
135. Saito, S. (1969). *Asahi Garasu Kogyo Gijutsu Shoreikai Kenkyu Hokoku*, **15**, 397
136. Hlousek, K. and Hala, E. (1970). *Collect. Czech. Chem. Commun.*, **35**, 1030
137. Markuzin, N. P. and Pavlova, L. M. (1971). *Zh. Prikl. Khim. (Leningrad)*, **44**, 311
138. Jain, D. V. S., Gupta, V. K. and Lark, B. S. (1970). *Indian J. Chem.*, **8**, 815
139. Harris, K. R. and Dunlop, P. J. (1970). *J. Chem. Thermodynamics*, **2**, 805
140. Smith, V. C. and Robinson, R. L., Jr. (1970). *J. Chem. Eng. Data*, **15**, 391
141. Treszczanowicz, T. and Kehiaian, H. V. (1971). *Bull. Acad. Polon. Sci., Ser. Sci. Chim.*, (in the press)
142. Paz Andrade, M. I. and Baluja, M. C. (1969). *Acta. Cient. Compostelana*, **6**, 245; Paz-Andrade, M. I., Jimenez, E. and Baluja, M. C. (1970). *An. Quim.*, **66**, 955; Paz-Andrade, M. I. Castromil, S. and Baluja, M. C. (1970). *J. Chem. Thermodynamics*, **2**, 775
143. Picquenard, E., Kehiaian, H. V., Abello, L. and Pannetier, G. (1972). *Bull. Soc. Chim. Fr.*, 120
144. Jain, D. V. S., Dewan, R. K. and Tewari, K. K. (1968). *Indian J. Chem.*, **6**, 511
145. Jain, D. V. S., Lark, B. S., Chamak, S. S. and Chander, P. (1970). *Indian J. Chem.*, **8**, 66
146. Harris, K. R. and Dunlop, P. J. (1970). *J. Chem. Thermodynamics*, **2**, 813
147. Sosnkowska-Kehiaian, K. and Kehiaian, H. V. (1965). *Bull. Acad. Polon. Sci., Ser. Sci. Chim.*, **13**, 659
148. Stokes, R. H., Marsh, K. N. and Tomlins, R. P. (1969). *J. Chem. Thermodynamics*, **1**, 211
149. Stokes, R. H., Levien, B. J. and Marsh, K. N. (1970). Ibid., **2**, 43
150. Powell, R. J. and Swinton, F. L. (1968). *J. Chem. Eng. Data*, **13**, 260
151. Diaz Pena, M. and Rodriguez Cheda, D. (1970). *An. Quim.*, **66**, 721
152. Rivenq, F. (1969). *Bull. Soc. Chim. Fr.*, 3034
153. Paz Andrade, M. I. and Amor, M. P. (1970). *An. Quim.*, **66**, 717
154. Wóycicki, W. and Sadowska, K. W. (1968). *Bull. Acad. Polon. Sci., Ser. Sci. Chim.*, **16**, 147
155. Vera, J. H. and Prausnitz, J. M. (1971). *J. Chem. Eng. Data*, **16**, 149
156. Rastogi, R. P., Nath, J. and Yadava, R. B. (1971). *Proc. 1st Intern. Conf. Calorimetry and Thermodynamics*, 791. (Warsaw: Polish Scientific Publishers-PWN); Nath, J. and Yadava, R. B. (1971). *Indian J. Chem.*, **9**, 45

157. Paz Andrade, M. I., Baluja, M. C. and Nunez, L. (1971). *An. Quim.*, **67**, 17
158. Lewis, G. and Johnson, A. F. (1969). *J. Chem. Eng. Data.* **14**, 484
159. Ocon, J., Tojo, G. and Espada, L. (1969). *An. Quim.*, **65**, 727
160. Hyder Khan, V. and Subrahmanyam, S. V. (1971). *Trans. Faraday Soc.*, **67**, 2282
161. Nigam, R. K. and Singh, P. P. (1969). *Trans. Faraday Soc.*, **65**, 950
162. Boublik, T. and Benson, G. C. (1969). *Can. J. Chem.*, **47**, 539
163. Kreglewski, A. (1967). *J. Phys. Chem.*, **71**, 2860
164. Jain, D. V. S., Gupta, V. and Lark, B. S. (1971). *Indian J. Chem.*, **9**, 465
165. Bissell, T. G., Okafor, G. E. and Williamson, A. G. (1971). *J. Chem. Thermodynamics*, **3**, 393
166. Jain, D. V. S., Yadav, O. P. and Gill, S. S. (1971). *Indian J. Chem.*, **9**, 339
167. Chang, E. T. and Westrum, E. F., Jr. (1970). *J. Phys. Chem.*, **74**, 2528
168. Rodger, A. J., Hsu, C. C. and Futer, W. F. (1969). *J. Chem. Eng. Data*, **14**, 362
169. Ocon, J., Tojo, G. and Espada, L. (1969). *An. Quim.*, **65**, 633
170. Boublik, T., Lam, V. T., Murakami, S. and Benson, G. C. (1970). *J. Phys. Chem.*, **73**, 2356
171. Boublik, T. and Benson, G. C. (1970). *J. Phys. Chem.*, **74**, 904
172. Young, C. L. (1969). *Trans. Faraday Soc.*, **65**, 2639
173. Sosnkowska-Kehiaian, K., Hryniewicz, R. and Kehiaian, H. V. (1969). *Bull. Acad. Polon. Sci., Ser. Sci. Chim.*, **17**, 185
174. Kehiaian, H. V. (1967). *The Chemical Society Anniversary Meetings Symposium on the Physical Chemistry of Weak Complexes* (Exeter, UK)
175. Letcher, T. M. and Bayles, J. W. (1971). *J. Chem. Eng. Data*, **16**, 266
176. Wóycicki, W. and Sadowska, K. W. (1968). *Bull. Acad. Polon. Sci., Ser. Sci. Chim.*, **16**, 531
177. Wolff, H. and Wurtz, R. (1968). *Ber. Bunsenges. Phys. Chem.*, **72**, 101
178. Siedler, R. and Bittrich, H.-J. (1969). *J. Prakt. Chem.*, **311**, 721
179. Andrews, A. W. and Morcom, K. W. (1971). *J. Chem. Thermodynamics*, **3**, 513; 519
180. Arm, H. and Bankay, D. (1969). *Helv. Chim. Acta*, **52**, 279
181. Murakami, S., Koyama, M. and Fujishiro, R. (1968). *Bull. Chem. Soc. Jap*, **41**, 1540
182. Tewari, Y. B., Sheridan, J. P. and Martire, D. E. (1970). *J. Phys. Chem.*, **74**, 3263
183. Heric, E. L. and Coursey, B. M. (1970). *Can. J. Chem.*, 48, 3911; (1971). *J. Chem. Eng. Data*, **16**, 185
184. Kehiaian, H. V., Sosnkowska-Kehiaian, K. and Grolier, J.-P. (1971). *2nd Intern. Conf. Calorimetry and Thermodynamics* (Orono, Maine, USA)
185. Wilhelm, E., Schano, R., Becker, G., Findenegg, G. H. and Kohler, F. (1969). *Trans. Faraday Soc.*, **65**, 1443
186. Miksch, G., Liebermann, E. and Kohler, F. (1969). *Monatsh. Chem.*, **100**, 1574
187. Liebermann, E. and Kohler, F. (1968). *Monatsh. Chem.*, **99**, 2514
188. Kreibich, U. T., Schaefer, K. and Hoepfner, A. (1970). *Ber. Bunsenges. Phys. Chem.*, **74**, 1020
189. Siedler, R., Grote, L., Kauer, E., Werner, U. and Bittrich, H.-J. (1969). *Z. Phys. Chem. (Leipzig)*, **241**, 203
190. Rastogi, R. P., Nath, J. and Yadava, R. B. (1970). *Indian J. Chem.*, **8**, 541
191. Booth, C. and Devoy, C. J. (1971). *Polymer (London)*, **12**, 309; 320
192. Chand, K. and Ramakrishna, V. (1969). *J. Phys. Soc. Jap.*, **26**, 239
193. Rastogi, R. P., Nath, J. and Misra, R. R. (1971). *J. Chem. Thermodynamics*, **3**, 307
194. Abello, L., Picquenard, E., Kern, M. and Pannetier, G. (1971). *Bull. Soc. Chim. Fr.* (in the press)
195. Turner, E. M., Anderson, D. W., Reich, L. A. and Vaughan, W. E. (1970). *J. Phys. Chem.*, **74**, 1275
196. Kiyohara, O. and Higasi, K. (1969). *Bull. Chem. Soc. Jap.*, **42**, 1158
197. Ewing, M. B., Marsh, K. N., Stokes, R. H. and Tomlins, R. P. (1970). *J. Chem. Thermodynamics*, **2**, 297
198. Levien, B. J. and Marsh, K. N. (1970). *J. Chem. Thermodynamics*, **2**, 227
199. Schmack, G. and Bittrich, H.-J. (1970). *J. Prakt. Chem.*, **312**, 730
200. Wang, J. L. H., Boublikova, L. and Lu, B. C. Y. (1970). *J. Appl. Chem.*, **20**, 172
201. Baur, M. E., Knobler, C. M., Horsma, D. A. and Perez, P. (1970). *J. Phys. Chem.*, **74**, 4594
202. Sosnkowska-Kehiaian, K. and Kehiaian, H. V. (1966). *Bull. Acad. Polon. Sci., Ser. Sci. Chim.*, **14**, 573

203. Murakami, T., Murakami, S. and Fujishiro, R. (1969). *Bull. Chem. Soc. Jap.*, **42**, 35
204. Bares, D., Metzger, J., Peneloux, A. and Laffitte, M. (1971). *Proc. 1st Intern. Conf. Calorimetry and Thermodynamics*, 809 (Warsaw: Polish Scientific Publishers-PWN)
205. Buchowski, H. and Jaroniowa, H. (1971). Ibid., 855
206. Powell, R. J. and Swinton, F. L. (1970). *J. Chem. Thermodynamics*, **2**, 87
207. Andrews, A., Morcom, K. W., Duncan, W. A., Swinton, F. L. and Pollock, J. M. (1970). *J. Chem. Thermodynamics*, **2**, 95
208. Wóycicki, W. and Sadowska, K. W. (1968). *Bull. Acad. Polon. Sci., Ser. Sci. Chim.*, **16**, 329
209. Garrett, P. R., Pollock, J. M. and Morcom, K. W. (1971). *J. Chem. Thermodynamics*, **3**, 135
210. Korchemskaya, K. M., Shakhparonov, M. I. and Antonova, N. M. (1970). *Termodin. Termokhim. Konstanty*, **83**, (Moscow: Izd. Nauka)
211. Chowdary, M. C., Naidu, P. R. and Krishman, V. R. (1969). *Indian J. Chem.*, **7**, 796
212. Kehiaian, H. V. (1963). *Bull. Acad. Polon. Sci., Ser. Sci. Chim.*, **11**, 153
213. Patrick, C. R. (1969). *Chem. Ind. (London)*, **28**, 940
214. Powell, R. J., Swinton, F. L. and Young, C. L. (1970). *J. Chem. Thermodynamics*, **2**, 105
215. Fenby, D. V. and Scott, R. L. (1967). *J. Phys. Chem.*, **71**, 4103
216. Bruce, G. R. and Malcolm, G. N. (1969). *J. Chem. Thermodynamics*, **1**, 183
217. Gray, D. F., Watson, I. D. and Williamson, A. G. (1968). *Aust. J. Chem.*, **21**, 379
218. Beath, L. A. and Williamson, A. G. (1969). *J. Chem. Thermodynamics*, **1**, 51
219. Dincer, S. and Van Ness, H. C. (1971). *J. Chem. Eng. Data*, **16**, 378
220. Fried, V., Franceschetti, D. R. and Schneider, G. B. (1968). *J. Chem. Eng. Data*, **13**, 415
221. Heric, E. L. and Coursey, B. W. (1970). Ibid., **15**, 536
222. Roveillo, J. and Gomel, M. (1968). *C. R. Acad. Sci. Paris*, **266**, 1655
223. Andrews, A. W., Hall, D. and Morcom, K. W. (1971). *J. Chem. Thermodynamics*, **3**, 527
224. Armitage, D. A. and Morkom, K. W. (1969). *Trans. Faraday Soc.*, **65**, 688
225. Boule, P. (1969). *C. R. Acad. Sci. Paris*, **268**, 5
226. Findlay, T. J. V. and Kenyon, R. S. (1969). *Aust. J. Chem.*, **22**, 865
227. Becker, F. and Kiefer, M. (1969). *Z. Naturforsch. A*, **24**, 7
228. Becker, F. and Kiefer, M. (1971). Ibid., **26**, 1040
229. Wolff, H., Hopfner, A. and Hopfner, H.–M. (1964). *Ber. Bunsenges. Phys. Chem.*, **68**, 410
230. Wolff, H. and Würtz, R. (1970). *J. Phys. Chem.*, **74**, 1600
231. Kern, M., Servais, B., Abello, L. and Pannetier, G. (1968). *Bull. Soc. Chim. Fr.*, 2763
232. Wóycicki, W. and Sadowska, K. W. (1968). *Bull. Acad. Polon. Sci., Ser. Sci. Chim.*, **16**, 365
233. Ramalho, R. S. and Ruel, M. (1968). *Can. J. Chem. Eng.*, **46**, 456
234. Diaz Pena, M. and Rodriguez Cheda, D. (1970). *An. Quim.*, **66**, 637
235. Brown, I., Fock, N. and Smith, F. (1969). *J. Chem. Thermodynamics*, **1**, 273
236. Smirnova, N. and Kurtynina, L. M. (1969). *Zh. Fiz. Khim.*, **43**, 1883
237. Vesely, F. and Pick, J. (1969). *Collect. Czech. Chem. Commun.*, **34**, 1854
238. Schmidt, R., Werner, G. and Schuberth, H. (1969). *Z. Phys. Chem. (Leipzig)*, **242**, 381
239. Tucker, E. E., Farnham, S. B. and Christian, S. D. (1969). *J. Phys. Chem.*, **73**, 3820
240. Kurtynina, L. M., Smirnova, N. A. and Andrukovich, P. F. (1968). *Khim. Thermodin. Rastvorov*, **2**, 43
241. Strubl, K., Svoboda, V., Holub, R. and Pick, J. (1970). *Collect. Czech. Chem. Commun.*, **35**, 3004
242. Ramalho, R. S. and Delmas, J. (1968). *Chem. Eng. Data*, **13**, 161
243. Diaz Pena, M. and Rodriguez Cheda, D. (1970). *An. Quim.*, **66**, 737; 747
244. Boublikova, L. and Lu, B. C. Y. (1969). *J. Appl. Chem. (London)*, **19**, 89
245. Raju, B. N. and Rao, D. P. (1969). *J. Chem. Eng. Data*, **14**, 283
246. Vonka, P., Svoboda, V., Strubl, K. and Holub, R. (1971). *Collect. Czech. Chem. Commun.*, **36**, 18
247. Seetharamaswamy, V., Subrahmanyan, V., Chiranjivi, C. and Dakshinamurty, P. (1969). *J. Appl. Chem.*, **19**, 258
248. Campbell, A. N. and Kartzmark, E. M. (1969). *Can. J. Chem.*, **47**, 619
249. Ramalho, R. S. and Ruel, M. (1969). *J. Chem. Eng. Data*, **14**, 20
250. Belousov, V. P., Kurtynina, L. M. and Kozulyaev, A. A. (1970). *Vestn. Leningrad. Univ., Fiz., Khim.*, 163

251. Recko, W. M., Sadowska, K. W. and Wóycicka, M. K. (1971). *Bull. Acad. Polon. Sci., Ser. Sci. Chim.,* **19,** 475
252. Geiseler, G., Quitzsch, K., Hesselbach, J. and Schmidt, K. (1968). *Z. Phys. Chem. (Frankfurt am Main),* **60,** 41
253. Iguchi, A. (1969). *Kagaku Sochi,* **11,** 76
254. Recko, W. M. (1968). *Bull. Acad. Polon. Sci., Ser. Sci. Chim.,* **16,** 549; 553
255. Wóycicki, W. and Sadowska, K. W. (1968). Ibid., **16,** 413
256. Ragaini, V., Santi, R. and Carra, S. (1968). *Atti Accad. Naz. Lincei, Cl. Sci. Fis., Mat. Natur., Rend.,* **45,** 540
257. Nagata, I. (1969). *J. Chem. Eng. Data,* **14,** 418
258. Butcher, K. L. and Medani, M. S. (1968). *J. Appl. Chem.,* **18,** 100
259. Ocon, J., Tojo, G. and Espada, L. (1969). *An. Quim.,* **65,** 641
260. Galska-Krajewska, A. (1971). *Rocz. Chem.,* **45,** 99
261. Ocon, J., Tojo, G. and Espada, L. (1969). *An. Quim.,* **65,** 735
262. Watson, I. D. and Williamson, A. G. (1965). *J. Sci. Ind. Res.,* **24,** 615
263. Sosnkowska-Kehiaian, K., Orzel, K. and Kehiaian, H. V. *Bull. Acad. Polon. Sci., Ser. Sci. Chim.,* (1966), **14,** 711
264. Abello, L. and Pannetier, G. (1971). *J. Phys. Chem.,* **75,** 1763
265. Brusset, H. and Duboc, C. (1968). *C. R. Acad. Sci. Paris,* **265,** 1; Duboc, C. (1969). *Bull. Soc. Chim. Fr.,* 2260; Brusset, H., Delvalle, P. and Philoche-Levisalles, M. (1969). Ibid., 3024
266. Wolff, H. and Hoeppel, H. E. (1968). *Ber. Bunsenges. Phys. Chem.,* **72,** 1173
267. Nagata, I. and Ohta, T. (1971). *J. Chem. Eng. Data,* **16,** 164
268. Brusset, H., Delvalle, P., Philoche-Levisalles, M. and Demanee, L. (1968). *Bull. Soc. Chim. Fr.,* 3113
269. Abello, L., Servais, B., Kern, M. and Pannetier, G. (1968). Ibid., 4360
270. Aarna, A., Kaps, T. and Malanowski, S. (1969). *Eesti NSV Tead. Akad. Tiom., Keem., Geol.,* **18,** 312; (1970). Ibid., **19,** 36
271. Raman, G. K., Naidu, P. R. and Krishnan, V. R. (1968). *Aust. J. Chem.,* **21,** 2717
272. Whetsel, K. B. and Lady, J. H. (1970). *Spectrometry of Fuels.* Chapter 20, p. 259 (New York: Plenum Press)
273. Murakami, S. and Fujishiro, R. (1967). *Bull. Chem. Soc. Jap.,* **40,** 1784
274. Langguth, U. and Bittrich, H.-J. (1970). *Z. Phys. Chem. (Leipzig),* **244,** 327
275. Benson, G. C. and Pflug, H. D. (1970). *J. Chem. Eng. Data,* **15,** 382; Pflug, H. D., Pope, A. E. and Benson, G. C. (1968). Ibid., **13,** 408
276. Singh, J. and Benson, G. C. (1968). *Can. J. Chem.,* **46,** 1249; 2065; Pflug, H. D. and Benson, G. C. (1968). Ibid., **46,** 287; Singh, J. Pflug, H. D. and Benson, G. C. (1969). Ibid., **47,** 543; Polak, J., Murakami, S., Lam., V. T., Pflug, H. D. and Benson, G. C. (1970). Ibid., **48,** 2457; Polak, J., Murakami, S., Benson, G. C. and Pflug, H. D. (1970). Ibid., **48,** 3782
277. Diaz Pena, M., Benitez de Soto, M. and Martin, C. (1970). *An. Quim.,* **66,** 447; Diaz Pena, M. and Sotomayor, C. P. (1971). Ibid., **67,** 233; 249; Diaz Pena, M. Trigueros, R. (1968). Ibid., **64,** 303; (1971). Ibid., **67,** 461; 467
278. Quitzsch, K., Koehler, S., Taubert, K. and Geiseler, G. (1969). *J. Prakt. Chem.,* **311,** 429
279. Doniec, A., Krauze, R., Michalowski, S. Serwinski, M. (1968). *Zesz. Nauk. Politech. Lodz. Chem.,* **19,** 171
280. Roveillo, J. and Gomel, M. (1969). *C. R. Acad. Sci. Paris,* **268,** 1560
281. Nakanishi, K. and Shirai, H. (1970). *Bull. Soc. Jap.,* **43,** 1634
282. Belousov, V. P. and Makarova, N. L. (1970). *Vestn. Leningrad. Univ., Fiz., Khim.,* 101
283. Martire, D. E. and Riedl, P. (1968). *J. Phys. Chem.,* **72,** 3478
284. Nakanishi, K., Toba, R. and Shirai, H. (1969). *J. Chem. Eng. Jap.,* **2,** 4
285. Nakanishi, K., Touhara, H. and Watanabe, N. (1970). *Bull. Chem. Soc. Jap.,* **43,** 2671
286. Krichevtsov, B. K. and Komarov, V. M. (1970). *Zh. Prikl. Khim. (Leningrad),* **43,** 703
287. Findlay, T. J. V. and Copp, J. L. (1969). *Trans. Faraday Soc.,* **65,** 1463
288. Nakanishi, K., Shirai, H. and Nakasato, K. (1968). *J. Chem. Eng. Data,* **13,** 188
289. Wóycicki, W. and Sadowska, K. W. (1968). *Bull. Acad. Polon. Sci., Ser. Sci. Chim.,* **16,** 537
290. Riccardi, R., Franzosini, P. and Rolla, M. (1968). *Z. Naturforsch. A,* **23,** 1816
291. Edwards, J., Rodriguez, J. I. and Soto, A. (1970). *Z. Naturforsch A.* Ibid., **25,** 1998

292. Edwards, J., Rodriguez, J. I. (1969). *Monatsh. Chem.*, **100,** 2066
293. Campbell, A. N. and Chatterjee, R. M. (1970). *Can. J. Chem.*, **48,** 277
294. Nagata, I., Tago, O. and Takahashi, T. (1970). *Kagaku Kogaku,* **34,** 1107
295. Paz-Andrade, M. I. and Casas, M. I. (1970). *An. Quim.*, **66,** 709
296. Franzosini, P., Geangu-Moisin, A. and Ferloni, P. (1970). *Z. Naturforsch. A,* **25,** 457
297. Scheller, W. A., Torres-Soto, A. R. and Daphtary, K. J. (1969). *J. Chem. Eng. Data,* **14,** 439
298. Grolier, J.–P. (1970). *Ph.D Thesis* (Univ. Clermont-Ferrand)
299. Grolier, J.-P. and Viallard, A. (1970). *J. Chim. Phys. Physicochim. Biol.*, **67,** 1582; (1971). Ibid, 1442; (1972). Ibid, 203
300. Miksch, G., Ratkovics, F. and Kohler, F. (1969). *J. Chem. Thermodynamics,* **1,** 257
301. Liszi, J. (1970). *Acta Chim. (Budapest),* **64,** 49
302. Affsprung, H. E., Findenegg, G. H. and Kohler, F. (1968). *J. Chem. Soc. A,* 1364
303. Nissema, A. (1970). *Ann. Acad. Sci. Fenn., Ser. A2,* **153,** 51
304. Lau, C. F., Wilson, P. T. and Fenby, D. V. (1970). *Aust. J. Chem.*, **23,** 1143
305. Quitzsch, K., Prinz, H. P., Suehnel, K., Pham Van Sun and Geiseler, G. (1969). *Z. Phys. Chem. (Leipzig),* **241,** 273
306. Chareyron, R. and Clechet, P. (1971). *Bull. Soc. Chim. Fr.* 2853
307. Benoit, R. L. and Charbonneau, J. (1969). *Can. J. Chem.*, **47,** 4195
308. Jannelli, L., Sciacovelli, O., Dell'Atti, A. and Della Monica, A. (1971). *Proc. 1st Intern. Conf. Calorimetry and Thermodynamics,* 907 (Warsaw: Polish Scientific Publishers-PWN)
309. Franzosini, P., Riccardi, R. and Rolla, M. (1968). *Ric. Sci.*, **38,** 3
310. Quitzsch, K., Strittmatter, D. and Geiseler, G. (1969). *Z. Phys. Chem. (Leipzig),* **240,** 107
311. Ababi, V. and Balba, N. (1968). *An. Stiint. Univ. Al. I. Cuza Iasi, Sect. 1c,* **14,** 155

6
Pulse Calorimetry of Solids at High Temperatures

A. CEZAIRLIYAN and C. W. BECKETT
National Bureau of Standards, Washington

6.1 INTRODUCTION 159

6.2 DESCRIPTION OF METHODS 160
6.2.1 *General description* 160
6.2.2 *Classification of methods* 162
6.2.3 *Formulation of relationships* 163

6.3 MEASUREMENT AND RECORDING OF QUANTITIES 164
6.3.1 *Measurement of power and temperature* 164
6.3.2 *Recording of quantities* 166

6.4 INVESTIGATIONS USING PULSE CALORIMETRY 167

6.5 AN ACCURATE MILLISECOND RESOLUTION PULSE CALORIMETER 168

6.6 AN EXAMPLE OF MEASUREMENTS AT HIGH TEMPERATURES 170

6.7 DISCUSSION AND CONCLUSIONS 173

6.1 INTRODUCTION

Steady-state and quasi-steady-state techniques have been developed for accurate measurements of thermodynamic properties at moderately high temperatures (up to 2500 K). In all these techniques, the specimen and its immediate environment are exposed to high temperatures for relatively long periods of time (minutes-to-hours). These relatively long exposure times in conjunction with the rapid increase with temperature of various phenomena, such as heat transfer, chemical reactions, evaporation, diffusion and loss of mechanical strength, limit the application of the conventional techniques to temperatures below 2500 K. An alternative for accurate measurements above this limit is the development of techniques in which the contribution of these undesirable phenomena can be made negligibly small by exposing the specimen to high temperatures for only a very short period of time (less than a second). It is in this context that most of the pulse techniques for the measurement of thermodynamic properties at high temperatures were developed.

The advantages of pulse techniques have been realised for over 50 years, and during this period a number of pulse calorimeters were constructed and used for measurements of specific heat at moderate and high temperatures. However, in spite of their distinct advantages, in almost all cases the progress did not extend beyond the 'preliminary' stage. This can be attributed largely to the lack of proper instrumentation and to the difficulties in accurate transient measurement techniques. Increasing demand for properties at high temperatures and rapid advances in the electronics field have stimulated new efforts in pulse calorimetry in recent years.

The objective of this chapter is to present a general description of pulse techniques for the measurement of specific heat of solid electrical conductors. The presentation is limited to methods that utilise rapid resistive self-heating of the specimen. Techniques that may yield the specific heat as a by-product, such as in some flash thermal diffusivity measurements, and methods that utilise cyclic heating (modulation methods) and free radiative cooling are not considered. Design and operational characteristics of pulse calorimetric systems reported in the literature are summarised. Emphasis is placed on sub-second pulse calorimetry for measurements above 2000 K. However, for the sake of completeness and continuity, investigations that utilised slower pulses and that were for measurements below 2000 K are also included.

6.2 DESCRIPTION OF METHODS

6.2.1 General description

The principle of pulse calorimetry is based on rapid resistive self-heating of the specimen by the passage of electrical currents through it, and on measuring the pertinent quantities with appropriate time resolution. The required quantities are power imparted to the specimen and specimen temperature, both as a function of time.

A pulse calorimeter consists of an electrical power pulsing circuit and associated high-speed measuring circuits. The pulsing circuit includes the specimen in series with a power source, a variable resistance, a standard resistance, and a fast-acting switch. Heavy duty batteries and, to a lesser extent, capacitors are used as power sources. The specimen is contained in a controlled-environment chamber. The high-speed measuring circuits include detectors and recording systems. A simplified block diagram of a generalised pulse calorimeter is shown in Figure 6.1.

Different techniques may be used for the measurement of pertinent quantities. In general, power imparted to the specimen is obtained from measurements of current through the specimen and potential difference

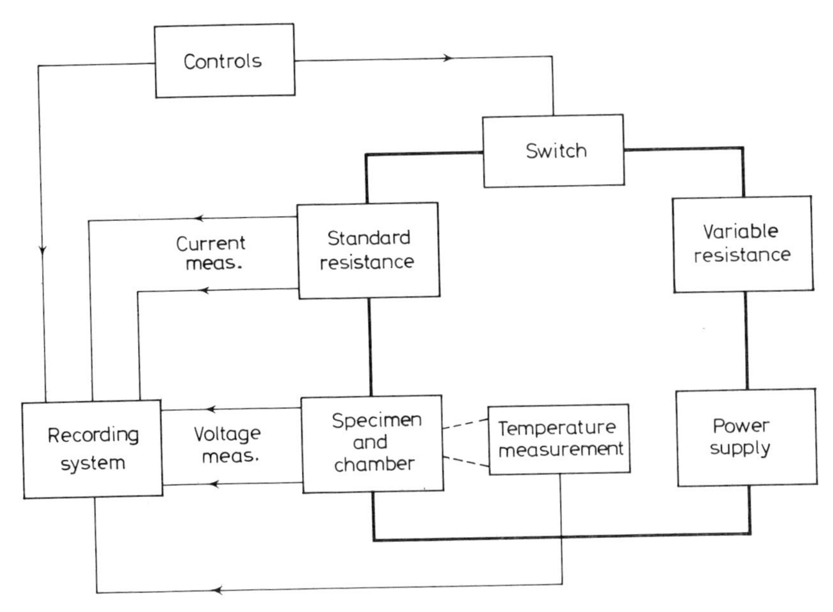

Figure 6.1 Block diagram of a generalised pulse calorimeter

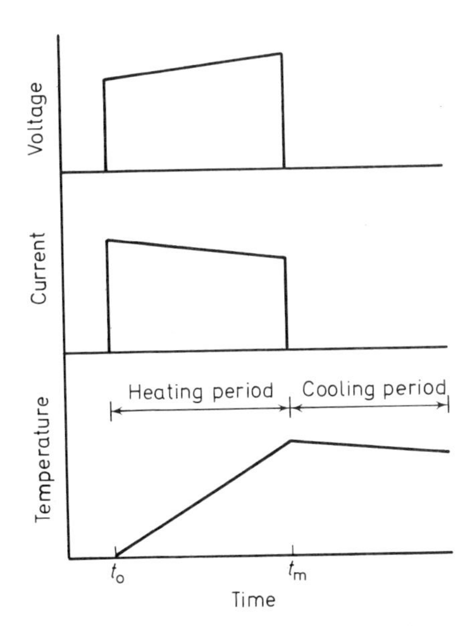

Figure 6.2 Simplified traces of voltage across the specimen, current through the specimen and specimen temperature in a typical pulse experiment. Time t_0 corresponds to closing of the switch and t_m to opening of the switch

across the specimen as a function of time. Temperature measurements are made using resistive, thermoelectric (thermocouples) or optical means. Simplified and smoothed traces of voltage, current and temperature obtained in a typical pulse experiment are shown in Figure 6.2.

Details on design and operational characteristics of an accurate millisecond resolution pulse calorimeter are given in Section 6.5.

6.2.2 Classification of methods

Considerable constructional and operational differences exist among pulse calorimeters used by various investigators. In order to be able to compare the techniques and to discuss their advantages as well as disadvantages it may be appropriate to classify them into the following categories:

(a) The specimen is initially at a temperature above room temperature under steady-state conditions. Then, a current pulse is sent through the specimen rapidly raising its temperature. The techniques in this category may be divided into two subgroups:

(i) The specimen is initially under steady-state conditions at a high temperature as the result of resistive self-heating.

(ii) The specimen is initially under steady-state conditions at a high temperature as the result of a high-temperature environment (furnace).

(b) The specimen is initially at or near room temperature. Then a current pulse experiments. The desired temperature level is achieved by steady-state desired level. The techniques in this category may be divided into two subgroups:

(i) During the pulse period, ambient temperature is maintained at or near room temperature.

(ii) During the pulse period, ambient temperature is raised rapidly and almost proportionately with the specimen temperature.

The methods under category (a) are a combination of steady-state and pulse experiments. The desired temperature level is achieved by steady-state power dissipation, which allows the measurement of initial temperature and other pertinent quantities with conventional equipment. Then, a current pulse is sent through the specimen to raise its temperature. Quantities required for the determination of power and temperature are measured during this time. Usually the temperature increase is of the order of a few degrees. Because of the small temperature rise, the magnitude of the pulse current is usually small, less than 10 A. Of course, this depends on the specimen geometry. In the case where the specimen is brought to a steady-state temperature by resistive self-heating (category (a)(i)), radiative heat losses during the pulse heating period become appreciable. This implies that a correction to the imparted power has to be made based either on an estimated emittance value for the specimen or on supplementary data which can be obtained in the course of the experiments. In the case where the specimen is initially in thermal equilibrium with its surroundings by being in a furnace (category (a)(ii)), the net temperature difference between the specimen and its surroundings during the pulse heating period is small; thus, radiative heat loss is small. This permits the performance of slow-pulse experiments, which have certain advantages from instrumentation and theoretical viewpoints.

The methods under category (b) represent truly high-speed experiments, in the sense that the entire measurement system is initially at or near room temperature, and only the specimen is heated to high temperatures during the very short pulse period. They do not have the limitations of those in category (a), which result from the exposure of the specimen and its surroundings to high temperatures for extended periods. However, the methods of category (b) are subject to other difficulties. Since the specimen must be heated rapidly from room temperature to the desired high temperature, greater pulse currents are required. Also, in the case of category (b)(i) radiation heat loss from the specimen becomes appreciable. In order to reduce this effect, high heating rates are required. In category (b) (ii) certain design and operational difficulties are encountered. In general, the speed of the experiments in category (b) is much greater than that in category (a). This presents difficulties in measuring and recording the pertinent quantities. High pulse currents pose additional problems from both design and operational viewpoints. For example, time-dependent high currents create varying electromagnetic fields which may affect the operation of the transducers and the entire measuring equipment.

6.2.3 Formulation of relationships

In the general case, power balance for the specimen during the heating period can be expressed as follows:

$$\text{power imparted} = \text{power stored} + \text{power losses}$$

which becomes

$$ei = c_p m \left(\frac{\mathrm{d}T}{\mathrm{d}t}\right)_\mathrm{h} + Q \tag{6.1}$$

Solving equation (6.1) for c_p, one obtains

$$c_p = \frac{ei - Q}{m\left(\dfrac{\mathrm{d}T}{\mathrm{d}t}\right)_\mathrm{h}} \tag{6.2}$$

where e is the potential difference across the specimen, i the current through the specimen, c_p the specific heat of specimen, m the effective mass of specimen, Q the total power loss from specimen, and $(\mathrm{d}T/\mathrm{d}t)_\mathrm{h}$ the heating rate of specimen. In pulse experiments in which the temperature rise is only a few degrees, the quantity $\mathrm{d}T/\mathrm{d}t$ in equation (6.2) may be replaced by $\Delta T/\Delta t$.

The quantity Q may be obtained from data during the initial cooling period. Power balance for this period gives

$$-c_p m \left(\frac{\mathrm{d}T}{\mathrm{d}t}\right)_\mathrm{c} = Q \tag{6.3}$$

where $(\mathrm{d}T/\mathrm{d}t)_\mathrm{c}$ is the cooling rate of the specimen. Substituting equation (6.3) in equation (6.2) for Q, one obtains

$$c_p = \frac{ei}{m\left(\frac{\mathrm{d}T}{\mathrm{d}t}\right)_\mathrm{h}\left(1+\frac{1}{M}\right)} \tag{6.4}$$

where

$$M = -\left(\frac{\mathrm{d}T}{\mathrm{d}t}\right)_\mathrm{h} \Big/ \left(\frac{\mathrm{d}T}{\mathrm{d}t}\right)_\mathrm{c} \tag{6.5}$$

Specific heat can be obtained from equation (6.4) provided that experimental data on voltage, current and temperature during the heating period and temperature during the initial cooling period, all as a function of time, are available.

At temperatures above approximately 1500 K, in pulse experiments of 0.001–1 s duration, thermal radiation is the major source of power loss. When data during the initial cooling period are not available, specific heat can be obtained from equation (6.2) after estimating power loss using the relation for thermal radiation

$$Q_r = \varepsilon\sigma A_\mathrm{s}(T^4 - T_0^4) \tag{6.6}$$

where ε is the hemispherical total emittance, σ the Stephan-Boltzmann constant, A_s the effective surface area, T the specimen temperature and T_0 the ambient temperature. Equation (6.6) requires a knowledge of hemispherical total emittance for the specimen. However, if power loss by thermal radiation is only a few per cent of imparted power, the effect of even an appreciable error (10 %) in estimated emittance on specific heat is much less than one per cent.

As may be seen from equations (6.2) or (6.4), pulse calorimetry permits the determination of specific heat from the data of one experiment.

6.3 MEASUREMENT AND RECORDING OF QUANTITIES

Various techniques that were reported in the literature for the measurement and recording of experimental quantities are summarised in the following paragraphs and in Table 6.1. To avoid duplications, references to specific investigators are omitted in the text.

6.3.1 Measurement of power and temperature

In general, power imparted to the specimen during the heating period is determined from the measurement of current through the specimen and potential difference across it as a function of time. This approach is straightforward in systems in which batteries are used as power sources. However, due to the existence of time-dependent high electro-magnetic fields, accurate measurements of current and especially voltage become very difficult in

Table 6.1 Summary of investigations for the measurement of specific heat of solid electrical conductors by pulse calorimetry

No.	Investigator	Ref.	Year	Category	Power source	Pulse current/A	Pulse length/s	Heating rate/Ks^{-1}	ΔT/K	Substance	Temperature range K	Specimen geometry	Power meas.	Temp. meas.	Record-ing	Inac-curacy
1	Worthing	3	1918	ai	B	5–7	1			W, C	1200–2400	W	EI	R	V	
2	Lapp	8	1929	aii	B				1–2	Ni	100–730	R	W	TC	V	2
3	Grew	9	1934	aii	B	3–5			2–4	Ni	90–720	R	EI	TC	V	2
4	Avramescu	21	1939	bi	B	2000	1–2			Al, Cu	400–1300	R	EI	R	G	2
5	Néel and Persoz	10	1939	aii	B	1000–5000	0.1		10–80	Cu, Ni, Pt	300–1300	W	W	R	V	2
6	Kurrelmeyer *et al.*	11	1943	aii	C		0.002–0.05			Pt	300	W	B	R	V	0.5
7	Baxter	22	1944	bi			0.05					W	EI	R	G	
8	Pallister	12	1949	aii	B	200		1	1–2	Fe	273–1500	R	EI	TC	V	2
9	Nathan	24	1951	bi	B			15–1000		Steel	770	R	EI	TC	S	
10	Khotkovich and Bagrov	23	1951	bi	B		0.01	10^4–5×10^4		Cu, W, Mo, Cd		W	EI	R	G	3
11	Pochapsky	13	1953	aii	C		0.001		1–3	Al, Pb	273–920	W	B	R	V	5
12	Pochapsky	14	1954	aii	C	4	0.001		1–3	Pt	273–900	W	B	R	V	5
13	Strittmater *et al.*	25	1957	bi	B		0.05–0.15	3000–9000		Pt, Ni	300–720	W	EI	R	S	5–10
14	Rasor and McClelland	16	1960	aii	B	300–1000		50	5	Mo, Ta, C	1300–3920	R	EI	O	G	5
15	Wallace *et al.*	17	1960	aii	B	10	0.04		100	Fe	300–1300	W	B	R	S	2
16	Wallace	18	1960	aii	B	10	0.03		100	Th	300–1300	W	B	R	S	2
17	Pasternak *et al.*	4	1962	ai	B	10	1–10	100–1000		Pt	300–1100	W	EI	R	R	
18	Pasternak *et al.*	5	1963	ai	B	10		35–2000		Pt	400–1300	W	EI	R	R	
19	Taylor and Finch	26	1964	bi	B			10^3–6×10^4		Mo, Ta	100–3200	W	EI	R	S	4–7
20	Cezairliyan	28	1965	bi	B	2000	0.25	6600		Mo	1300–1600	T	EI	O	S	5
21	Parker	6	1965	ai	C		10^{-5}	10^4–10^9	60–160	Ti	300–1100	S	C	TC	S	
22	Kollie	7	1967	ai	R	< 50	4–60	10–20		Fe	300–1200	R	EI	TC	D	1
23	Affortit and Lallement	29	1968	bi	B		0.1–1			Nb, W	300–3600	W	EI	TC, O	S	3–5
24	Affortit	30	1969	bi	B		< 1			UN, UC, UO_2	700–3100	R	EI	TC, O	S	5
25	Finch and Taylor	27	1969	bi	B	1000–5000		7.10^3–1.6×10^5		$ZrU_{0.04}H_{0.5}$	300–800	R	EI	R, O	S	5
26	Jura and Stark	31	1969	bi	B		0.1			Fe	80–300	W	EI	TC	S	5
27	Kollie *et al.*	19	1969	aii	R	< 100	10–35	20	100–300	Fe	300–1500	R	EI	TC	D	1–2
28	Cezairliyan *et al.*	33	1970	bi	B	1500–2200	0.3–0.7	3300–8000		Mo	1900–2800	T	EI	O	D	2–3
29	Dikhter and Lebedev	32	1970	bi	C		10^{-5}			W	2600–3600	W	EI	O	S	10
30	Cezairliyan *et al.*	34	1971	bi	B	1300–1500	0.3–0.5	3700–5700		Ta	1900–3200	T	EI	O	D	2–3
31	Cezairliyan and McClure	35	1971	bi	B	1500–1900	0.4–0.6	5600–7100		W	2000–3600	T	EI	O	D	2–3
32	Cezairliyan	36	1971	bi	B	1300–1500	0.4–0.5	5200		Nb	1500–2700	T	EI	O	D	2–3

Abbreviations and Notes
Power source: B = Battery, C = Capacitor, R = Regulated d.c. Power Supply
ΔT: Specimen Temperature Rise Due to Pulse Current
Temperature Range: Range between two extreme temperatures covered by the investigator regardless of particular substance
Specimen Geometry: R = Rod (diam. > 1 mm), S = Strip, T = Tube, W = Wire (diam. ⩽ 1 mm)
Power Meas.: B = Bridge, C = Capacitor energy, EI = Voltage-current, W = Wattmeter
Temperature Meas.: O = Optical, R = Resistive, TC = Thermocouple
Recording: D = Digital, G = Oscillographic, R = Chart recording, S = Oscilloscopic, V = Visual-manual
Inaccuracy: Total error in specific heat as reported in the literature by the investigators

the case where capacitors are used. Techniques of measuring current and voltage in capacitor discharge experiments were reviewed in an earlier publication (Beckett and Cezairliyan[1]). In some slow experiments power was measured by specially designed wattmeters; in others, the specimen was placed in one of the arms of a bridge (Wheatstone or Kelvin) and sudden deflection of the galvanometer, resulting from the passage of pulse current, was used to determine power input to the specimen.

Most investigators determined specimen temperature either with thermocouples or by measuring specimen resistance as a function of time. In general, thermocouples were formed by spot welding the wires individually to the specimen at a plane perpendicular to the current flow. In the resistance method, a separate steady-state experiment was required to obtain the resistance–temperature relation for the specimen, which is used to convert the experimentally obtained resistance–time relationship to a temperature–time relationship. Although during the last decade a few attempts were made to measure specimen temperature optically, only one method (with millisecond resolution) has yielded results comparable in accuracy to those obtained by conventional photoelectric pyrometers (with one second resolution).

6.3.2 Recording of quantities

During the early years of high-speed experiments, recording of electrical quantities under transient conditions was achieved by relatively crude methods. These methods were based on the observation or recording of the deflection of a ballistic instrument (galvanometer) placed either in a bridge (Wheatstone or Kelvin) or a potentiometric circuit. Several variations of this method were used by different investigators. In most cases, the magnitude of galvanometer deflection gave the magnitude of the quantity to be measured. In other cases, the bridge or the potentiometer circuit was pre-adjusted, after several trial experiments, to give zero galvanometer deflection under pulse conditions. The pre-adjusted value gave the magnitude of the quantity to be measured. In two cases, voltage and current were recorded using a fast two-channel recorder. A few investigators have reported the use of oscillographs as a means of recording the electrical quantities.

It may be seen that all of these recording techniques are limited in speed; thus, they are applicable only in the case of slow experiments, in which specimen temperature rise, as the result of pulse currents, is small.

The first truly high-speed recording started with the use of oscilloscopic techniques. Because of the very fast response characteristics (as low as nanoseconds), oscilloscopes were used for a wide range of experimental conditions, from millisecond to nanosecond resolution. Since 1950 most investigators have used oscilloscopic recording techniques. However, they too have limitations, which stem primarily from their relatively poor recording resolution.

During the last few years, advances in high-speed digital recording techniques and their application to high-speed measurements of millisecond resolution (0.01% full-scale signal resolution) have improved the recording resolution by approximately two orders of magnitude.

6.4 INVESTIGATIONS USING PULSE CALORIMETRY

Historical developments of pulse calorimetry were given in detail recently by Cezairliyan[2]. Therefore, in this section only a summary of early investigations, with proper updating, is presented. The salient features of these investigations are given in Table 6.1.

Historically, the earliest attempts in using pulse techniques were confined to those in category (a) (see Section 6.2.2). The choice was most probably dictated by the fact that smaller (in amplitude) and longer pulses could have been used for the methods in this category which, in turn, would have allowed the measurement of pertinent quantities with instruments then available. During the last decade, the development of instruments with faster response times and increased accuracy, along with growing interest in the measurement of specific heat at temperatures above the reliable operation of steady-state systems, have encouraged a concentration of effort in methods belonging to category (b).

The earliest successful attempt to use a pulse technique for the measurement of specific heat of electrical conductors may be attributed to Worthing[3]. It was based on the measurement of absorbed power by the specimen (tungsten and carbon filaments) while going from one steady-state to another as the result of the passage of electric current through it in a short time (1 s). An initial high-temperature steady-state condition was achieved by resistive self-heating of the specimen. Batteries were used as the pulse power source. Temperature of the specimen was determined from the measurement of specimen resistance during the pulse experiment and from the knowledge of the resistance-temperature relation, which was obtained separately under steady-state conditions. This general concept was used, after considerable modifications and refinements, by Pasternak *et al.*[4,5], Parker[6] and Kollie[7].

Because of certain disadvantages in establishing initial steady-state high-temperature conditions by resistive self-heating, some investigators placed the specimens in a furnace and then pulsed the specimen to a higher temperature. This approach was first adopted by Lapp[8] and was used, after considerable modifications and refinements, by Grew[9], Néel and Persoz[10], Kurrelmeyer *et al.*[11], Pallister[12], Pochapsky[13,14], Rasor and McClelland[15,16], Wallace *et al.*[17], Wallace[18], Kollie *et al.*[19] and Kollie[20].

The technique in which the specimen is heated from room temperature to high temperatures in a short time was tried successfully for the first time by Avramescu[21]. The specimens (aluminium and copper) in the form of rods were heated to their respective melting points in a few seconds by passing high currents (2000 A) through them. The principle of this technique, after considerable modifications and refinements, was used by Baxter[22], Khotkovich and Bagrov[23], Nathan[24], Strittmater *et al.*[25], Taylor and Finch[26], Finch and Taylor[27], Cezairliyan[28], Affortit and Lallement[29], Affortit[30], Jura and Stark[31],

Dikhter and Lebedev[32], Cezairliyan *et al.*[33, 34], Cezairliyan and McClure[35], and Cezairliyan[36].

6.5 AN ACCURATE MILLISECOND RESOLUTION PULSE CALORIMETER

In this section, an accurate sub-second pulse calorimeter (Cezairliyan[37]) for the measurement of specific heat of solid electrical conductors in the temperature range from 1500 K to the melting point of the specimen is described. In this calorimeter, the specimen is initially at room temperature and is heated rapidly to the desired high temperature in less than one second by the passage of high currents (2000 A).

The functional diagram of the pulse calorimeter is shown in Figure 6.3. The power-pulsing circuit includes the specimen in series with a 28 V battery

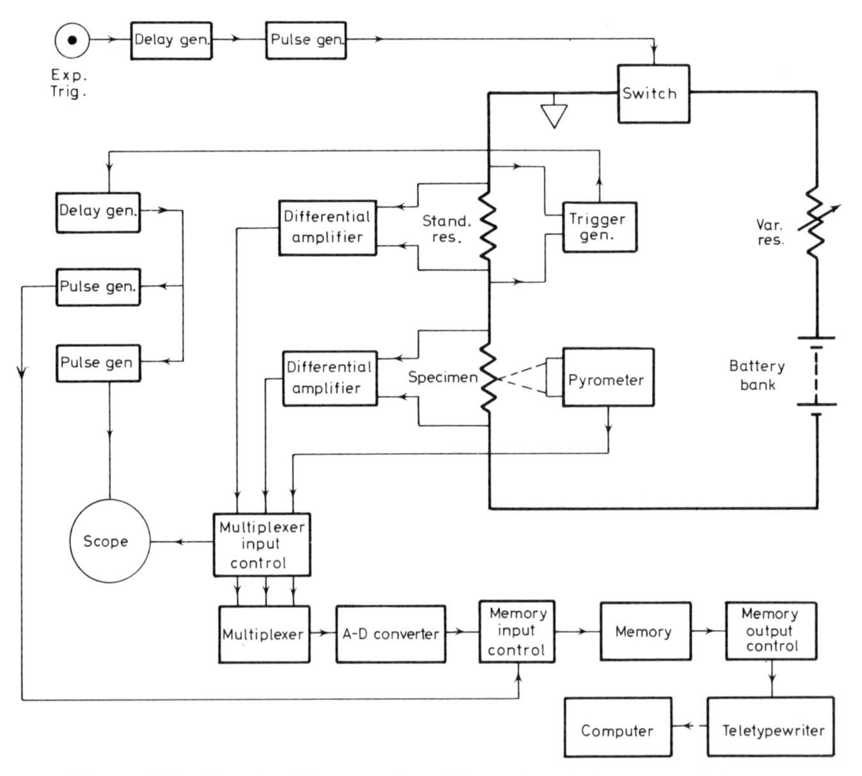

Figure 6.3 Functional diagram of a millisecond-resolution pulse calorimeter

bank, a 0.001 Ω standard resistance, a variable resistance and a switch. The specimen is a tube of the following nominal dimensions: length, 100 mm, outside diameter, 6 mm, and wall thickness, 0.5 mm. A small rectangular hole (1×0.5 mm) in the wall at the middle of the specimen provides an approximation to black-body conditions for optical temperature measurements.

The experiment chamber (Figure 6.4) is designed for conducting experiments with the specimen either *in vacuo* or in a controlled atmosphere.

Current through the specimen is determined from the measurement of the potential difference across the standard resistance placed in series with the specimen. Potential difference across the middle two-thirds of the specimen is measured using spring-loaded, knife-edge probes. Specimen temperature is measured with a high-speed photoelectric pyrometer (Foley[38]) which permits 1200 evaluations of the specimen temperature per second. A unique

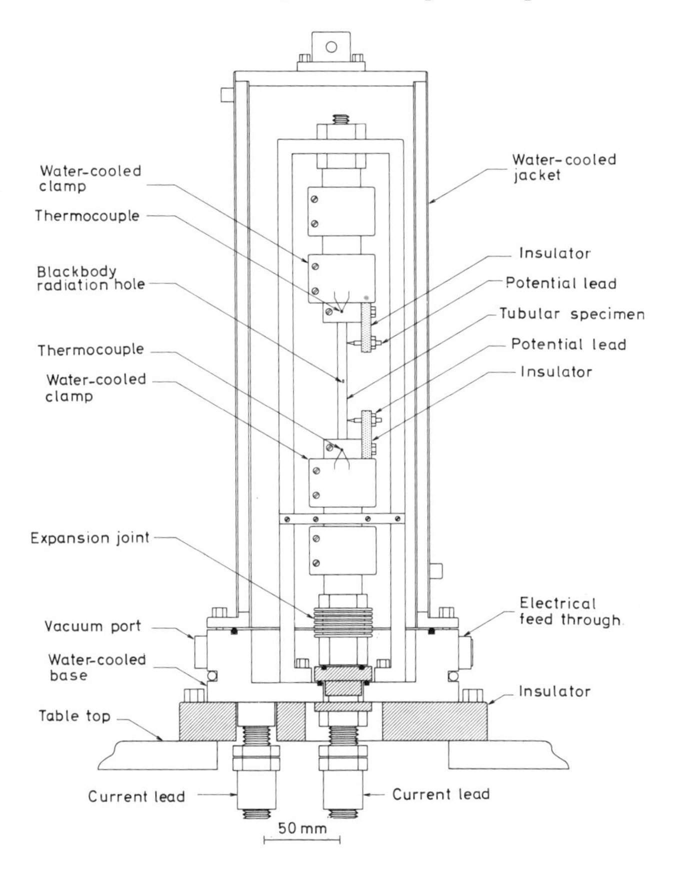

Figure 6.4 Experiment chamber of a millisecond-resolution pulse calorimeter

feature of this pyrometer is that it alternately samples three reference radiances in addition to the specimen radiance. This scheme eliminates possible errors that result from detector instabilities.

Experimental quantities are recorded with a high-speed digital data acquisition system consisting of a multiplexer, analogue-to-digital converter, core memory and various control and interfacing equipment. The system is capable of recording sets of data corresponding to temperature, voltage, and current approximately every 0.4 ms with a full-scale signal resolution of 1 in 8192 ($8192 = 2^{13}$). In a sub-second duration experiment, up to 2048 data points can be digitised and stored in the memory. At the end of the experiment, data are retrieved via a teletypewriter and are processed immediately using a time-shared computer. Oscilloscopes are used for monitoring purposes only.

Oscilloscope-trace photographs of typical voltage and current pulses are shown in Figure 6.5. Radiances from the blackbody hole in the specimen as well as the reference radiances as seen by the pyrometer are shown in Figure 6.6.

Typical characteristics of the pulse calorimeter are as follows: current, 1300–2200 A; power, 7000–15000 W; heating rate, 3000–8000 Ks^{-1}; heating duration, 0.3–0.7 s; and heating rate/cooling rate, 10–100. These figures are the results of typical experiments and do not represent the full capabilities of the measurement system.

In the experiments performed with this calorimeter, heat loss by thermal radiation is the only significant power loss from the specimen. In the computations of specific heat, this loss is accounted for by measuring the initial cooling rate of the specimen after opening the switch (at the end of the heating period) according to equation (6.4). The experimental quantities, voltage, current, and temperature in equation (6.4) are expressed by third degree polynomial functions of time obtained by the least squares approximation of the data. Typical standard deviation of an individual point from the pertinent function is 0.02% for voltage, 0.03% for current and 0.5 K for temperature.

An advantage of the pulse calorimeter described above is that it permits the determination of several other properties in addition to specific heat. Data taken during the rapid-heating period and the subsequent initial-cooling period yield electrical resistivity and hemispherical total emittance as a by-product. By performing an additional pulse experiment in which the pyrometer is aimed at the surface of the specimen, normal spectral emittance (at the effective wavelength of the pyrometer, $\lambda = 650$ nm) is obtained. During the rapid-heating period if the current through the specimen is not interrupted the specimen reaches its melting point. The plateau in measured temperature yields the melting point.

Estimated inaccuracies in measured properties in a typical experiment are: specific heat, 2% at 2000 K, 3% at 3000 K; electrical resistivity, 0.5%; hemispherical total emittance, 3%; normal spectral emittance, 3%; and melting point, 10 K at 3000 K. The details regarding estimation of errors are given by Cezairliyan *et al.*[33].

6.6 AN EXAMPLE OF MEASUREMENTS AT HIGH TEMPERATURES

The pulse-calorimeter described above has been used for the measurement of specific heat, electrical resistivity, hemispherical total emittance, and normal spectral emittance of molybdenum[33], tantalum[34], tungsten[35] and niobium[36]. The experimental details are given in Table 6.1. It was also used for the measurements of the melting points of molybdenum[39] and tungsten[40].

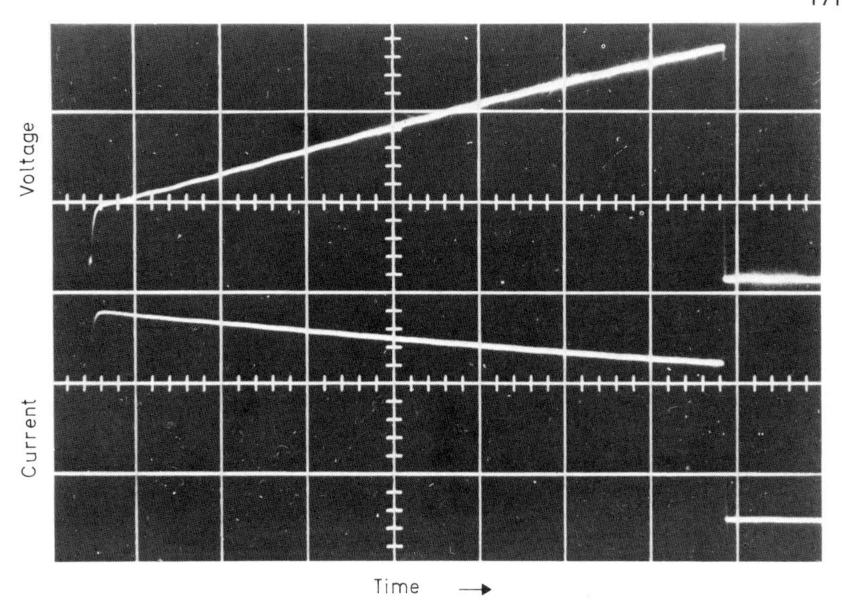

Figure 6.5 Oscilloscope-trace photographs of typical voltage and current pulses. Equivalance of each major division is, time = 50 ms, voltage = 2 V, current = 1000 A

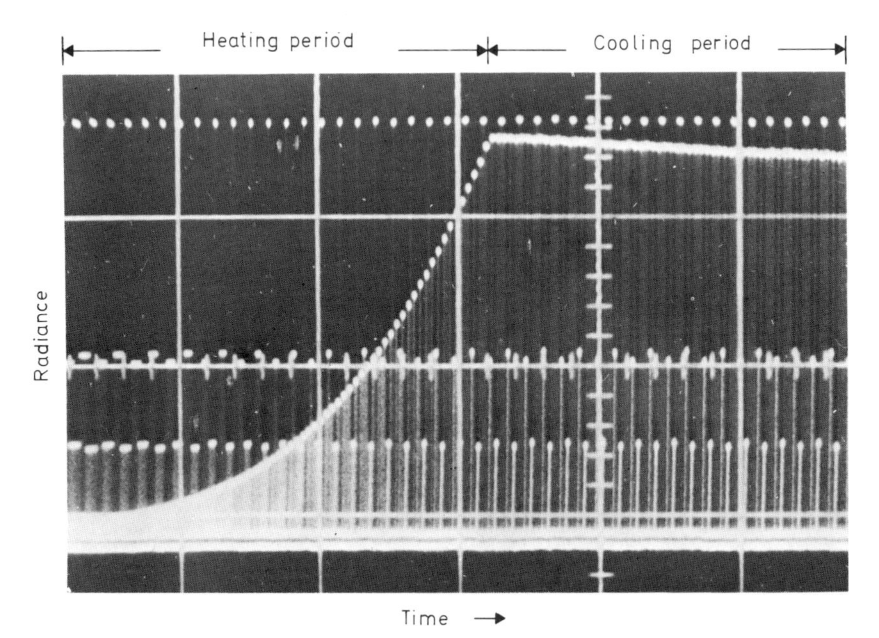

Figure 6.6 Oscilloscope-trace photograph of radiance of a rapidly heating specimen as seen by the pyrometer. Dots forming the long horizontal lines correspond to the reference radiances. Equivalence of each major division is, time = 20 ms, radiance = arbitrary unit

As an example, results for specific heat of tungsten obtained using the pulse calorimeter, in addition to those obtained by other investigators, are shown in Figure 6.7. The results represent measurements using three different techniques, namely 'drop', 'modulation', and 'pulse'. There is no evidence for any bias with respect to a particular technique.

From Figure 6.7 it may be noted that specific heat results of tungsten at high temperatures are considerably higher than the Dulong and Petit value

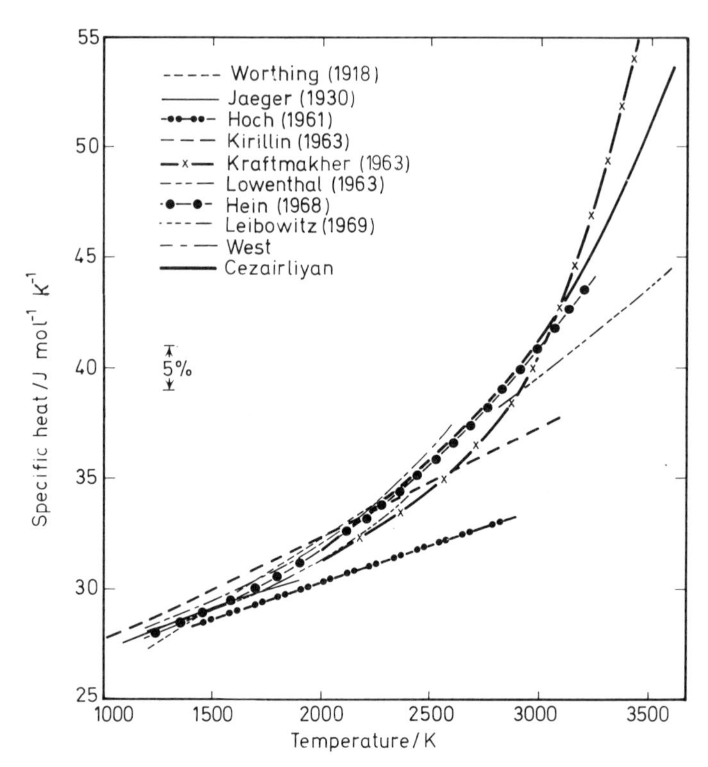

Figure 6.7 Specific heat of tungsten reported in the literature. For brevity, the names of only the first authors are given in the figure. References to investigations presented in the figure may be found in the publication by Cezairliyan and McClure[35]. The techniques that were used are as follows. Pulse: Worthing, Cezairliyan; modulation: Kraftmakher, Lowenthal; drop: remainder

of $3R$. Some of this departure is due to $c_p - c_v$ and the electronic terms. However, they do not account for the entire departure.

Specific heat at high temperatures may be expressed by

$$c_p = A - \frac{B}{T^2} + CT + \Delta c \tag{6.7}$$

where the constant term A is $3R$ (24.943 J mol^{-1} K^{-1}), the term in T^{-2} is the first term in the expansion of the Debye function, the term in T represents $c_p - c_v$ and first-order anharmonic and electronic contributions, and the quantity Δc represents excess in measured heat capacity at high temperatures, which is not accounted for by the first three terms. The coefficients B and C can be obtained from heat capacity data at low and moderate temperatures (below 1000 K).

Using equation (6.7) and specific heat of tungsten reported by Cezairliyan and McClure[35] the quantity Δc (in J mol^{-1} K^{-1}) is computed to be 2 at 2000 K, 9 at 3000 K and 22 at the melting point (3695 K). The estimated uncertainty in the computed Δc may be as high as 1 J mol^{-1} K^{-1}. This is obtained from the combined uncertainties in the coefficients in equation (6.7) and the measured specific heats. The trend of specific heat results for other refractory metals, niobium, molybdenum and tantalum, is similar to that of tungsten.

Although vacancy generation becomes appreciable at high temperatures, it is not likely that the high values are due entirely to vacancies. Higher order terms in electronic specific heat and in lattice anharmonicity probably account for most of the increment Δc at high temperatures. More accurate theoretical work may produce a quantitative explanation of this phenomena.

6.7 DISCUSSION AND CONCLUSIONS

From Table 6.1 it may be seen that reported errors in specific heat measurements range from 1 to 10%. It is likely that, in most cases, the actual errors are higher than the reported values. Until recently, because of crude instrumentation and difficulties in measurement techniques, most pulse calorimeters did not extend beyond the 'preliminary' stage. Recent advances in the electronics fields and concentrated efforts have contributed to the development of millisecond-resolution calorimeters for accurate measurement of specific heat of solid electrical conductors from room temperature to approximately 4000 K. In this connection, two pulse calorimeters may be cited; (a) for the range from room temperature to 1500 K (Kollie *et al.*[19]), and (b) for the range from 1500 to 4000 K (Cezairliyan[37]). Estimated errors in specific heat measurements with the above calorimeters are: 1% at 1000 K, 2% at 2000 K, and 3% above 3000 K.

Pulse calorimetry below 2000 K is useful primarily for measurements in which high resolution in temperature is required, such as in measurements at or near solid–solid transition points. However, the main advantage of pulse calorimetry is at temperatures above 2000 K. In the range 2000 to 2500 K, accuracy of specific heat measurements by pulse and conventional calorimetry (assuming the best ones in each category) is approximately the same. Above 2500 K, it becomes very difficult if not impossible to conduct accurate steady-state experiments, thus pulse calorimetry becomes a unique and indispensable tool for measurements at high temperatures. An additional advantage of pulse calorimetry is that it permits the simultaneous measurement of several electrical and thermal properties.

The next challenge is likely to be in the extension of pulse calorimetry to the measurement of specific heat of liquid metals at temperatures above 2500 K and possibly up to the critical points. However, a number of difficulties, such as specimen containment, in addition to high temperatures present serious problems. Very rapid (microsecond or sub-microsecond resolution) accurate pulse calorimetry may be an answer to these. However, at the present time, no serious and systematic effort has been made in this direction. Equipment and instrumentation required for the construction and operation of an accurate microsecond calorimeter is considerably different than those used in millisecond calorimeters. Some of these items are independently developed in connection with other high-speed work, such as exploding conductors and electrical discharges. Integration of existing techniques with some modifications and further advances in the areas of high-speed measurement of temperature and electrical quantities may result in the development of accurate pulse calorimeters of microsecond or even sub-microsecond resolution.

References

1. Beckett, C. W. and Cezairliyan, A. (1968). High-Speed Thermodynamic Measurements and Related Techniques, in *Calorimetry on Non-Reacting Systems,* J. P. McCullough and D. W. Scott. (London: Butterworths)
2. Cezairliyan, A. (1969). *High Temp.-High Pres.,* **1,** 517
3. Worthing, A. G. (1918). *Phys. Rev.,* **12,** 199
4. Pasternak, R. A., Fraser, E. C., Hansen, B. B. and Wiesendanger, H. U. D. (1962). *Rev. Sci. Instr.,* **33,** 1320
5. Pasternak, R. A., Wiesendanger, H. U. D. and Hansen, B. B. (1963). *J. Appl. Phys.,* **34,** 3416
6. Parker, R. (1965). *Trans. Met. Soc. AIME,* **233,** 1545
7. Kollie, T. G. (1967). *Rev. Sci. Instr.,* **38,** 1452
8. Lapp, E. (1929). *Ann. de Physique,* **12,** 442
9. Grew, K. E. (1934). *Proc. Roy. Soc. (London),* **145,** 509
10. Néel, L. and Persoz, B. (1939). *Compt. Rend.,* **208,** 642
11. Kurrelmeyer, B., Mais, W. H. and Green, E. H. (1943). *Rev. Sci. Instr.,* **14,** 349
12. Pallister, P. R. (1949). *J. Iron Steel Inst.,* **161,** 87
13. Pochapsky, T. E. (1953). *Acta Met.,* **1,** 747
14. Pochapsky, T. E. (1954). *Rev. Sci. Instr.,* **25,** 238
15. Rasor, N. S. and McClelland, J. D. (1960). *Rev. Sci. Instr.,* **31,** 595
16. Rasor, N. S. and McClelland, J. D. (1960). *J. Phys. Chem. Solids,* **15,** 17
17. Wallace, D. C., Sidles, P. H. and Danielson, G. C. (1960). *J. Appl. Phys..* **31,** 168
18. Wallace, D. C. (1960). *Phys. Rev.,* **120,** 84
19. Kollie, T. G., Barisoni, M., McElroy, D. L. and Brooks, C. R. (1969). *High Temp.-High Pres.,* **1,** 167
20. Kollie, T. G. (1969). Contributions to the Specific Heat Capacity of Nickel, Iron and the Alloy Ni_3Fe, ORNL-TM-2649
21. Avramescu, A. (1939). *Z. Tech. Physik,* **20,** 213
22. Baxter, H. W. (1944). *Nature (London),* **153,** 316
23. Khotkovich, V. I. and Bagrov, N. N. (1951). *Doklady Acad. Nauk,* **81,** 1055. (English Translation, U.S. AEC (Oak Ridge), Tr. No. 1817)
24. Nathan, A. M. (1951). *J. Appl. Phys.,* **22,** 234
25. Strittmater, R. C., Pearson, G. J. and Danielson, G. C. (1957). *Proc. Iowa Acad. Sci.,* **64,** 466
26. Taylor, R. E. and Finch, R. A. (1964). *J. Less-Common Metals,* **6,** 283
27. Finch, R. A. and Taylor, R. E. (1969). *Rev. Sci. Instr.,* **40,** 1195
28. Cezairliyan, A. (1965). 'A High-Speed (Millisecond) Method for the Simultaneous Measurement of Enthalpy, Specific Heat and Resistivity of Electrical Conductors at High Temperatures' in *Advances in Thermophysical Properties at Extreme Temperatures and Pressures,* 253. (New York: A.S.M.E.)
29. Affortit, C. and Lallement, R. (1968). *Rev. Int. Hautes Tempér. et Réfract.,* **5,** 19
30. Affortit, C. (1969). *High Temp.–High Pres.,* **1,** 27
31. Jura, G. and Stark, W. A. (1969). *Rev. Sci. Instr.,* **40,** 656
32. Dikhter, I. Ya. and Lebedev, S. V. (1970). *High Temp.–High Pres.,* **2,** 55
33. Cezairliyan, A., Morse, M. S., Berman, H. A. and Beckett, C. W. (1970). *J. Res. Nat. Bur. Stand. (U.S.),* **74A,** 65

34. Cezairliyan, A., McClure, J. L. and Beckett, C. W. (1971). *J. Res. Nat. Bur. Stand. (U.S.)*, **75A,** 1
35. Cezairliyan, A. and McClure, J. L. (1971). *J. Res. Nat. Bur. Stand. (U.S.)*, **75A,** 283
36. Cezairliyan, A. (1971). *J. Res. Nat. Bur. Stand. (U.S.)*, **75A,** 565
37. Cezairliyan, A. (1971). *J. Res. Nat. Bur. Stand. (U.S.)*, **75C,** 7
38. Foley, G. M. (1970). *Rev. Sci. Instr.*, **41,** 827
39. Cezairliyan, A., Morse, M. S. and Beckett, C. W. (1970). *Rev. Int. Hautes Hautes Tempér. et Réfract.*, **7,** 382
40. Cezairliyan, A. (1972). *High Temperature Science*, in press

7
Thermodynamics of Electrolyte Solutions

K. P. MISHCHENKO
Leningrad Institute of Pulp and Paper Technology

7.1 GENERAL INTRODUCTION 178

7.2 PHYSICAL THEORY OF THE ENERGETICS OF ELECTROLYTE SOLUTIONS 180
 7.2.1 *Physical theories* 180
 7.2.2 *Physico-chemical theory of Mikulin* 183

7.3 EXPERIMENTAL TECHNIQUES 184
 7.3.1 *General principles* 184
 7.3.2 *Calorimeters for electrolyte solutions. Measurement errors* 185
 7.3.3 *Standard reference substances in solution calorimetry* 185

7.4 STANDARD ENTHALPIES OF ION FORMATION IN SOLUTION 186
 7.4.1 *Selection of standard states* 186
 7.4.2 *The calculation of the enthalpies of dilution of electrolyte solutions up to infinite dilution* 187
 7.4.3 *Standard enthalpies of ion formation in solution* 187

7.5 THERMODYNAMIC PROPERTIES OF ELECTROLYTE SOLUTIONS OVER A WIDE RANGE OF CONCENTRATION AND TEMPERATURE 188
 7.5.1 *Aqueous electrolyte solutions* 188
  7.5.1.1 *Dependence of solution enthalpies and excess relative partial molar (r.p.m.) solution entropies on concentration* 188
  7.5.1.2 *Complete hydration (solvation) limit* 191
  7.5.1.3 *The temperature dependence of thermodynamic properties of electrolyte solutions* 191
  7.5.1.4 *Isotopic effects: the thermodynamics of electrolyte solutions in* H_2O *and* D_2O 193

7.5.2 *Non-aqueous solutions of electrolytes* 194
7.5.2.1 *Concentration and temperature dependence of thermodynamic properties of electrolyte solutions in non-aqueous and mixed (aqueous–non-aqueous) solvents* 194
7.5.2.2 *Solutions of tetra-alkylammonium salts (TAS)* 197
7.5.2.3 *Heat capacities of electrolyte solutions in water and in non-aqueous solvents* 198

7.6 STANDARD THERMODYNAMIC PROPERTIES OF SOLVATED IONS AT INFINITE DILUTION 199

7.7 CONCLUSION 202

ACKNOWLEDGEMENTS 202

7.1 GENERAL INTRODUCTION

The study of Nature is an unabating process. Every new achievement has its origin in a preceding one, and it is seldom that this continuous step-wise progress is interrupted by the discovery of a hitherto unknown phenomenon or fact which makes it imperative radically to reconsider accepted and seemingly obvious concepts. A survey of basic achievements in any field of science over a two or even five year period is not necessarily one which will reveal sensational discoveries or even fundamental generalisations.

This is largely the case for recent studies on electrolyte solutions, and on the thermodynamics of electrolyte systems. The general trend in this field became sufficiently clear as far back as the first decade after World War 2.

The five-year period considered in this survey is mainly characterised by the obtaining of new, and the refining of old, experimental data, and by the verification and development of hypotheses and ideas of earlier origin.

It is natural that any review and assessment of the whole bulk of material for a specified period will be to some degree subjective. Moreover, because of the limited size of this survey it may appear that special attention has been given to current research work in the Soviet Union. It is the case that reviews of studies on electrolyte solutions published in English and German from 1968–1971 have not given Soviet contributions their due coverage[1–8]. As for the similar Soviet reviews[9–12] over the same periods these are not always easy to come by, and are available only in the Russian language.

Common to all the basic trends in the thermodynamic studies of solutions in recent years has been a renaissance of some ideas which date back to the classical era of the theory of solutions, and have since been most undeservedly ignored. This renaissance arose from the impact of many new data resulting from the wider scope of modern research and the higher precision of modern measurements.

Of particular significance here are Mendelejev's classical works published in 1862–1865[13]. It is only today, given the background of modern data, that the genuine content of his fundamental ideas can be appreciated and it seems to me that they should be adopted as guiding principles by those doing research in solutions. Mendelejev makes it clear that *all* components of a solution have equal rights, so that it is not possible to neglect, for example, a solvent by treating it as an inert medium. He particularly emphasises the need to consider all types of interactions between all sorts of particles forming a solution. It seems appropriate to quote here from Debye, another classic name in the theory of solutions. His introductory lecture to the International Symposium on Electrolytes in Trieste (1959) was entitled 'Intermolecular Forces'. Its conclusion runs as follows: 'It seems to me that essential progress in the theory of concentrated solutions of electrolytes can only be achieved if sufficient attention is paid to the existence of short-range molecular forces on top of the long-range Coulombic interactions. It is for this reason that I have given preference to the general subject of molecular forces as an introduction to the deliberations of this symposium'[14].

Evidently a better understanding of the nature of solutions over a wide range of concentrations and temperatures requires that the sum total of evidence on the properties of the particles forming the system be taken into account, and that the significance of solvent structure be fully recognised.

The true solution is the state of thermodynamic equilibrium characterised by the continuous formation and dissociation of the products of interparticle interactions. This dynamic picture of solution interactions is another idea which Mendelejev propounded, and is acceptable today. However, some modern contributors neglect the possibility of stoichiometrical compounds being formed which differ for different concentrations and temperatures.

Unfortunately, the possibility of making an accurate estimation of the properties of electrolyte solutions by theoretical calculation is limited to solutions of concentration less than 1 mol per kg solvent. This means that the thermodynamic approach is the only effective one for all compositions. Given that the thermodynamic properties are available over a wide range of concentration and temperature, systems of practicable interest can be quantitatively handled. An adequate interpretation of the mechanism of a given phenomenon cannot depend exclusively on thermodynamic data and requires other information, e.g. spectral, electrochemical, diffusion, paramagnetic resonance, nuclear magnetic resonance, etc. But a precise thermodynamic analysis can be relied upon in terms of verifying hypotheses and models resulting from different methods. It is for this reason that recent contributions to the thermodynamics of electrolyte solutions have steadily increased in number and in scope.

Special attention is paid in recent work to obtaining such basic thermodynamic values as standard enthalpies of ion formation in solutions, heats of solution, heat capacities etc. in infinite dilution. To these ends, much effort has been given towards improving the methods of both calorimetric and tensimetric measurements.

Particular interest has been given to the investigation of the thermodynamics of electrolyte solutions in non-aqueous and in mixed media, and especially in aqueous–non-aqueous solvents. The investigation of solutions over a wide range of temperature has also proved fruitful. The correlation of thermal coefficients of thermodynamic quantities has proved advantageous, and so has appreciation of the function of structural factors, especially when dealing with aqueous and non-aqueous systems. This involves extensive use of excess partial molar functions of which excess relative partial molar entropy is of prime importance. The problem of ion hydration and solvation remains a focus of attention. Some progress has been made in interpreting the mechanisms of these phenomena, and several empirical and semi-empirical relationships have been proposed enabling applied calculations to be performed.

7.2 PHYSICAL THEORY OF THE ENERGETICS OF ELECTROLYTE SOLUTIONS

The first part of this section is concerned with some achievements of purely physical theories; the second part discusses the little-known theory of Mikulin, which can be styled a physico-chemical theory.

7.2.1 Physical theories

It should be noted that even the most successful attempts at extending the quantitative purely physical theories to high concentrations of ionic solutions have so far been limited to concentrations below 1.5 mol dm^{-3}. Practical requirements demand at least a formal thermodynamic interpretation for more concentrated solutions. Evidently, further progress of non-thermodynamic methods of quantitative description of such systems is to be welcomed, inasmuch as a quantitative estimation (at any concentration) of an effective potential of inter-ion interaction, which takes into account ion solvation and desolvation (when ions are drawn together) as well as structural changes in the solvent, is a prerequisite to theoretical calculation of all thermodynamic properties, and their excess values in particular. However, as regards higher concentrations, this approach remains difficult.

New theories combine both the classical electrostatic approach and the statistical investigation of ionic solutions. Using microscopic models of such solutions (i.e. certain assumptions are made concerning the interactions and laws of motion of ions and neutral particles), thermodynamic functions and transport values can then be calculated by means of statistical mechanics. The choice of a model depends on the contributor, but recent experimental work favours a transition from the system of point charges to that of hard spheres, taking into account short-range potentials.

In the opinion of some theoreticians (e.g. Falkenhagen)[7] it is of considerable importance that a statistical analysis shows that for very high dilutions of strong electrolytes, the Coulombic forces are paramount in inter-ion interactions. In this event, an ionic solution can justly be considered to be a system of point charges in a continuous medium of given dielectric constant and viscosity. Although it is now roughly half a century since 1923, the classical era of Debye and Hückel, their conception still remains valid for very dilute solutions.

But for higher concentrations, the theoretician is faced with many obstacles, and so far there does not exist any general irrefutable theory which accounts for all the available experimental data. For example, the statistical handling of ion–solvent interactions, the significance of solvent structure, and the mechanisms and energetics of ion solvation and desolvation remain as difficulties to be solved.

The theoretical physics of solutions is concerned with the general problems of modern electrochemistry, and with transport processes in particular (electroconductivity, etc.), and detailed consideration of them all in this chapter is not possible. I shall dwell only upon the most important contributions which seek new ways of solving the problems involved. Falkenhagen's *Theorie der Elektrolyte* (1971) gives a good historical survey of the science, and provides a thorough analysis of its present-day activities[7].

Kelbg and Ebeling's work can justly be referred to as a major contribution promoting further progress in the physical theories of electrolyte solutions[15]. Using an expression for the free energy in virial form, these authors have obtained a specific solution for electrolytes at low concentrations for systems having a high Bjerrum parameter, r_b.

It was found that the correction to the Debye–Hückel free energy value which Bjerrum denoted by the ion-association constant, should in fact be expressed as a function, $f(r_b)$. This correction gives a positive contribution to the free energy for small r_b, and a negative one for high r_b. For very high r_b this function changes to $K_a = a^3 e^{r_b/akT}$ which is similar to K_a in Fuoss' papers (1958). Thus a correct statistical account of all details of ion partition leads to a considerable improvement in electrostatic theories by enabling proper construction of partition functions, even though based on the Coulombic potential of hard charged spheres.

The statistical theory of ionic association demands that the term 'ion pairs' be clearly defined. But here some difficulties arise, since the common definition as the persistent state of negative relative energy proves to be invalid for the statistical handling of systems with Coulombic interactions. Since 1965 several papers have dealt with reformulating the atomic partition function, but it is only recently that a satisfactory solution to this problem was found (Ebeling)[16]. The end result is very similar to Bjerrum's K_a, becoming identical with it for high values of the Bjerrum parameter. But Ebeling's method can be said to have some considerable advantages: (i) the theory has no arbitrary limitations, (ii) the formulae for the thermodynamic functions are relatively simple, and, (iii) the theory is internally consistent. Even so, there remains an arbitrary assumption, i.e. the Debye expression for osmotic pressure. The author, however, considers it to be well substantiated from the physical point of view.

It was shown[17] that the effects of short-range forces between ions can be described in terms of chemical equilibrium. Some authors believe that the values a (the centre distance of two ions) and $K_{(PT)}$ in the case of non-electrostatic interactions should be treated as independent parameters. In the case of pure electrostatics they should be dependent. Comparison of this theory of ionic association with practice shows that the effects of association are described quite reasonably.

Another statistical approach to this problem has been given by Stillinger and Lovett[18], but their results require experimental verification. The most recent treatment is due to Kessler[19]. He considers two basic problems in the thermodynamics of strong electrolyte solutions: (i) the search for an effective inter-ionic interaction potential which takes account of the peculiarities of the solvent in a quasi-rigorous treatment whereby the solution is divided into a sub-system of ions and that of the solvent, (ii) the formulation of partition functions for the ion sub-system, using an arbitrary potential for the interaction of ions separated from one another by at least the diameter of the solvent molecules. Based on this procedure, a method of estimating the inter-ionic potential curves, $U_{ab(r)}$, is described which makes use both of experimental activity coefficients, and other properties derived from the free energy equation for the electrolyte solution given by the theory of Levitch and Kiryanov[20]. Eighty-eight electrolytes were treated in this way, in water, hydrogen cyanide, formamide, formamide–acetamide mixtures, and methylformamide solvents at several temperatures.

Kessler has also shown that the potential $U_{ab(r)}$, apart from the Coulombic interaction, is dependent on ion desolvation resulting from overlap of the solvated spheres as they come together, as well as on the solvation of the ion pair as a whole. The latter effect is closely connected with the lability of the solvent structure. The acceptability of the theory[20], the concentration limits of its applicability, and the reliability of its conclusions have been successfully tested on EDC, and by comparison with results obtained by use of the Monte Carlo method[21].

EDC can be said to be the only means of obtaining up-to-date calculations according to the equations of statistical thermodynamics. For example it was on EDC that Rosaiah, Friedman *et al.*[22–23] investigated in detail the so-called HNC (hypernetted chain equation) with the correlation function

$$g(r) = (1+\tau_{ij}+j_{ij})\mathrm{e}^{-\psi_{ij}/kT} \,.\, \exp\left(-\frac{r_{\mathrm{b}}}{r}\,.\,\mathrm{e}^{-xr}\right)$$

where τ_{ij} and j_{ij} are composite functions depending on the type of diagrams and the method of their summation, ψ_{ij} is the cut-off potential, and x is the reciprocal $(1/r_{\mathrm{d}})$ Debye radius. The HNC equation has been correlated with the Monte Carlo method[23]. The results closely approximate one another, and indicate the validity of the HNC equation. These studies have shown that the charged hard-sphere potential does not correspond to physical reality, which requires an additional contribution due to overlap of the ion–solvate spheres, as was discussed above with reference to Kessler's approach.

In other papers[24–25], Rosaiah and Friedman solve the integral equations of type HNC, derived in 1964 by Allnatt[26], in a computer experiment by introducing a square-well potential with two parameters (width and depth of the well). It appears that an adequate choice of parameters enables the entire thermodynamics of electrolyte solutions (activity coefficients, excess enthalpy and excess entropy of dilution) to be described up to concentrations as high as 1.5 mol l^{-1}. It should be noted that the significance of overlap of the solvate spheres of ions on close approach was revealed by Kessler and Gorbanev[27] as early as 1958; their equation is only applicable up to concentrations 0.3 mol l^{-1}. Evidently the HNC equation with proper choice of parameters covers a much wider concentration range.

7.2.2 Physico-chemical theory of Mikulin

Of another pattern is Mikulin's theory, unfortunately not widely known as yet. After refraining from publishing articles in journals for a number of years, Mikulin with his co-workers in 1968 issued a collection of papers as a separate book[28] which appeared in limited circulation. For this reason, we shall dwell on his results at some length.

Mikulin's studies associate electrostatic concepts with the idea pioneered by Mendelejev[13] on the formation in solution of complex compounds of ions with water, having a stoichiometric composition and able to dissociate (reference 28 pp. 5–43, 116–125). Starting from experimental data on solution properties, Mikulin derived a real function of charge-density distribution in the ion atmosphere, explicable on the assumption that water in the 'secondary hydration' zone, beyond the first hydrate layer, and lying within a radius of from 3–4 to 10–12 Å from the centre of the ion, has a dielectric constant of the order of 105–110.

As is evident from Section 7.2.1, neither statistical-mechanical methods, nor methods which take account of the finite size of the ion, or the effects of ionic association, have enabled quantitative calculation of thermodynamic properties in highly concentrated solutions, or provided an interpretation of a number of observed empirical regularities in these systems. Mikulin proposed to evaluate the electrostatic contribution to the Gibbs energy, G^{El}, in concentrated solutions by reference to an empirical observation by Messon[29], who established that for a fairly high region of concentrations, there is a linear relationship between the apparent molar volume, ϕ, of the electrolyte and concentration, $\phi = \phi^* + K\sqrt{c}$, where ϕ^* and K are constants which depend on temperature and pressure. A similar dependence was found by Randall and Rossini[30] for partial molar heat capacities.

Starting from his charge-density distribution function, Mikulin succeeded in deriving theoretically the Messon formula, and showed, further, that a linear dependence of molar properties of a solution upon $\sqrt{c}$ or $\sqrt{m}$ will hold true, provided that the electrostatic part of the thermodynamic potential, G^{El}, in a concentrated aqueous solution of a strong electrolyte also depends linearly upon $\sqrt{c}$ and is of the form

$$G^{El} = -mRT(a+b\sqrt{c})$$

where a and b are specific constants for a given electrolyte, which depend upon temperature and pressure.

Combining this proposition with the concept of ion hydration as a process of formation in solution of stoichiometric complex compounds (similar to ions of aqua-salts and characterised by integral hydrate numbers and by stability constants), and, using the laws of chemical equilibrium, Mikulin was able to calculate activity coefficients of the electrolyte and evaluate the water vapour pressure (reference 28, pp. 44–72). The theoretical analysis shows that the higher the electrolyte concentration, the higher the concentration of liquid hydrates. The latter reaches a certain maximum and thereafter remains constant. In terms of this theory the ions K^+, Rb^+, Cs^+, Ag^+, NO_3^-, $H_2PO_4^-$ do not form complex compounds with water, whereas the ions

H_3O^+, Na^+, Li^+, Cl^-, Br^-, I^- form liquid hydrates which are partly dissociated (not more than 20%). This hydrate theory is well confirmed experimentally in the thermodynamic aspects and of great significance for technological processes involving the use of concentrated electrolytic solutions.

Of interest also are studies due to Mikulin dealing with the theory of concentrated mixed electrolyte solutions. He derived (reference 28, pp. 202–221) the principal differential equations of the theory of mixed solutions, corrected the formula of McKay and Perring[31] and gave analytical expressions to give approximate values of log a_w and log γ for mixed solutions. A formula was proposed to calculate the density of mixed solutions (reference 28, pp. 401–416).

More recently, Soviet scientists have focused their attention on the additivity rule for iso-piestic solution properties suggested by Zdanovskii[32] as early as 1936. Mikulin (reference 28, pp. 304–360) comprehensively developed the theory of mixed-electrolyte solutions governed by this rule. He derived formulae for the activity of water and electrolyte in mixed solutions, and evolved a method of constructing diagrams for the simultaneous solubility of salts which agree remarkably with experiment. He also suggested methods of calculating the activities of water and electrolyte in 'supersaturated' binary solutions, iso-piestic to a mixed solution. According to Mikulin a mixed solution of electrolytes should be considered as a nearly perfect mixture of 'quasi-particles' of iso-piestic binary solutions, each of which consists of one dissociated molecule of electrolyte and $55.5/m$ molecules of water. Such an approach, in Mikulin's opinion, imparts a real physical sense to 'supersaturated' solutions and provides a straightforward approach towards a theory of concentrated electrolyte solutions.

In the same treatise[28] the reader will find comprehensive tables of activities of water, activity coefficients for electrolytes, and osmotic coefficients for 150 aqueous solutions of various salts at 25 °C for the range of concentrations up to saturation.

7.3 EXPERIMENTAL TECHNIQUES

7.3.1 General principles

The thermodynamic properties of electrolyte solutions are determined experimentally mainly by the methods of thermochemistry, by measuring the pressure of saturated vapours over solutions (tensimetry), and by means of potentiometric measurements. Calorimetry, in conjunction with component activities from tensimetric data, makes it possible to measure changes in enthalpy and entropy of the system over a wide range of temperature and concentration.

It should be noted that the enthalpy and entropy properties of solutions, as a rule, more clearly depict the specific properties of solutions and their change with temperature and concentration. The effect of concentration and temperature changes on the Gibbs energy, G, is usually less pronounced due to mutual compensation of ΔH by $T\Delta S$. Furthermore, electrochemical measurements, in non-aqueous solutions in particular, encounter serious difficulties, e.g., on extrapolation to $m = 0$. Thus, the most reliable data are obtained calorimetrically or tensimetrically.

The present requirements demand experimental results of high accuracy. As pointed out by one of the prominent experts in the field of solution calorimetry, Vorobyev[33], the quoted errors in reported thermochemical properties of solutions are often insufficiently objective or incompletely assessed. By using such initial values, and treating them, e.g. by the method of partial molal values, there is a risk of arriving at totally unjustified conclusions.

Of no less significance is the need for pure substances. For example, the results of experiments with non-aqueous solvents, which are not quite water free, are more likely to mislead than assist; as a rule, trace water enters the solvated sphere surrounding an ion at the expense of the true solvent[9].

7.3.2 Calorimeters for electrolyte solutions. Measurements errors

A number of monographs has appeared on precision calorimetry in general, and on the methods of solution thermochemistry in particular. Of the latest, references 34–37 should be noted. Special refinements of apparatus for precision measurements on the thermochemistry of electrolyte solutions have been described by Vorobyev (1965–1971) at the Luginin Thermochemical Laboratory, Moscow University. In order to measure the standard enthalpies of ion formation in infinitely dilute solutions (see Section 7.4.1 and 7.4.2) several different types of calorimeter were designed, including a block calorimeter to measure enthalpies of reactions between solid and gaseous substances[38]; a sealed rocking calorimeter to measure enthalpies of reactions in solution[39], and two calorimeters of high sensitivity ($>10^{-5}$ degrees) to determine enthalpies of solution and of dilution. One of these calorimeters made use of a germanium resistance thermometer and of semiconductor thermocouples[40]; another made use of thermistors[41].

At the 5th All-Union Calorimetric Conference in Moscow (1971) calorimeters designed to study solutions at high dilution were described by Emelin and Kessler[42] and by Gurevich and Sokolov[437].

In order to measure the heat capacity of electrolyte solutions as accurately as possible, calorimeters have been developed in recent years with a precision of $\pm 0.05\,\%$. For aqueous solutions, these include a twin adiabatic calorimeter designed by Latisheva and Kozhevnikov[44] and, for non-aqueous solutions, an adiabatic liquid sealed calorimeter by Novoselov and Mishchenko[45].

For heat of solution measurements in water a precision of ± 0.1–$0.4\,\%$ is now usually attained, and in non-aqueous solvents (and for volatile media in particular), the precision is of the order ± 0.4–$1\,\%$.

7.3.3 Standard reference substances in solution calorimetry

Potassium chloride has been widely used as a standard reference substance in solution calorimetry for calibration purposes, although no official international agreement has yet been reached about this. As early as 1949,

Mishchenko and Kaganovich[46], made a comprehensive examination of the data then available, including their own measurements and recommended $\Delta H_m = 4194 \pm 3$ cal mol^{-1} for the integral heat of solution at 25 °C up to $m = 0.278$ mol kg^{-1} of H_2O (KCl·200 H_2O). In 1963 Somsen *et al.*[47] presented a summary of results obtained by various authors during 1950–1961, but did not make a critical analysis. Parker[48] published a table of values for the heat of solution of potassium chloride at 25 °C and at infinite dilution, covering all available data from 1873 to 1962. In 1968, Mishchenko and Poltoratskii critically revised and extended this table[9] giving $\Delta H_m = 4201 \pm 8$ cal mol^{-1} as the best value at 25 °C for $m = 0.278$.

Quite recently (1971) Emelin and Kessler[42] as well as Gurevich and Sokolov[43] made a new series of careful measurements on the heat of solution of potassium chloride in water. Their results, referred to infinite dilution, give 4123 and 4101 cal mol^{-1} respectively, within error limits of ± 5 cal mol^{-1}. Taking the average, $\Delta H_0 = 4112$ cal mol^{-1}, which corresponds, for $m =$ 0.278, to $\Delta H_m = 4198 \pm 5$ cal mol^{-1}, in close agreement with that recommended in reference 9.

Of standards proposed recently, the greatest attention has been given to tris(hydroxymethyl)aminomethane $(HOCH_2)_3CNH_2$, (Tris) suggested by Irving and Wadsö[49]. In principle, this preparation has a number of advantages; dissolution of Tris may be used as standard for both endothermic (dissolution in water) and exothermic (dissolution of 5 g of Tris in 1 dm^3 of 0.100 N hydrochloric acid) processes. In addition, the temperature coefficient of the heat of solution is equal to only 0.6 % degree^{-1} whereas for potassium chloride it is markedly higher, i.e. 0.9 % degree^{-1}.

The exothermic solution process has been studied quite carefully since 1964 and is now an internationally recommended procedure. Further details are given in the article by Pilcher in this volume (Chapter 3).

No substantial innovations have been made recently in the methods of measuring the pressure of saturated vapours. The best results are provided by iso-piestic methods, and good data are obtainable by use of static techniques when sufficient care is taken.

7.4 STANDARD ENTHALPIES OF ION FORMATION IN SOLUTION

7.4.1 Selection of standard states

Particular attention continues to be paid to refining the concepts of 'standard state' and 'standard change' in respect of the thermodynamic functions for electrolyte solutions. The thermodynamics literature is inconsistent on these concepts and definitions sometimes lack rigour. The standard state for a solution is most commonly thought of as the hypothetical ideal solution wherein the partial molal enthalpy and heat capacity of a dissolved substance are the same as in an infinitely dilute solution and the entropy and Gibbs energy are the same as in a solution of unit molality.

Whereas the enthalpy of formation of a strong electrolyte in an infinitely dilute solution truly relates to the totally dissociated ions in solution, a weak electrolyte even at infinite dilution relates to but slight dissociation into ions. This problem also arises for complex associated molecules and for complex ions which may not exist as such in an infinitely dilute solution because they may be totally dissociated into simple ions. In this connection Vorobyev[33, 50] was the first to suggest that for thermodynamic purposes it might be advantageous to define a hypothetical state in infinitely dilute solution in which the molecules are totally undissociated. The numerical difference between the enthalpies of formation of such substances in infinitely dilute solution and in the hypothetical undissociated state with $m = 0$ is equal to the enthalpy of dissociation (ionisation). The introduction of this hypothetical state made it possible to extend the thermodynamic tables for complex ions, weak electrolytes and associated complex molecules. This standard state has been included in the fundamental tables *Thermodynamic Constants of Substances*[51] edited by Academician Glushkov and published in the USSR (1965–1971). The more recent tables produced by the National Bureau of Standards (Washington) also take note of this state.

7.4.2 The calculation of the enthalpies of dilution of electrolyte solutions up to infinite dilution

Another intensively discussed problem has been that of the extrapolation of experimental thermodynamic data to infinite dilution. Even though present techniques of measurement allow reliable points to be obtained experimentally at low dilutions, a substantial interval remains for extrapolation to $m = 0$, in which the course of concentration dependence may change in an unexpected way. When extrapolating experimental enthalpies and heat capacity curves it is usual to calculate the limiting slope using the improved equations of Debye–Hückel. This procedure is arbitrary, since even if a fairly accurate determination of the limiting slope has been made the cross-over point of this straight line with the zero ordinate is unknown. Several authors have made suggestions aiming at a more satisfactory solution to this problem.

Most authors have previously used the Debye–Hückel theory of long-range forces[9, 52–55] extrapolating linearly throughout the interval between the extreme experimental points and $m = 0$. However, the improved accuracy of recent thermochemical measurements on highly-dilute solutions has revealed deviations from rectilinearity in heats of solution[56–60] and in heat capacities[61] for concentrations between the undeniably Debye region and those traditionally amenable to calorimetric study. Theoretical considerations suggest that such deviations may be genuine, and in the papers of Kessler[19, 21] mentioned above, as well as in his reports to the All-Union Conference on Solutions at Ivanovo[62], it is argued that the interaction of ions with the solvent structure persists even at very small concentrations.

7.4.3 Standard enthalpies of ion formation in solution

Several new data has been obtained, and existing enthalpies of formation of ions in infinitely dilute, mainly aqueous, solution have been critically revised in recent years. The revisions are in constant progress by the compilation team in the U.S.A. at the National Bureau of Standards, under the supervision of Wagman[63], and in the Soviet Union by the team preparing the reference tables, edited by Glushkov[51]. New measurements of $\Delta H_{f,0}$ of ions in aqueous solutions include those by Vorobyev and co-workers using the calorimeters mentioned in Section 7.3.2. Among these are: iodide and tri-iodide ions (from the enthalpies of reactions of HI with H_2O_2 and I_2, cryst.)[64], Ba^{2+} ion (from the enthalpy of reaction of barium with water and of barium and barium hydroxide with HCl)[65–66]; Rb^+ and Cs^+ ions (from the enthalpy of reaction of the metals with water)[67]; Zn^{2+} ion (from the enthalpy of reaction of zinc with HCl)[68]; Cd^{2+} ion (from the enthalpy of reaction of cadmium oxide with sulphuric acid)[69].

Measurements of the standard enthalpies of formation of the ions TiO^{2+}, Bi^{3+} and Pb^{2+} in aqueous solutions have been reported[70] by Vasilyev at the Ivanovo Institute of Chemical Technology.

Values for the standard thermodynamic properties of ions in infinitely dilute non-aqueous solution remain very scarce. It is only in recent years that efforts to measure these have begun, and so far only half-a-dozen or so laboratories are involved throughout the world. A survey on the subject was made by Vorobyev[33] in 1970. By the end of 1970 only 20 solvents had been examined, and in most of these only the enthalpies of solution of a few salts had been measured. The accuracy of results with organic solvents is markedly lower than for aqueous systems. This is mainly due to the special difficulties encountered in such experiments, including the problem of absolute dehydration of the solvents, and sealing of the calorimeters. Finally, the extrapolation of the solution heat to infinite dilution is especially difficult in this case, since in non-aqueous media (with low dielectric constants) nearly all electrolytes are associated.

The results examined critically by Vorobyev are included in the reference book mentioned above[51].

7.5 THERMODYNAMICAL PROPERTIES OF ELECTROLYTE SOLUTIONS OVER A WIDE RANGE OF CONCENTRATION AND TEMPERATURE

Special attention has been given over the period 1966–1971 to relatively dilute solutions at 25 °C, but in some laboratories systematic measurements over a wide range of concentration, up to saturation, and at varying temperatures have been made.

7.5.1 Aqueous electrolyte solutions

Studies at high dilution and the extrapolation of results to $m = 0$ have already been dealt with. Special attention is now paid to more concentrated solutions.

7.5.1.1 Dependence of solution enthalpies and excess relative partial molar (r.p.m.) solution entropies on concentration

Mishchenko[9] has analysed typical curves of the enthalpy of integral solution against concentration, i.e. $\Delta H_m = f(m)$, and found that these can be divided into two main types, I and II. Excluding very high dilutions, where the curves are unambiguously determined by the Debye–Hückel limiting law, type I curves (given by electrolytes with highly hydrophilic ions such as Li^+, Mg^{2+}, Ca^{2+}, Co^{2+}, H^+) are characterised by a continuous reduction of exothermicity of solution with increasing concentration. Type II curves, on the contrary, show increasing exothermicity (or decreasing endothermicity) of solution with increasing concentration and are given by electrolytes with less hydrophilic ions, such as Na^+, K^+, Rb^+, Cs^+. Semi-quantitatively, type I curves result when the effect of diminishing degree of hydration of the ions with increasing concentration is energetically predominant. On the contrary, type II curves indicate the predominance of exothermic contributions with increasing concentration, which are caused by closer ion approach and increasing ease of destruction of the water structure as more electrolyte enters into solution. The latter implies a diminution in the expenditure of energy needed to extract water molecules from the solvent structure in order to introduce them into the hydrated sphere of the ions. As shown later Section (7.5.1.3), these considerations are corroborated by the temperature dependence of ΔH_m, and have also proved useful when dealing with the thermodynamics of electrolyte solutions in non-aqueous and mixed solutions.

Frank and Robinson[71] in 1940 were the first to correlate relationships between excess r.p.m. entropies of water, $\Delta \bar{S}_1^E$, with electrolyte concentration at 25 °C. They have shown that with the rare exception of highly hydrophilic ions, values of $\Delta \bar{S}_1^E$ in aqueous solutions are positive, i.e. the basic integral effect of electrolyte solution in water at 25 °C is the destruction of the water structure. Extensive investigations by Mishchenko and his co-workers at the Leningrad Institute of Pulp and Paper Technology[9], fully support Frank and Robinson's views.

Of particular interest are the investigations by Krestov at the Ivanovo Institute of Chemical Technology, who has measured the heat of solution of salts in water[76] and in organic solvents over the dilution range $m = 0.3$ to $m = 1.10^{-3}$ at varying temperatures, using solution calorimeters of several different types[72–75]. The results in organic solvents are discussed in Section 7.5.2; in aqueous solution, measurements were made on the integral heats of solution of 17 nitrates and chlorides of the elements of Groups I–III, and of 7 rare-earth chlorides at concentrations below $m = 0.3$, and at temperatures from 10 to 95 °C [60]. A new method of calculation of the thermodynamic properties of monatomic aqueous ions at varying temperatures has been proposed by Krestov, and the entropies and heat capacities of monatomic ions in aqueous solution for a number of elements were calculated over the temperature range 0–100 °C [77–80]. Consistent values for the thermodynamic properties of the proton in aqueous solution at varying temperatures were found[81].

Mishchenko and Shpigel[82] reported on the thermodynamic properties of aqueous sodium and potassium nitrate solutions over a wide range of concentrations at 1, 25, 50 and 70 °C; their results indicated the special role of NO_3^- ion as a water-structure breaker. Measurements were also made on the integral heats of solution of strontium chloride at 25 °C in the concentration range $m = 0.050$–2.53 at temperatures from -5 to 65 °C [83]. The r.p.m. of water and the electrolyte enthalpies were calculated. The temperature limits for the change of the curves $\Delta H_m = f(m)$ from type I into II (or *vice versa*) were established.

At Leningrad University, Lilich *et al.*[84, 85] have classified triple aqueous electrolyte solutions in terms of 'the function of enthalpy inter-influence', $\Delta H^b = f(m)$, where $\Delta H^b = \Delta H_{exp.} - (\Delta H_I + \Delta H_{II})$. Here $\Delta H_{exp.}$ is the enthalpy of triple-salt-solution formation from the components I, II and water in quantities m_I, m_{II} with 55.51 mol of H_2O, and ΔH_I, ΔH_{II} are the enthalpies of formation of binary solutions ($m_I + 55.51$ H_2O) and ($m_{II} + 55.51$ H_2O) respectively.

The heats of the mixing in iso-molar binary solutions at 25 °C of $NaClO_4$–H_2O, NaH_2PO_4–H_2O and $HClO_4$–H_2O[86], and of $(CH_3)_4NBr$–H_2O, $NaBr$–H_2O and $CaBr_2$–H_2O [87] were also measured experimentally. Negative values of ΔH^b correspond to an increase in stability of the system, and positive values imply a decrease.

The potentiometric method has been used in thermodynamic investigations of the hydrolysis reactions of metal ions in aqueous solutions, in a constant ionic background medium (3 mol dm^{-3}) of $NaClO_4$, NaCl, or $NaNO_3$ at varying temperatures, and values were obtained for the hydrolysis constants β_{pq} for the reaction[88–92]

$$q\mathrm{Me}^{2+} + p\mathrm{H_2O} \rightleftharpoons \mathrm{Me}_q(\mathrm{OH})_p^{(qz-p)+} + p\mathrm{H}^+$$

Measurements of the temperature dependence of the hydrolysis constants, indicated that both ΔH and ΔS of hydrolysis remain virtually constant over the whole temperature range studied.

Several thermodynamic investigations have been made at Leningrad University of heat-capacity changes, volumes and activities on complex-compound formation in aqueous solutions. One of the more recent of these relates to the formation of zinc and cadmium chloro-complexes[93–95]. The deviations from additivity on the formation of triple aqueous solutions by mixing the corresponding iso-molar binary solutions have been studied over a wide range of concentration at 25 °C. Although the interpretation of the behaviour of such systems is complicated, certain features may enable definite conclusions to be drawn as to the structure of the solutions formed. In the case of the zinc and cadmium chloro-complexes, the investigation made it clear that the complexes formed in solution are not identical, and have compositions $ZnCl_4^{2-}$, and $Cd(H_2O)_2Cl_4^{2-}$ respectively. The zinc complex is tetrahedral, and the cadmium complex is likely to be a distorted octahedron.

At the Department of General and Inorganic Chemistry, Moscow Mendelejev Institute of Chemistry and Technology, measurements of heats of mixing (ΔH_{mix}) at 25 °C of various aqueous salt solutions (iso-molal, iso-ionic and iso-piestic) from electrolytes of similar and different valence types have been made. Studies on complexing ions (Co^{2+}, Ni^{2+}, Cu^{2+}) are also under investigation. A method of estimating the energy contributions to ΔH_{mix} arising from the redistribution of water among the ions, from structural changes in the system, and from the interaction of ions[96–97] has been proposed. The effect of the interaction of the electrolytes has been shown to predominate.

At the same department, Drakin, Karapetyants *et al.* have measured the heat capacities of aqueous solutions of the nitrates of cadmium, zinc, manganese, magnesium, aluminium, chromium, iron, lanthanum, cerium, strontium, gallium, indium and yttrium[98–100] and of the chlorides of rubidium, caesium and zinc[101] at 25 °C, at concentrations from $m = 0.2$ to saturation. For singly- and doubly-charged ions it has been found empirically that $\Delta C_{hyd} = -Azn$, where ΔC_{hyd} is the change of the heat capacity of the ion on hydration, z is the ion charge and n is the coordination number of the ion.

7.5.1.2 Complete hydration (solvation) limit

As early as 1951, Mishchenko and Sukhotin introduced the term 'Complete (Hydration) Solvation Limit'—symbolised by CHL[9,102]. Accepting certain coordination numbers of hydration (n_h) for the simplest ions (having values between 4 and 8) and assuming that the average values of n_h are retained, with increasing electrolyte concentration, in the first hydrate layer as long as there is sufficient water, then on reaching the concentration when for each mole of completely dissociated electrolyte there are Σn_h (cation and anion number) moles of water, the system is in a peculiar state wherein the ions are separated from one another only by the initial hydrate shells surrounding each of them. It was found that the average interionic separation at the CHL, as determined from the solution density, differs from the sum of the crystal ionic radii by a constant amount—e.g. $r' - (r_c + r_a) = 4.67 + 0.08$ Å for all alkali halides (r_c = cation, r_a = anion radius). Beyond the CHL there is a 'lack' of water, and redistribution occurs in favour of the more hydrophilic ion-partner.

Since the introduction of the term CHL, the existence of this limit has been proved by a number of authors, and characteristic breaks have been frequently noted in the region of the CHL[9]. Drakin and Karapetyants[98–101], e.g. observed maxima near the CHL in curves of the concentration-dependence of apparent heat capacities (ϕ_{Cp}) of electrolyte solutions and in relation to heats of mixing, ΔH_{mix}, and $(\Delta C_p)_{mix}$, for mixtures of lithium and sodium chloride solutions with cobalt, nickel and copper chlorides[103]. The conception of CHL was used to analyse the processes of water redistribution, dehydration and complexing taking place on mixing the solutions. The CHL concept has also proved helpful when dealing with the thermodynamic characteristics of electrolyte solutions in mixed (aqueous–non-aqueous) solvents (see Section 7.5.3). When considering the properties of the solutions over a wide range of concentration it is undoubtedly advantageous to treat the data up to the CHL in a different manner from that for more concentrated solutions.

7.5.1.3 The temperature dependence of thermodynamic properties of electrolyte solutions

Those structural factors that influence the properties of electrolyte solutions have an even greater effect on the temperature dependence of thermodynamic properties. Until recently, measurements have generally been restricted to

25 °C or to room temperatures, but nowadays more studies are being made on temperature dependency, particularly in aqueous solutions. The two types of the $\Delta H_m = f(m)$ isotherm observed at 25 °C (see Section 7.5.1.1) have already been mentioned; type I is characterised by a steady decrease in exothermicity with increasing concentration, whereas type II is characterised by increasing exothermicity.

From studies by Mishchenko *et al.*[9] over the temperature range −6 to 70 °C, it was noted that the general trend of type I curves accentuates with rising temperature, and that type II curves gradually change their form towards type I, and may even become identical with type I curves. On the other hand, type I curves with falling temperature tend to take on the form of type II curves. The transition temperature for I ↔ II is qualitatively related to the heat of hydration of the ions concerned; the greater the enthalpy of hydration, the lower the transition temperature. Thus for NH_4Cl, which is relatively weakly hydrated ($\Delta H_{hyd}(NH_4^+) = -78$ kcal mol^{-1}, $\Delta H_{hyd}(Cl^-) = -84$ kcal mol^{-1}), the change II → I occurs from 35–37 °C, whereas for the strongly hydrated $MgCl_2$ (ΔH_{hyd} (Mg^{2+}) = −467 kcal mol^{-1}), the change I → II occurs near to −6 °C. The higher the transition temperature, the less the endothermic contribution to ΔH_m, since the energy needed to remove water molecules from the quasi-crystalline water-structure for hydration purposes is smaller at higher temperatures. Thus, the temperature coefficients $\delta\Delta H_m/\delta T$ for salt solution at low m are negative. On the other hand, the second derivatives $\delta^2(\Delta H_m)/\delta T^2$ are >0, as are the derivatives $(\delta(\Delta H_m)/\delta m)_T$, since the water-structure is weakened both by rising temperature and by dissolved electrolyte. Hence it follows from Kirchhof's law, that the heat capacity of an electrolyte solution is less than the additive sum of the heat capacities of salt plus water, and often is even less than that of the water alone. The effect of temperature change on water structure is even more evident in the temperature dependence of excess r.p.m. water entropies. With increasing temperature, $\Delta\bar{S}_1^E$ becomes less positive, so that a substance such as NH_4Cl, a typical 'structure-breaker' at −6, 18 and 25 °C, becomes a 'structure-maker' at 50 and 75 °C, whereas $CoCl_2$, a typical 'structure-maker' in the range 25–50 °C, begins to act as a 'structure-breaker' over the interval −6 to +2 °C[9]. In Section 7.5.2 we shall see that in non-aqueous solutions, where the solvent does not possess the characteristic features of the water-structure, there is totally different behaviour.

In recent years the above considerations have been fully examined by Mashovets and Puchkov at the Leningrad Lensoviet Institute of Technology[104–109]. Systematic investigations of the thermodynamic properties of aqueous electrolyte solutions over the temperature range 25–350 °C and a wide range of concentration included measurements on densities, vapour pressure, heat capacities and viscosity of aqueous solutions of lithium, sodium and potassium hydroxides, nitrates and aluminates. Integral and partial enthalpies, Gibbs potential ΔG_m, solution entropy ΔS_m, and other properties were calculated from the experimental data. Equilibrium thermodynamic functions of solutions under saturated vapour pressure were also derived.

It has become clear that at 200–250 °C all ions act as 'structure-makers', and the peculiarity of water disappears. But the exothermicity of the dissolution process with the rise of temperature increases as before $(\delta\Delta H_m/\delta T)_m < 0$. The second derivative increases $(\delta^2\Delta H_m/\delta T^2)_m > 0$, up to 70 °C, whereafter it decreases more rapidly than does the second derivative with temperature of the evaporation heat of water from the solution.

The authors have also found that above 200 °C, the activity of water increases and that of the electrolyte decreases, due to ionic association. The apparent molar volumes reach a maximum with rising temperature, and then decrease rapidly, becoming negative in diluted solutions. The solution entropies, ΔS_m, become more negative at high temperatures, the more so the lower the solution concentration. Evidently under such conditions dissolution is mainly governed by the enthalpy factor. In concentrated solutions the qualitative change in the character of thermodynamic functions has not been observed, but in systems with stable crystalline hydrates (for example $NaOH{\cdot}H_2O$, $KOH{\cdot}H_2O$, $LiNO_3{\cdot}H_2O$ and others) the concentration dependence of r.p.m. enthalpies and entropies showed that distinct breaks occur at compositions corresponding approximately to the crystalline hydrates. The solvated structures formed during the melting of crystalline hydrates thus seem to retain their stability at high temperatures.

High-temperature investigations of electrolyte solutions in other laboratories, include studies by Gardner, Jekel and Cobble on the standard partial molal heat capacities of sodium sulphate and sulphuric acid over the range 0–100 °C [110], and of the osmotic coefficient of aqueous solutions of sodium chloride over the range 125–275 °C in terms of vapour-pressure measurements on salt concentrations from $m = 0.5$–3 by Gardner[111]. The results have been analysed on the basis of the competition between hydration and ionic association. Also noteworthy are the mass-spectroscopic studies of Searles and Kebarle[112] from which they determined the enthalpy and entropy of potassium-ion hydration in the gaseous phase at temperatures up to 390 °C.

Investigations of this type are of great interest to geochemists, thermotechnicians and other specialists dealing with aqueous salt solutions, in equilibrium with vapour, at high temperatures. One can mention, for example, the work of Rafalskii[113] on ion activity coefficients in aqueous solutions at elevated temperatures (100–350 °C) and on high-temperature studies by Khodukovskii[114] on the entropy of ions in aqueous solutions.

7.5.1.4 *Isotopic effects: the thermodynamics of electrolyte solutions in H_2O and D_2O*

Following the discovery of heavy water in 1932, many investigations were made on its properties up to the beginning of World War 2, but the only studies in this period on the thermodynamic properties of electrolyte solutions in D_2O were those of Lange[115]. Relatively little further work has been done until recently. A review was given by Gold in 1967 at the symposium on 'Physico-chemical processes in mixed aqueous solutions' held at Bradford University[116]. Gold paid special attention to the behaviour of acids in H_2O–D_2O solvent.

Systematic studies of isotopic effects on the thermodynamics of solutions of both electrolytes and non-electrolytes are being carried out at Lobachevsky

University in Gorky by Rabinovitch[117] and co-workers. These have considerably extended the earlier work of Lange and heats of solution in D_2O have now been measured for 30 different salts[118]. The solution of a salt in water changes the number of hydrogen bonds in the solvent. The energy needed to break $2N$ (N = Avagadro number) hydrogen bonds in $D_2{}^{16}O$ at 25 °C is greater than in $H_2{}^{16}O$. The difference in the solution heats of a salt in $H_2{}^{16}O$ and in $D_2{}^{16}O$ can thus be discussed in terms of the changes in the number of hydrogen bonds on solution. This change is, however, dependent on the nature of the ions involved, their charge, and their size. Thus, e.g. solution of the alkali metal iodides results in the loss of $4N$ hydrogen bonds per mole of salt dissolved, whereas on solution of an alkali metal fluoride there is virtually no change in the number of hydrogen bonds.

The differences in the heats of solution of a salt in $H_2{}^{18}O$ compared with $H_2{}^{16}O$ are less by a factor of 10 than resulting from solution in $D_2{}^{16}O$. This implies that there is very little energy difference[119] involved in breaking hydrogen bonds in $H_2{}^{18}O$ compared with $H_2{}^{16}O$.

7.5.2 Non-aqueous solutions of electrolytes

The structure of liquid water at moderate temperatures is unique so that the formation of a general theory based only on aqueous systems is in principle impossible. Accordingly, recent research has paid more attention to the thermodynamic and physico-chemical properties of electrolyte solutions both in non-aqueous media and in mixed aqueous–non-aqueous solvents. This interest has also been directed to aqueous–non-aqueous binary non-electrolyte systems, and in fact studies of the latter seem to have been more numerous. In 1967, a special symposium was arranged by Franks and held in Bradford on physico-chemical processes in mixed aqueous solvents[120]. Summing up the proceedings, Franks wrote: '...it almost seems that, as the volume of experimental results grows, so does the number of unsolved problems'.

7.5.2.1 Concentration and temperature dependence of thermodynamic properties of electrolyte solutions in non-aqueous and mixed (aqueous–non-aqueous) solvents

Systematic studies using an increasing number of solvents of differing physical and chemical nature are now being made, although only a limited range of electrolytes has so far been used. The measurements of solution enthalpies by Somsen and Coops deserve special attention[121–123]. These, made on dilute solution, were extrapolated to $m = 0$ to obtain for the first time the solution enthalpies (ΔH_0) at 25 °C for alkali halides in formamide, N-methylformamide, N-methylacetamide and N,N-dimethylformamide. Similar and reliable results were obtained in N-methylformamide and N-dimethylformamide by Held and Criss[124,125]. The values obtained agree satisfactorily. Interpretation of measured values is assisted by the results of measurements on the electroconductivity, viscosity and density of the NaCl and CsCl solutions in *N*-methylformamide at 5 and 25 °C by Mostkova, Kessler and Semenov[126], which show that alkali metal chlorides in this solvent have the free ions solvated to no less an extent than in water, and that solvation increases with rise in temperature, probably due to solvation of the ion pairs in the solution.

The conclusions reached are significant to the understanding of the thermodynamics of solutions in media of high dielectric constant. Krestov and Zverev have measured ΔH_m for 11 salts in dimethylformamide at 25 °C at high dilutions[127]. Certain regularities in the heats of solvation of singly- and doubly-charged ions in aliphatic alcohols have been noted by Drakin, Karapetjants and others[128]. The temperature dependence (10–32 °C) of the heats of solution of KCl, $AgNO_3$, $NaNO_3$, RbCl and CsCl in methanol has been studied[129]. Studies on non-aqueous systems by Mishchenko up to 1967 were surveyed in reference 9, and others include solutions of NaI in methanol[130], ethanol[131] *n*-propyl and *n*-amyl alcohols[132] over a wide range of concentrations at two or three different temperatures; of NaSCN and NH_4SCN in methanol[133]; and of NaI in acetonitrile[134], ethyleneglycol[135], acetone[136, 137] and formamide at four different temperatures[138].ΔH_m, ΔG_m and $\Delta \bar{S}_1^E$ were obtained by combining these calorimetric measurements with measurements of vapour pressures.

Comparison of results from a variety of solvents allows several interesting general conclusions to be drawn. In all solvents studied, the integral enthalpies of solution (ΔH_m) are found to be more exothermic than they are in water. The Gibbs' potentials (ΔG_m) on the contrary, are somewhat less exothermic because the solution entropy is much more negative. The lack of orderly water-like structures in the organic media studied results in a marked increase of order when ions are introduced, so that the values $\Delta \bar{S}_1^E < 0$. All curves of $\Delta H_m = f(m)$ belong to type I, i.e. 'energetic desolvation' and the effects of ion approach prevail, and the exothermicity of dissolution diminishes with increasing concentration. The lack of a specific structure as in water is clearly shown by the temperature coefficients of ΔH_m. These are invariably positive, i.e. the exothermicity of solution diminishes with temperature. For solvents with low dielectric constant the steepness of slope of the isotherms $\Delta H_m = f(m)$ is considerably greater than in water, which is due to ionic association. The steepness is least in formamide ($\varepsilon = 109.5$). In this solvent the tendency to association seems to be relatively small.

Whereas in aqueous solution the temperature coefficients of ΔH_m are maximal at $m = 0$, the isotherms coming closer together with increasing concentration, in non-aqueous solvents the situation is reversed, and the isotherms are closest together at $m = 0$ and diverge fan-like with increasing concentration. This is a characteristic feature. It seems clear that the idea advanced by Frank[139] of the 'co-operation' of hydrogen bonds in liquid water has been confirmed. The introduction of the first few ions into a solvent gives rise to a peculiar 'explosive effect'.

A series of measurements on the enthalpies of solution of benzoic acid in methyl, ethyl, propyl and isopropyl alcohols by Tanevska–Osinka are worthy of note[140–142]; only a few investigations have been made on weak electrolytes in organic solvents.

Ivanova and co-workers at Kharkov University have thoroughly investigated the isotherms $\Delta H_m = f(m)$ (25 °C) given on solution of a number of electrolytes in formic acid, acetic acid and acetic anhydride over a wide range of concentration. Sodium, potassium, rubidium and caesium halides were dissolved in formic acid, and LiBr, NaI, $NaClO_4$, CH_3COOK, CH_3COOCs and H_2SO_4 in acetic anhydride, the last at three temperatures[143,144]. In these systems, ΔS_{solv} is greater than ΔS_{hydr}. Systematic studies have been made by Stern on the enthalpies of electrolyte transfer from water to mixed aqueous–non-aqueous media, including transfer of hydrochloric acid from water into aqueous solutions of acetic acid at 25° [145], of perchloric and hydrochloric acids from pure water into aqueous solutions of ethylene glycol and acetic acid[146] and of sodium chloride from water into aqueous solutions of urea[147].

The behaviour of HCl in pure CH_3COOH appears to be very similar to that of weak 1:1 acids in water, and the role of urea as a characteristic 'structure-breaker' in water has been confirmed.

Complete thermodynamic data for electrolyte solutions in mixed aqueous–non-aqueous media over all compositions of solvent from $x_1 = 1$ to $x_2 = 1$, and over a wide range of electrolyte concentration and temperature, are few and far between, due no doubt to the complexity of studies of this type. A few systems studied in this way are mentioned below, and further work is being pursued very intensively.

NaI solutions at 25 °C in the mixed solvent methanol–water[148], and NaBr in methanol–water at 25 and 40 °C have been studied in some detail[149]. A similar investigation of the LiCl–methanol–water system at 25 °C has been carried out by Woycicka and Minc[150]. The complete enthalpy and entropy properties reveal a regular transformation of the isotherms $\Delta H_m = f(m)$ and $\Delta \bar{S}_1^E = f(m)$ from typical 'aqueous' to typical 'non-aqueous' patterns. Ionic hydration is clearly seen to predominate over solvation, and methanol enters into the solvate spheres only below the CHL where there is a lack of water. The minimum exothermicity of solution occurred at composition of the solvent of *c*. 17 mol %. In this system Samilov's hypothesis[151,152] in respect of the filling of holes in the open structure of water with alcohol molecules and the attainment of maximum stabilisation upon filling all vacancies was confirmed. Similar results were obtained by Krestov and co-workers for mixtures of water with other alcohols[66,153] and for the water–methanol–ethanol system[154].

However, this interpretation although satisfactory in some cases, is not general. In a number of other solutions of electrolyte–water–organic-component a minimum of exothermicity has not been observed although the size of molecules of the organic component would allow them to penetrate into vacancies in the water structure. Recent studies by Stennikova, Poltoratskii and Mishchenko[155] on solutions of NaI in acetonitrile–water showed no minimum, but on the contrary, at 90 mol % CH_3CN, a maximum of solution exothermicity was noted, which was explained assuming the monomolecular state of water in such a solution, and by the formation of stoichiometric solvates of the electrolyte–solvent. Specific interactions were observed also in the temperature coefficients and the r.p.m. enthalpies of all three components of the solution. Poltoratskii has developed a special method for the calculation of r.p.m. values of each component in terms of heats of solution[156].

The calculation of $\bar{L}_i$ for any component of the electrolyte–mixed-solvent system is possible given the following experimental values:

(i) The heat of solution of the electrolyte (ΔH_m) in a mixed solvent of a given composition with varying molalities (m).

(ii) The heats of mixing of the solvent components according to composition.

$\bar{L}_i$ for triple systems of the type water–organic liquid–electrolyte for which sufficient experimental data are available were computed. The final equations for the particular ease of a triple system are as follows:

$$\bar{L}_1 = -10^{-3}m^2M_1\left[\frac{\delta(\Delta H_m)}{\delta m}\right]_{T,p} + \bar{\mathscr{L}}_1$$

$$\bar{L}_2 = -10^{-3}m^2M_2\left[\frac{\delta(\Delta H_m)}{\delta m}\right]_{T,p} + \bar{\mathscr{L}}_2$$

$$\bar{L}_3 = m\left[\frac{\delta(\Delta H_m)}{\delta m}\right]_{T,p} + \Delta H_m - \Delta\mathscr{H}_0$$

Here $\bar{L}_1, \bar{L}_2, \bar{L}_3$ are r.p.m. enthalpies of water (1), organic liquid (2) and electrolyte (3), m is the molality of an electrolyte (mol salt kg^{-1}) in a mixed binary solvent; M_1 and M_2 are molecular weights of components 1 and 2; $\bar{\mathscr{L}}_1$ and $\bar{\mathscr{L}}_2$ are their r.p.m. enthalpies in the double solution *without* electrolyte; $\Delta\mathscr{H}_0$ is the heat of solution of the third component in the first.

The above treatment revealed interesting details which are hidden by the conventional thermodynamic treatment of solutions consisting of many components.

7.5.2.2 Solutions of tetra-alkylammonium salts (TAS)

Considerable attention has been given recently to solutions of TAS in water and in organic solvents, since these solutions show extremely unusual properties. Lindenbaum[157–159], Frank[160], Franks and Smith[161], and Wen[162] have made systematic investigations on the thermodynamics of these solutions, and densities, vapour pressure, heats of dilution and heat capacities of aqueous solutions for the series of tetra-alkylammonium halides from alkyl = CH_3 to C_4H_9 have been measured. Apparent and partial molal properties of the components have been computed. The results show that there is a difference in the properties of salts with short and long hydrocarbon chains. The salts of methyl- and ethyl-ammonium behave like ordinary inorganic electrolytes, but the salts having longer hydrocarbon chains have anomalous properties. They are characterised by large positive deviations from ideal behaviour, with activity coefficients of water almost unity in concentrated solution, unusually high heat capacities, heats of dilution, and apparent salt volumes. Positive temperature coefficients of ΔH_m for aqueous solutions of tetrabutyl ammonium nitrate have been found[163].

These phenomena are generally explained by three hypotheses:

(i) association of the ions into large conglomerates,

(ii) formation of hydrates of clathrate type,

(iii) colloidal or semi-colloidal behaviour of the salts.

The second hypothesis accounts for positive temperature coefficients of ΔH_m provided that the formation of hydrates of clathrate type occurs without the destruction of elements of the water-structure. However, measurements of refractive index, rate of sound propagation and optical density in solutions of Bu_4NNO_3 have indicated that above the critical micelle concentration this system displays colloidal character[164]. Further investigation is needed to establish the connection between clathrate and micelle formation in such extremely complex systems.

The thermodynamic properties of TAS in non-aqueous solvents have been investigated[9, 165–167] and some data are available on the extraction properties of these solutions. Like aqueous solutions of TAS, the non-aqueous solutions show large positive deviations from the ideal, and some of them form complex adducts with the solvent[9]. The properties of such systems are usually explained in terms of 'the polarisation' of ionic pairs, but their colloidal properties have not yet been analysed. The thermodynamic data available so far do not confirm the existence of triple ions in TAS solutions in media of low dielectric constant[168]. The degree of 'polymerisation' as a function of solution composition does not show any maximum[165] and there is a profound effect from the addition of small quantities of water on the structure of such solutions.

7.5.2.3 *Heat capacities of electrolyte solutions in water and in non-aqueous solvents*

The difficulties of extrapolation of apparent molal heat capacities to infinite dilution have already been mentioned (see Sections 7.4.2 and 7.5.1). The heat capacities at 25 °C, and the densities of aqueous two- and three-component solutions of electrolytes have been investigated in detail at the Moscow Institute of Chemical Technology. Apparent molal heat capacities were computed, and the volume deviations from additivity on the formation of three-component systems from binary electrolyte–water solutions have been considered on the basis of solvation and structure concepts[169–171]. Methods of comparative computation for the approximate estimation of the C_p (25 °C) and $d_{4°}^{25°}$ values of electrolyte aqueous solutions have been described[172–175].

These studies, however, have not solved the problem of the correct extrapolation of apparent values to $m = 0$. The problem has been examined experimentally by Mishchenko and Novoselov on the basis of precision measurements of the heat capacities of aqueous and non- aqueous electrolytes at high dilution. Sodium iodide solutions in water, methanol and acetone at 0.8, 25 and 50 °C were studied at concentrations from $m = 0.01$ up to saturation, and the respective partial molal values were calculated[9, 61, 176–178].

The results of these experiments have shown that the widely-used method of extrapolation to $m = 0$ of the isotherms $\phi_{C_p} = f(\sqrt{m})$ from the region obeying the so-called Randall–Rossini rule[179, 180] is open to question. Linearity is observed commonly enough in aqueous solutions above $m = 0.4$. At lower concentrations, there is a deviation towards more negative values, which was indeed noticed by Randall and Rossini. But in the case of sodium iodide solutions in methanol, the deviation is towards positive values, and the curve becomes non-linear at concentrations as high as $m = 0.622$. In acetone solutions deviations were observed below $m = 0.57$, but as in water, these tend towards more negative values. A comparison of the limiting values, $\phi^0_{C_p}$, obtained by rectilinear extrapolation with experimental points, shows considerable discrepancies, and one must be sceptical of many generally-accepted tabulated values of $\phi^0_{C_p}$. The temperature dependence of the isotherms is even more complex. Since the quantitative theory of the heat capacities of electrolyte solutions is insufficiently advanced, the main task at present is to improve the accuracy of C_p measurements to better than $\pm 0.01\,\%$ uncertainty.

7.6 STANDARD THERMODYNAMIC PROPERTIES OF SOLVATED IONS AT INFINITE DILUTION

For the purpose of calculation of thermodynamic properties, widespread usage is made of tabulated data on the standard enthalpies, energies and entropies of individual ions in solution. These are often regarded as 'experimental' data, but they are subject not only to the errors in experimental measurement of the solution heats upon which they are based, but also on the extrapolation errors in approaching $m = 0$. The *total* enthalpy of solvation of the ions composing an electrolyte, $\Delta H^{\pm}_{solv}$ at $m = 0$ is related to the enthalpy of solution of the salt, ΔH_0, and to the lattice energy of the crystal, U, through the equation $\Delta H^{\pm}_{solv} = \Delta H_0 - U$. The division of $\Delta H^{\pm}_{solv}$ into its anion and cation contributions, ΔH^{+}_{solv}, ΔH^{-}_{solv}, requires the acceptance of an arbitrary assumption.

Total enthalpies, $\Delta H^{\pm}_{solv}$ can be fairly reliably assessed. Lattice energies, U, are perhaps more likely to contain serious error than the solutions heats, ΔH_0, although in favourable cases (e.g. alkali halides) it is thought[33] that the calculated lattice energies are correct within fractions of 1 kcal mol^{-1}. In non-aqueous solvents, ΔH_0 values are less reliable than in aqueous solutions, and may be the chief source of error.

However, when comparing results obtained by different authors the main difference is often in the value selected for the lattice energy. The comparison of $\Delta H^{\pm}_{solv}$ values obtained by Somsen[181, 182], by Vorobyev[33] and at our laboratory for sodium iodide solutions in water, methanol, formamide, and methylformamide at 25 °C may serve as an illustration (see Table 7.1).

The differences in the $\Delta H^{\pm}_{solv}$ values of Vorobyev and of Somsen and ourselves amount to 3–4 kcal mol^{-1}, and arise from the difference in the values selected for the lattice energies. In so far as $\Delta H^{\pm}_{solv}$ values carry the uncertainty in lattice energies, the solvation enthalpies of single ions are still less accurate, since the problem of separating $\Delta H^{\pm}_{solv}$ into its single ionic components has not yet been solved. The assumption commonly used is that of equality of the enthalpies of hydration of caesium and iodine ions, suggested by Lange and Mishchenko[183] as far back as 1930. Many of the widely-used data tables have been constructed on this basis.

Recently Krestov and Zverev[59] proposed a new method of division which takes into account both the ion–dipole interactions with nearest molecules

Table 7.1 Comparison of U, ΔH_0 and $\Delta H^{\pm}_{solv}$ values (kcal mol^{-1}) obtained by different workers

	U	ΔH_0 (H_2O)	ΔH_0 (CH_3OH)	ΔH_0 (*formamide*)	ΔH_0 (*methylformamide*)	$\Delta H^{\pm}_{solv}$ (H_2O)	$\Delta H^{\pm}_{solv}$ (CH_3OH)	$\Delta H^{\pm}_{solv}$ (*formamide*)	$\Delta H^{\pm}_{solv}$ (*methylformamide*)
Somsen	164.5	−1.79	−7.89	−7.43	−8.27	−166.3	−172.3	−171.9	−172.8
Mishchenko[9, 138]	165.7	−1.80	−7.66	−7.45	—	−168	−173.4	−173.1	—
Vorobyev	168.69 ±0.23	−1.80	−8.10	−7.42	−8.26	−170.49 ±0.23	−176.79 ±0.25	−176.11 ±0.23	−176.95 ±0.25

of solvent, and the geometrical characteristics of these molecules. This approach made possible an extension of the partition of $\Delta H^{\pm}_{solv}$ values to cover six non-aqueous solvents in addition to water. In water the results obtained agree quite well with those based on the equality of enthalpies of hydration of Cs^+ and Cl^-.

Somsen's[123, 181, 182] approach to this problem takes into consideration ion–quadrupole interactions in a solution. He concludes that the most accurate method is based on the discontinuous model of a solution and he has a preference for the calculation method of Halliwell and Nyburg[184] rather than for that of Verwey[185]. Compared with results obtained by division on the caesium–iodine scale Somsen's calculations require anion solvation to be more exothermic and cation solution to be less exothermic, i.e. $\Delta H^{Cs+}_{solv} < \Delta H^{I-}_{solv}$. At present it seems difficult to draw a final conclusion on the merits of the two methods, and the enthalpies of single ion solvation remain uncertain in the absolute sense to several kcal mol^{-1}.

Results obtained lately by Sukhotin and Kazankina are of special interest to the further development of our concepts of the mechanism and energetics of ionic solvation. A physical property of solutions closely associated with hydration is the transfer of water by ions during electrolysis, but, up to now such transport measurements have presented difficulties and the data obtained were not trustworthy. Sukhotin and Kazankina[186] have developed a unique method free from the usual defects enabling accurate measurements of the translational coefficients of ion hydration (h_i), which determine the number of molecules of H_2O associated with an ion during its translational motion in an electric field. The chief advantage of the method is that it is entirely independent of the processes at the electrodes, since the observed volume changes occur in the middle region of the electrolyser. The method can be used for electrolytes at concentrations $\geqslant 1$ gm equiv. dm^{-3}.

The authors have studied concentration and temperature dependence in aqueous solutions of KCl, NaCl, LiCl, $MgCl_2$, $CaCl_2$, $SrCl_2$, $BaCl_2$, KNO_3, $NaNO_3$, NH_4NO_3, $(NH_4)_2SO_4$ and $NaClO_4$[187, 188]. Relative to h_i for ClO_4^- equal to zero, h_i values for other ions at 25 °C are: $K^+ = 1.03$, $Na^+ = 4.02$, $Li^+ = 6.71$, $NH_4^+ = 0.73$, $Ca^{2+} = 8.23$, $Sr^{2+} = 7.40$, $Ba^{2+} = 1.18$, $NO_3^- = 0.0$, $Cl^- = 0.96$, $SO_4^{2-} = 0.54$.

Assuming that the simultaneous motion of ions and water molecules is feasible only for particular orientations of the water in the vicinity of the ion, the authors[189] have developed a theory of translational coefficients based on the equilibrium between various hydrate forms which differ from one another by the 'degree of disorientation' of the H_2O molecules in the solvate sphere.

For the ions Li^+, Na^+, K^+, Mg^{2+}, Ca^{2+}, Sr^{2+}, Ba^{2+}, and Cl^- the measured h_i values were used to determine the equilibrium constants (K) and disorientation heats were derived. It was shown that the disorientation enthalpy of water molecules constitutes no more than 10–20% of the ionic hydration enthalpy. The values of K give a quantitative measure of 'positive' and 'negative' hydration (using Samilov's terminology), and of the effect of ions on the orientational transport of water. It appears that this value might also serve as a quantitative measure of 'energetic dehydration'[9], mentioned in Section 7.5.2.1.

The link between thermodynamic and transport properties is relevant to the problems of extraction and salting-out from solutions. Samoilov and co-workers who have developed a molecular-kinetic theory of salting-out in aqueous solutions, the thermodynamical aspects of which relate to the dependence of nearest-ion hydration in a solution on concentration and composition of the solution[190, 191]. The theory shows the dependence of salting-out on the initial hydration of the cation salted-out with the hydration of cations and anions of the salting-out agent. The details of the phenomena can be studied by a thermochemical method, and by n.m.r. studies[192–194]. It has been established that the maximum separation of highly-hydrated cations salted-out in aqueous solutions results by using a salting-out agent the cation of which does not specifically affect the structure of water. Experimental results obtained from the chromatographic separation of rare earth elements support this conclusion[195]. The stabilisation of the solvent structure results in a decrease of ion solvation in a solution; on the contrary, destruction of solvent structure is accompanied by the growth of solvation[196, 197]. This generalisation can be made equally from the interpretation of phenomena observed in thermodynamic studies of ion solution in mixed aqueous–non-aqueous solvents.

7.7 CONCLUSION

Generalising, we would like to stress once more that the principal problems for future study in the thermodynamics of solution are the attainment of high precision in experimental measurements, and the further accumulation of experimental data on the thermodynamic properties of ions in different solvents, over a wide range of concentration and of temperature. Measurements at low temperatures, when the structure of the solvent is most pronounced, would be of special interest.

Acknowledgements

The author expresses his sincerest thanks to his colleagues who have kindly provided reprints of their articles, and have given valuable advice in the course of preparing the manuscript: Professors Falkenhagen, H. Frank, F. Franks, Karapetjants, Kessler, Krestov, Lilich, Rabinovich, Samoilov, Somsen, Sukhotin, Drs Ivanova, Poltoratskii and Puchkov, and also his pupils who have helped in its preparation.

References

1. Guggenheim, E. A. and Stokes, R. (1969). *Equilibrium Properties of Aqueous Solutions of Single Strong Electrolytes.* (Oxford: Pergamon Press)
2. Harned, H. S. and Robinson, R. A. (1969). *Multicomponent Electrolyte Solutions.* (Oxford, Pergamon Press)
3. Desnoyer, I. E. and Jolicoeur, C. (1969). *Hydration Effect and Thermodynamic Properties of Ions* in *Modern Aspects of Electrochemistry,* Vol. 5, 1-89. (London: Butterworths)
4. Blandamer, M. J. (1970). *Quart. Rev. Chem. Soc.,* **24,** 169
5. Hertz, H. G. (1970). *Angew. Chem.,* **82,** 91

6. Falkenhagen, H., Ebeling, W. and Kraeft, W. D. (1970). *Ionic Interactions* (S. Petrucci, editor). (New York: Academic Press)
7. Falkenhagen, H. (1971). *Theorie der Elektrolyte.* (Leipzig: Hirsel Verlag)
8. Wood, R. U. and Reilly, P. I. (1970). *Ann. Rev. Phys. Chem.*, **21**, 387
9. Mishchenko, K. P. and Poltoratskii, G. M. (1968). *Voprosy Termodinamiki i Stroenija vodnykh i nevodnykh rastvorov Elektrolitov.* (Leningrad: Khimia)
10. Mishchenko, K. P. and Samoilov O. Ya. (1967). *Rastvory. Sbornik-Rasvitie fizicheskoi Khimii v SSSR,* pp. 177–222. (Moskva: Nauka)
11. Lilich, L. S. and Mishchenko, K. P. (1969). *Periodicheskii Zakon i Rastvory. Sbornik-100 let Periodicheskogo Zakona Khimicheskikh Elementov,* pp. 302–313. (Moskva: Nauka)
12. Yermakov, V. I. and Atanasjants, A. G. (1970). *Assotsiatsia Ionov i Struktura wodnykh Rastvorov Elektrolitov. Elektrokhimia, Serii Itogi Nauki,* pp. 65–95. (Moskva: Isd. VINITI AN SSSR)
13. Mendelejev, D. I. (1959). *Rastvory.* (Moskva: Izd. AN SSSR)
14. Debye, P. (1962). *Molecular Forces, Proceedings International Symposium on Electrolytes, Trieste,* 1. (Oxford: Pergamon Press)
15. Ebeling, W. and Kelbg, G. (1966). *Z. Phys. Chem.*, **233,** 209
16. Ebeling, W. (1967). *Ann. Phys.*, **19,** 104; (1968). *Physica,* **38,** 478; (1969). **43,** 298; (1968). *Z. Phys. Chem.*, **238,** 400
17. Ulbricht, H. and Ebeling, W. (1970). *Z. Phys. Chem.*, **249,** 129
18. Stillinger, F. H. and Lovett, R. (1968). *J. Chem. Phys.*, **48,** 3858, 3869; **49,** 1991
19. Kessler, Y. M. (a) (1968). *Zh. Strukt. Khim.*, **9,** 133; (b) (1967). *Ukr. Phys. Zh.*, **12,** 1279; (c) (1967). *Zh. Strukt. Khim.*, **8,** 692; (1968). **9,** 950; (1967). *Elektrochimiya,* **3,** 373, 695; (1968). *ibid.*, **4,** 92; (1969). *ibid.*, **5,** 623
20. Levich, V. G. and Kirjakov, V. A. (1962). *Zh. Fiz. Khim.*, **36,** 1646
21. Kessler, Y. M., Kirjakov, V. A. and Jemelin, V. P. (1970). *Electrokhimija,* **6,** 742
22. Rosaiah, J. C. and Friedman, H. L. (1968). *J. Chem. Phys.*, **48,** 2742; (1969). **50,** 3965
23. Vorontsov-Veliaminov, P. N., Eliashevich, A. M., Rosaiah, J. C. and Friedman, H. L. (1970). *J. Chem. Phys.*, **52,** 1013
24. Rosaiah, J. C. and Friedman, H. L. (1968). *J. Phys. Chem.*, **72,** 3352
25. Rosaiah, J. C. (1970). *J. Chem. Phys.*, **52,** 704
26. Allnatt, A. R. (1964). *Molec. Phys.*, **8,** 533
27. Kessler, Y. M. and Gorbanjev, A. I. (1957). *Dokl. Acad. Nauk SSSR,* **117,** 437
28. Mikulin, G. I. (1968). *Voprosy fizicheskoy khimii rastvorov elektrolytov.* Leningrad: Isdatelstvo Khimiia)
29. Messon, D. O. (1929). *Phil. Mag.*, **8,** 218
30. Randall, M. and Rossini, F. D. (1929). *J. Amer. Chem. Soc.*, **51,** 323
31. McKay, H. A. C. and Perring, J. K. (1953). *Trans. Faraday Soc.*, **49,** 163
32. Zdanovskii, A. B. (1936). *Zakonomernosti v izmenenijakh svoistv smeshannykh rastvorov, Trudy soljanoi laboratorii Akad. Nauk SSSR,* vyp. **6**
33. Vorobjev, A. F. (1971). *Materialy vsesojusnogo simposiuma po termokhimii rastvorov elektrolytov i neelectrolytov,* 5. (Ivanovo: Ivancvskii khimiko-tekhnologicheskii institut)
34. McCullough, J. P. and Scott, D. W. (editors) (1968). *Experimental Thermodynamics,* Vol. I, *Calorimetry of Non-reacting systems.* (London: Butterworths)
35. Skuratov, S. M., Kolesov, V. P. and Vorobjev, A. F. (1964). *Thermokhimiia,* Vol. I; (1966). *ibid.*, Vol. II. (Izd. Moskovskogo Gosud. Universiteta)
36. Oleinik, B. N. (1964). *Tochnaja Kalorimetrija,* Isdatelstvo Gosudarstvennogo Komiteta Standartov Mer i Ismeritelnykh priborov SSSR
37. Calvet, E. and Prat, H. (1956). *Microcalorimétrie.* (Paris: Masson et Cie)
38. Vorobyev, A. F., Monaenkova, A. S. and Skuratov, S. M. (1965). *Zh. Fiz. Khim.*, **39,** 2068
39. Vorobyev, A. F., Broijer, A. F. and Skuratov, S. M. (1967). *Zh. Fiz. Khim.*, **41,** 922
40. Kostjuk, B. G. and Vorobjev, A. F. (1971). *Zh. Fiz. Khim.*, **45,**
41. Privalova, N. M., Bunatjan, R. A. and Vorobyev, A. F. (1971). *Zh. Fiz. Khim.*, **45,**
42. Emelin, V. P. and Kessler, Y. M. (1971). *V. Vsesojusnaja Konferentsija po kalorimetrii.* (Isdat. Moskovskogo Universisteta, p. 443)
43. Gurevich, V. M. and Sokolov, V. A. (1971). *V. Vsesojusnaja Konferentsija po kalorimetrii.* (Isdat. Moskovskogo Universisteta, p. 435)
44. Latysheva, V. A. and Kozhevnikov, O. A. (1965). *Vestnik Leningradskogo Gosud. Universiteta,* N 22, vyp. 4, 109

45. Novoselov, N. P. and Mishchenko, K. P. (1968). *Zh. Fiz. Khim.*, **42,** 535
46. Mishchenko, K. P. and Kaganovich, Y. Ja. (1949). *Zh. Prikl. Khim.*, **22,** 1078
47. Somsen, G., Coops, J. and Tolk, M. W. (1963). *Rec. Trav. Chim.*, **82,** 231
48. Parker, V. B. (1965). *Thermal Properties of Aqueous Uni-univalent Electrolytes.* National Standards Reference Data, Ser. NBS-2, Category 5-T (Thermodynamic and Transport Properties), April 1
49. Irving, R. J. and Wadsö, I. (1964). *Acta Chem. Scand.*, **18,** 195
50. Vorobyev, A. F. (1971). *Teoretich. i. Experiment. Khimija,* **7,**
51. Glushkov, V. P., editor, *Termicheskie konstanty veshchestv.* (Isdanie AN SSSR, Vol. 1 (1965); Vol. 2 (1966); Vol. 3 (1968); Vol. 4 (1971))
52. Criss, C. M. and Cobble, J. W. (1961). *J. Amer. Chem. Soc.*, **83,** 3223
53. Mitchell, K. E. and Cobble, J. W. (1964). *J. Amer. Chem. Soc.*, **86,** 5401
54. Cross, C. M. and Cobble, J. W. (1964). *J. Phys. Chem.*, **68,** 5404
55. Krestov, G. A. and Jegorova, I. V. (1968). *Teor. i Experim. Khimija,* **4,** 623
56. Somsen, G. and Coops, J. (1965). *Rec. Trav. Chim.*, **84,** 965
57. Held, R. P. and Criss, C. M. (1965). *J. Phys. Chem.*, **69,** 2611
58. Finch, A., Gardner, P. J. and Steadman, G. J. (1967). *J. Phys. Chem.*, **71,** 1966
59. Krestov, G. A. and Zverev, V. A. (1969). *Isv. Vyssh. Ucheb. Zaved., khim. i khim. tekhnol.,* **12,** 25
60. Krestov, G. A. (1971). *Proc. Intern. Conf. Calorimetry and Thermodynamics,* 949. (Warsaw: Polish Scientific Publishers)
61. Novoselov, N. P. and Mishchenko, K. P. (1971). *Proc. Intern. Conf. Calorimetry and Thermodynamics,* 943. (Warsaw: Polish Scientific Publishers)
62. Kessler, J. M. (1969). *Materialy Vsesojusnogo Simposiuma po Termokhimii rastvorov elektrolitov i neelektrolitov.* (Ivanovo, IKhTI, 1971, p. 145)
63. Wagman, D. D., Evans, W. H., Parker, V. B., Halow, J., Bailey, S. M., Schumm, R. H. and Churney, K. L. (1965–1971). *Selected Values of Chemical Thermodynamic Properties,* NBS Technical Notes, 270-1, 2, 3, 4, 5
64. Vorobyev, A. F., Brojer, A. F. and Skuratov, S. M. (1967). *Dokl. Acad. Nauk SSSR,* **173,** 385
65. Vorobyev, A. F., Monaenkova, A. S. and Skuratov, S. M. (1968). *Dokl. Acad. Nauk SSSR,* **179,** 1129
66. Vorobyev, A. F., Monaenkova, A. S. and Skuratov, S. M. (1967). *Vestnik. Mosk. Univ., Ser. Khim.,* N 6, 3
67. Vorobyev, A. F. and Monaenkova, A. S. (1972). *Vestnik Mosk. Univ., Ser. Khim.,* N 1
68. Vorobyev, A. F. and Brojer, A. F. (1971). *Zh. Fiz. Khim.,* **45,** 1307
69. Vorobyev, A. F. and Brojer, A. F. (1971). *Zh. Fiz. Khim.,* **45,** 2390
70. Vasiljev, V. P., Vorobjev, P. N., Ikonnikov, A. A. and Shitova, V. V. (1969). *Preprint, E.4. Intern. Conf. Calorimetry and Thermodynamics,* Warsaw, Aug. 31–Sept. 4
71. Frank, H. S. and Robinson, A. L. (1940). *J. Chem. Phys.,* **8,** 933
72. Krestov, G. A. (1968). *Trudy Ivanovskogo khimiko-technologicheskogo Instituta* (Jubileinyi vypusk). (Isd. IKhTI, Ivanovo)
73. Krestov, G. A. and Klapov, V. I. (1966). *Isv. Vysch. Ucheb. Zaved., Khim. khim. Tekhnol.,* **9,** 34
74. Abrosimov, V. K. and Krestov, G. A. (1967). *Zh. Fiz. Khim.,* **41,** 3150
75. Zverev, V. A. and Krestov, G. A. (1968). *Zh. Fiz. Khim.,* **42,** 540
76. Krestov, G. A. and Abrosimov, V. K. (1967). *Isv. Vyssh. Ucheb. Zaved., Khim. khim. Tekhnol.,* **10,** 1005
77. Krestov, G. A. and Kobenin, V. A. (1969). *Trudy Ivanovskogo Khimikotekhnologicheskogo Instituta,* **11,** 5
78. Kestov, G. A. and Kobenin, V. A. (1971). *Materialy Vsesojusnogo Simposiuma po khimii rastvorov.* (Isd. IKhTI, Ivanovo, 165)
79. Krestov, G. A. and Abrosimov, V. K. (1969). *Teor. i Experim. Khimija,* **5,** 415
80. Krestov, G. A. and Jegorova, I. V. (1970). *Radiokhimija,* **12,** 903
81. Krestov, G. A. and Kobenin, V. A. (1970). *Isv. Vyssh. Ucheb. Zaved., Khim. khim. Tekhnol.,* **13,** 1722
82. Mishchenko, K. P. and Shpigel, L. P. (1967). *Zh. Obshch. khim.,* **37,** 2145
83. Mishchenko, K. P. and Stagis, A. Ja. (1970). *Zh. Obshch. Khim.,* **40,** 2537
84. Lilich, L. S., Mogilev, M. E. and Chernykh, L. V. (1967). *Zh. Strukt. Khim.,* **8,** 197
85. Aleksejeva, E. A., Bart, T. Ja. and Lilich, L. S. (1969). *Zh. Strukt. Khim.,* **10,** 977
86. Aleksejeva, E. A. and Lilich, L. S. (1970). *Zh. Fiz. Khim.,* **44,** 2041

87. Lilich, L. S. and Bart, T. Ja. (1970). *Zh. Fiz. Khim.*, **44,** 2409
88. Burkov, K. A., Lilich, L. S. and Sillen, L. L. (1965). *Acta Chem. Scand.*, **19,** 14
89. Burkov, K. A. and Lilich, L. S. (1965). *Vestn. Leningr. Univ.*, vyp. 1, N 10, 103
90. Burkov, K. A. and Lilich, L. S. (1966). *Vestn. Leningr. Univ.*, vyp. 3, N 16, 120
91. Burkov, K. A., Zinevich, N. J., Nikolaeva, N. M. and Lilich, L. S. (1970). *VINITI,* N 2488-71, Novosibirksk
92. Burkov, K. A., Zinevich, N. J. and Lilich, L. S. (1970). *Isv. Vyssh. Uchebn. zawed, khim. khim. Tekhnol.*, **13,** 1250
93. Latysheva, V. A. and Andrejeva, I. N. (1968). *Zh. Neorg. Khim.*, **13,** 2080
94. Latysheva, V. A. and Andrejeva, I. N. (1969). *Zh. Neorg. Khim.*, **14,** 1180
95. Latysheva, V. A. and Andrejeva, I. N. (1969). *Zh. Fiz. Khim.*, **43,** 483
96. Stakhanova, M. S., Mikulin, G. I., Karapetjants, M. Kh., Vlasenko, K. K. and Bazlova, I. V. (1971). *Fizicheskaja khimija Rastvorov.* (Nauka: Moskva)
97. Karapetjants, M. Kh. and Vlasenko, K. K. (1970). *Zh. Fiz. Khim.*, **44,** 2333
98. Drakin, S. I., Lantukhova, L. V. and Karapetjants, M. Kh. (1967). *Zh. Fiz. Khim.*, **41,** 98
99. Mjasnikova, V. F., Drakin, S. I. and Karapetjants, M. Kh. (1968). *Zh. Fiz. Khim.*, **42,** 2055
100. Majsnikova, V. F., Drakin, S. I. and Karapetjants, M. Kh. (1969). *Zh. Fiz. Khim.*, **43,** 2141
101. Karapetjants, M. Kh., Drakin, S. I. and Lantukhova, L. V. (1967). *Zh. Fiz. Khim.*, **41,** 2653
102. Mishchenko, K. P. and Sukhotin, A. M. (1951). *Isvestija Sektora Platiny,* IONKh, AN SSSR, **26,** 203
103. Vlasenko, K. K. (1971). *Materialy Vsesojusnogo Simposiuma po Termokhimii Vodnykh i nevodnykh rastovorov elektrolitov.* (Isd. IKhTI, Ivanovo, p. 146)
104. Mashovets, V. P., Dibrov, I. A. and Matvejeva, R. P. (1964). *Zh. Prikl. Khim.*, **37,** 29
105. Mashovets, V. P., Dibrov, I. A. and Krumhalz, B. S. (1965). *Zh. Fiz. Khim.*, **39,** 1723
106. Mashovets, V. P., Baron, N. M. and Zavodnaja, G. E. (1966). *Zh. Strukt. Khim.*, **7,** 825
107. Puchkov, L. V. and Matvejeva, R. P. (1967). *Zh. Prikl. Khim.*, **40,** 2588
108. Mashovets, V. P., Baron, N. M. and Zavodnaja, G. E. (1969). *Zh. Fiz. Khim.*, **43,** 1737
109. Puchkov, L. V. and Matvejeva, R. P. (1970). *Zh. Fiz. Khim.*, **44,** 1970
110. Gardner, W. L., Jekel, E. C. and Cobble, J. W. (1969). *J. Phys. Chem.*, **73,** 2017 .
111. Gardner, E. R. (1969). *Trans. Faraday Soc.*, **65,** 91
112. Searles, S. K. and Kebarle, P. (1969). *Can. J. Chem.*, **47,** 2619
113. Rafalskii, R. P. (1970). *Geog. Rudnykh Mestorozhdenii,* **12,** 111
114. Khodakovskii, I. L. (1969). *Geokhimia,* N 1, 57
115. Lange, E. and Martin, W. (1937). *Z. Phys. Chem.*, **A180,** 233; Birnthaler, W. and Lange, E. (1938). *Z. Elektrochem.*, **44,** 679
116. Gold, V. (1967). *The H_2O–D_2O Solvent System* in *Physico-Chemical Processes in Mixed Aqueous Systems,* 1 (F. Franks, editor). (Heinemann: London)
117. Rabinovich, I. B. (1964). *Vlijanie Isotopii na Fiziko-khimicheskije Svojstva Zhidkostei.* (Isd. Nauka: Moskva)
118. Cvetkov, V. G. and Rabinovich, I. B. (1968). *Zh. Fiz. Khim.*, **42,** 1213
119. Cvetkov, V. G., Nikolajev, P. N., Moseeva, E. M., Martynenko, L. Ja. and Borisov, L. V. (1971). *V Vsesojusnaja Konferentsija po Kalorimetrii,* 203–224. Isd. Moskovskogo Universiteta
120. Franks, F. (editor). (1967). *Physico-chemical Processes in Mixed Aqueous Solvents* (Heinemann: London)
121. Somsen, G. and Coops, J. (1965). *Rec. Trav. Chim.*, **84,** 985
122. Somsen, G. (1966). *Rec. Trav. Chim.*, **85,** 159, 517, 526; (1967). *ibid.*, **86,** 893
123. Somsen, G. (1971). *Proc. Intern. Conf. Calorimetry and Thermodynamics,* 959 (Warsaw: Polish Scientific Publishers)
124. Held, R. P. and Criss, C. M. (1965). *J. Phys. Chem.*, **69,** 2611
125. Held, R. P. and Criss, C. M. (1967). *J. Phys. Chem.*, **71,** 2487. Kriss, C. M. and Luksha, E. (1968). *J. Phys. Chem.*, **72,** 2966
126. Mostkova, R. J., Kessler, J. M. and Semenov, V. N. (1971). *Electrokhimija,* **7,** 642
127. Krestov, G. A. and Zverev, V. A. (1968). *Isv. Vyssh. Ucheb. Zaved. khim. khim. Tekhnol.*, **11,** 990. (1969). *ibid.*, **12,** 25; (1971). *ibid.*, **14,** 528
128. Drakin, S. I., Karapetjants, M. Kh., Erbanova, L. N., Kurmalieva, R. Kh., Lantukhova, L. V. and Sokolov, V. B. (1968). *Trudy I Konfer. po analit. khim. nevodnikh rastvorov,* p.II, Isd. MKhTI, p.52

129. Djakonova, M. E., Drakin, S. I. and Karapetjants, M. Kh. (1970). *Trudy Moskovskogo Khim.-Tekhnolog. Inst. im. D. Mendelejeva,* **67,** 12
130. Mishchenko, K. P. and Kljujeva, M. L. (1965). *Teor. i Exper. Khim.,* **1,** 20
131. Mischenko, K. P., Subbotina, V. V. and Krumhalz, B. S. (1969). *Teor. i Experim. Khim.,* **5,** 268
132. Lonkevich, I. I., Mishchenko, K. P. and Shadskii, S. V. (1970). *Zh. Obshch. Khim.,* **40,** 723
133. Prosvirjakova, E. P., Mishchenko, K. P. and Poltoratskii, G. M. (1969). *Teori. i Experim. Khim.,* **5,** 129
134. Kushchenko, V. V. and Mishchenko, K. P. (1968). *Zh. Prikl. Khim.,* **41,** 646; *Teor. i Experim. Khim.,* **4,** 403
135. Mishchenko, K. P. and Tungusov, V. P. (1965). *Teor. i Experim. Khim.,* **1,** 55; (1967). *Zh. Strukt. khim.,* **8,** 1005
136. Mishchenko, K. P. and Sokolov, V. V. (1965). *Zh. Strukt. Khim.,* **6,** 819
137. Kushchenko, V. V. and Mishchenko, K. P. (1970). *Zh. Strukt. Khim.,* **11,** 140
138. Bogdanov, V. G., Kliujeva, M. L., Mishchenko, K. P. (1970). *Zh. Obshch. Khim.,* **60,** 721
139. Frank, H. S. (1965). *Federation Proceedings, vol. 24,* N 2, Part III, March-April; (1958). *Proc. Roy. Soc. (London),* **A247,** 481
140. Osińska-Tanijewska, S. M. (1967). *Z. Phys. Chem.,* **235,** 272
141. Taniewska-Osińska, S. M. and Mishchenko, K. P. (1968). *Zh. Fiz. Khim.,* **42,** 487
142. Taniewska-Osińska, S., Grohowski, R. and Piekarski, H. (1971). *Proc. Intern. Conf. Calorimetry and Thermodynamics,* 875. (Warsaw: Polish Scientific Publishers)
143. Ivanova, E. F. and Kotljarova, G. P. (1966). *Zh. Strukt. Khim.,* **7,** 455; (1964). *Zh. Fiz. Khim.,* **38,** 422; (1966). *Zh. Fiz. Khim.,* **40,** 996
144. Ivanova, E. F. (1971). *Materialy Vsesojusn. Symposiuma po Termokhimii rastvorov Elektrolitov i Neelektrolitov,* p.154. (Ivanovo: Isd. IKhTI)
145. Stern, J. H. and Nobilione, J. (1968). *J. Phys. Chem.,* **72,** 1064
146. Stern, J. H. (1969). *J. Phys. Chem.,* **73,** 928
147. Stern, J. H. and Kulluk, J. K. (1969). *J. Phys. Chem.,* **73,** 2795
148. Karpenko, G. V., Mishchenko, K. P. and Poltoratskii, G. M. (1967). *Zh. Strukt. Khim.,* **8,** 413
149. Karpenko, G. V., Mischchenko, K. P. and Poltoratskii, G. M. (1970). *Teor. i Experim. Khim.,* **6,** 107
150. Wóycicka, M. and Minc, S. (1971). *Proc. Intern. Conf. Calorimetry and Thermodynamics,* 967. (Warsaw: Polish Scientific Publishers)
151. Samojlov, O. Ja. (1946). *Zh. Fiz. Khim.,* **20,** 1411; (1951). *Zh. Strukt. Khim.,* **2,** 551; (1960). *Radiokhimija,* **2,** 183; (1966). *Zh. Strukt. Khim.,* **7,** 15, 175
152. Buslajeva, M. N. and Samojlov, O. Ja. (1963). *Zh. Strukt. Khim.,* **4,** 4
153. Klopov, V. I., Pirogov, A. I., Krestov, G. A. (1971). *Zh. Fiz. Khim.,* **45,** 1349; (1971). *Zh. Strukt. Khim.,* **12,** 419
154. Jegorova, I. V., Sorokin, V. D. and Krestov, G. A. (1971). *Zh. Strukt. Khim.,* **12,**
155. Stennikova, M. F., Poltoratskii, G. M. and Mishchenko, K. P. (1971). *Zh. Obschch. Khim.,* **41,** 2588
156a. Poltoratskii, G. M. (1971). *VINITI,* N 2984–71
156b. Poltoratskii, G. M. and Karpenko, G. V. (1971). *VINITI,* N 2985–71
156c. Poltoratskii, G. M. and Stennikova, M. F. (1971). *VINITI,* N 2987–71
157. Lindenbaum, S. and Boyd, G. E. (1964). *J. Phys. Chem.,* **68,** 911
158. Lindenbaum, S. (1966). *J. Phys. Chem.,* **70,** 814
159. Lindenbaum, S. (1968). *J. Phys. Chem.,* **72,** 212
160. Frank, H. S. (1965). *Z. Phys. Chem. (Leipzig),* **228,** 384; *Federation Proceeding,* **24,** N 2, part III, March-April, S–1
161. Franks, F. and Smith, H. T. (1967). *Trans. Faraday Soc.,* **63,** 2586
162. Wen, W.-Y. and Nara, K. (1967). *J. Phys. Chem.,* **71,** 3907; (1968). *ibid.,* **72,** 1137; (1968). *ibid.,* **72,** 3048
163. Kudrjavtseva, I. V., Mishchenko, K. P. and Poltoratskii, G. M. (1968). *Teor. i Experim. Khim.,* **4,** 468
164. Shpenzer, N. P., Mishchenko, K. P., Poltoratskii, G. M. and Talmud, S. L. (1971). *Materjaly Vsesojusnogo Simposiuma po Termokhimii rastvorov elektrolytov,* p.80. (Ivanovo: Isd. IKhTI)
165. Mishchenko, K. P., Poltoratskii, G. M. and Fedotova, G. V. (1969). *Zh. Strukt. Khim.,* **10,** 3

166. Fedotova, G. V., Mishchenko, K. P. and Poltoratskii, G. M. (1969). *Teor. i Exper. Khim.*, **5,** 838
167. Poltoratskii, G. M. and Fedotova, G. V. (1971). *Materialy Vsesojusnogo Simposiuma po termokhimii rastvorov electrolitov.* (Ivanovo: Isd.IKhTI)
168. Sukhotin, A. M. (1959). *Voprosy Teorii Rastvorov electrolitov v sredakh s nizkoi Dielektricheskoi Pronitsaemostju.* (Leningrad: Goskhimisdat)
169. Stakhanova, M. S., Karapetjants, M. Kh., Vasil'ev, V. A., Byzlova, I. V. and Vlasenko, K. K. (1969). *Redkie schelochnye metally,* p.185. (Perm: Isdat. Permskogo Politekhnicheskogo Instituta)
170. Vasil'ev, V. A., Fedjainov, N. V., Karapetjants, M. Kh. and Serafimova, T. I. (1971). *Zh. Fiz. Khim.*, **45,** 1867
171. Vasil'ev, V. A. (1971). *Zh. Fiz. Khim.*, **45,** 1480
172. Karapetjants, M. Kh., Vasil'ev., V. A. and Fedjainov, N. V. (1970). *Zh. Fiz. Khim.*, **44,** 1822
173. Vasil'ev, V. A. and Shevchenko, E. Ja. (1970). *Isv. Vyssh. Uhebn. Zaved., Khim. khim. Tekhnol.*, **13,** 789
174. Karapetjants, M. Kh. (1965). *Metody sravnitelnogo Rascheta fiziko-khimich. svojstv.* (Moskva: Isd. Nauka)
175. Karapetjants, M. Kh. (1969). *Sbornik Sto let Periodicheskogo Zakona khimicheskikh elementov,* 256 (Moskva: Isd. Nauka)
176. Novoselov, N. P. and Mishchenko, K.P. (1968). *Zh. Obshch. Khim.*, **38,** 2129
177. Novoselov, N. P. and Mishchenko, K. P. (1971). *Zh. Fiz. Khim.*, **45,** 1254
178. Novoselov, N. P. and Mishchenko, K. P. (1971). *Zh. Obshch. Khim.*, **41,** 255, 489
179. Randall, M. and Rossini, F. D. (1929). *J. Amer. Chem. Soc.*, **51,** 323
180. Rossini, F. D. (1930). *J. Res. Nat. Bur. Stand.*, **4,** 313; (1931). *ibid.*, **6,** 791; *ibid.*, **7,** 47
181. Somsen, G. and Weeda, L. (1971). *Rec. Trav. Chim.*, **90,** 81
182. Somsen, G. and Weeda, L. (1971). *J. Electroanal. Chem.*, **29,** 375
183. Lange, E. and Mishchenko, K. P. (1930). *Z. Phys. Chem.*, **A149,** 1
184. Halliwell, H. F. and Nyburg, S. C. (1963). *Trans. Faraday Soc.*, **59,** 1126
185. Verwey, E. J. W. (1942). *Rec. Trav. Chim.*, **61,** 127
186. Sukhotin, A. M. and Kazankina, A. F. (1970). *Elektrokhimija,* **6,** 1323
187. Sukhotin, A. M. and Kazankina, A. F. (1971). *Elektrokhimija,* **7,** 932
188. Sukhotin, A. M. and Kazankina, A. F. (1970). *Elektrokhimija,* **6,** 1530
189. Sukhotin, A. M. and Kazankina, A. F. (1970). *Elektrokhimija,* **6,** 1514
190. Samojlov, O. Ja. (1966). *Zh. Strukt. Khim.*, **7,** 15
191. Samojlova, O. Ja. (1966). *Zh. Strukt. Khim.*, **7,** 175
192. Samojlov, O. Ja. (1970). *Zh. Strukt. Khim.*, **11,** 990
193. Samojlov, O. Ja., Aleshko-Ozhevskii, Y. P., Buslajeva, M. N. and Ojari, P. U. (1971). *Zh. Fiz. Khim.*, **45,** 974
194. Ojari, P. U. and Samojlov, O. Ja. (1971). *Zh. Strukt. Khim.*, **12,** 708
195. Varshal, G. M. and Senjavin, M. M. (1964). *Zh. Strukt. Khim.*, **5,** 681
196. Samojlov, O. Ja. (1967). *Sbornik Sostojanie i rol vody v biologicheskikh objektakh,* 31 (Moskva: Isd. Nauka)
197. Samojlov, O. Ja. *Structure and Transport Processes in Water and Aqueous Solutions.* In the press (New York: Wiley-Interscience)

8
Equilibrium Studies at High Temperatures

GIOVANNI DE MARIA and GIOVANNI BALDUCCI
Università di Roma, Roma

8.1 INTRODUCTION 209

8.2 HOMONUCLEAR DIATOMIC MOLECULES 210

8.3 HETERONUCLEAR DIATOMIC MOLECULES 212
8.3.1 *Intermetallic molecules* 213
8.3.2 *Other molecules* 213

8.4 HALIDES 213
8.4.1 *Fluorides* 214
8.4.2 *Chlorides* 214
8.4.3 *Bromides and iodides* 215

8.5 OXIDES 216

8.6 CARBIDES 218

8.7 SULPHIDES, SELENIDES AND TELLURIDES 222

8.8 NITRIDES AND BORIDES 223

8.9 MISCELLANEOUS 224

8.10 CONCLUSIONS 225

8.1 INTRODUCTION

The study of chemical equilibria at high temperatures has encountered a sharp increase in activity in the last decade not only because of interest of a theoretical nature but also for some particular applications of importance in modern technology. The field covered by this subject, which has been treated more or less extensively in articles and review papers published in the last 15 years[1–7], is extremely broad. Here, we shall restrict ourselves mainly to the condensed phase–gas phase equilibria of inorganic materials with particular emphasis on the description of the gaseous species originating from high-temperature equilibrium reactions. Correlations among the properties of the molecular species will be reported because of their importance, both theoretical and practical, in the study of new chemical compounds.

It is well known to high-temperature chemists that the gaseous phase in thermodynamic equilibrium with the condensed system becomes more complex as the temperature increases. This at first sight surprising generalisation is known as 'Brewer's paradox'[9]. The complexity of the gas phase at high temperature is due not only to the great variety of compounds but also to the presence of compounds of unusual oxidation states.

The thermodynamic treatment of equilibrium reactions is usually based on two independent methods, known as the second- and third-law methods, which utilise the following expressions:

$$\Delta H^\circ_T = -R \, \mathrm{d} \ln K / \mathrm{d}(1/T) \tag{8.1}$$

$$\Delta H^\circ_0 = -RT \ln K - T\Delta (G^\circ_T - H^\circ_0)/T \tag{8.2}$$

where K is the equilibrium constant, ΔH° the reaction enthalpy, and $(G^\circ_T - H^\circ_0/T$ the free-energy function.

Due to the lack of spectroscopic information for most high temperature molecules, the thermodynamic functions (in particular the free energy functions) are usually estimated on the basis of analogy. The high-temperature reactions considered in this review were mainly investigated by the Knudsen-cell molecular source, utilising effusion, torsion and mass spectrometric techniques. Because of their importance for the calculation of the thermodynamic functions of high temperature molecules, mention of some recent optical spectroscopic studies is also made.

8.2 HOMONUCLEAR DIATOMIC MOLECULES

Of all possible homonuclear diatomic molecules, about half have so far been identified mainly through optical spectroscopic investigation and through the application of mass spectrometry to the thermochemical study of vapours in equilibrium with elements and compounds at high temperature. Values of the dissociation energies of homonuclear diatomic molecules have been reviewed by Drowart[10], Drowart and Goldfinger[11], and De Maria[8]. Since these reviews were written, other diatomic molecules of the principal elements, particularly of the transition metals, have been studied. In some cases the dissociation energies of known molecules have been redetermined. Table 8.1 contains the original compilation by Drowart and Goldfinger[11] updated to include the new values.

The dissociation energies of homonuclear diatomic molecules can be correlated with the lattice energies of the corresponding elements. This correlation was extended to a number of molecules of different groups, and it was shown[12] that when the values of the ratio of the lattice energy of an element to the dissociation energy of the corresponding diatomic molecule, $\Delta H^\circ_0/D^\circ_0 = \alpha$, is plotted as a function of the atomic number, the molecules can be grouped into 'categories' characterised by a given value of α. The problem of the detection of unknown homonuclear diatomic molecules in terms of the relation of their chemical stability to the α parameter is discussed in Ref. 12 and reported also in Ref. 8. On the basis of this correlation, Balducci *et al.*[13] observed the first reported homonuclear diatomic molecule of the rare-earth elements with unfilled f shells and calculated a third law $D_0^\circ(Ce_2) = 65 \pm 5$ kcal mol^{-1} from the reaction: $Ce(l) + Ce(g) = Ce_2(g)$. This

Table 8.1 Dissociation energies of homonuclear diatomic molecules (in kcal mol^{-1}) [From Ref. 11, integrated with new values]

Molecule	D_0°	*Ref.*	*Molecule*	D_0°	*Ref.*
Li_2	25.0 ±1.2		Co_2	39 ±6	
Na_2	17.3 ±0.7		Ni_2	52.5±5	34
K_2	11.8 ±1.2		Pd_2	26 ±5	21
Rb_2	10.8 ±1.2		Cu_2	45.5±2.2	
Cs_2	10.4 ±1.0		Ag_2	37.6±2.2	
Be_2	(14)		Au_2	51.5±2.2	
Mg_2	(6)	23	Zn_2	(6)	
Ca_2	(7)	23	Cd_2	2.1±0.5	
Sc_2	38.0 ±5		Hg_2	1.4±0.07	
Y_2	37.3 ±5		B_2	65.5±5.5	
La_2	57.6 ±5		Al_2	41.1±4.3	32
Ce_2	65 ±5	13	Ga_2	32 ±5	
	57 ±4	14	In_2	19.5±2.5	
Nd_2	<39	17	Tl_2	(14)	
Tb_2	37	16	C_2	142.3±3.4	
Ho_2	19 ±4	15	Si_2	74 ±5	
Lu_2	25	16	Ge_2	64 ±5	
Th_2	⩽68 ±8	192	Sn_2	45.8±4	
Ti_2	32 ±5	19	Pb_2	23 ±5	
V_2	56.9 ±4	19	N_2	225.1±0.5	
Cr_2	36 ±7		P_2	116.0	
Mn_2	7.5 ±6	18	As_2	90.8	
Fe_2	25 ±5	20	Sb_2	70.6±1.5	
Bi_2	47.0 ±2		Te_2	61.3±1.1	25
O_2	117.96±0.04		F_2	36.8	
				37.5±2.3	31
S_2	101.0 ±0.2	26	Cl_2	57.4	
Se_2	75.7 ±2.5	25	Br_2	45.5	
			I_2	35.6	

*1 cal = 4.184 J

value is to be compared with the value $D_0^\circ(Ce_2) = 57 \pm 4$ kcal mol^{-1} recently found by Gingerich and Finkbeiner[14] from studies of the gaseous equilibria over the Au–Ce–CeS–BN–C system.

For other rare earth elements it proved possible quite recently to determine the dissociation energies of the diatomic molecules of Ho[15], Tb, and Lu[16], and an upper limit has been obtained for the Nd_2 molecule: $D_0^\circ(Nd_2) < 39$ kcal mol^{-1} (Ref. 17). The remaining homonuclear diatomic molecules of the first transition period Mn_2 [18], Ti_2 [19], V_2 [19], and Fe_2 [20] have been detected and their dissociation energies determined from studies at high temperatures of equilibria of the type: $M(s,l) + M(g) = M_2(g)$ and $2M(g) = M_2(g)$, using a combination of Knudsen effusion and mass spectrometric techniques.

The dissociation energy of the Pd_2 molecules has also been determined[21]. The discrepancy between the second- and third-law values of $D_0^\circ(Pd_2)$, 16.9 ± 6 and 26 ± 5 kcal mol^{-1}, respectively, although large, is within the uncertainties reported. However, more credence should be given to the third-law value. The uncertainties in the third-law computation for these molecules depend mainly on the ground state configurations assumed. The relatively low reported dissociation energies of Fe_2 and Mn_2 are consistent with a correlation that has been noted between dissociation energies of transition element molecules of the first period and the atomic valence-state energies of the corresponding atoms[22].

In the case of Mn_2, the dissociation energies calculated by the third-law method using estimated covalent parameters, and on the assumption that the bonding is of the van der Waals type, are 12.9 ± 6 and 7.5 ± 6 kcal mol^{-1} respectively. The agreement of the latter value with the second-law value (3 ± 3 kcal mol^{-1}) suggests that Mn_2 is a van der Waals type molecule.

Low dissociation energies are also found for other diatomic molecules of Groups IIA and IIB. Values of 6 and 7 kcal mol^{-1} were estimated for the bond energies of Mg_2 and Ca_2 [23].

The dissociation energies of the Group VIB molecules S_2, Se_2, Te_2 obtained from Knudsen cell effusion and mass spectrometry were previously quoted in a review[24]. Since then, equilibrium constants between atoms and molecules in the vapours of S_2, Se_2 and Te_2 have been measured using a combination Knudsen-torsion-effusion apparatus[25], from which the derived dissociation energies were $D_0^\circ(S_2) = 101.7 \pm 2.9$, $D_0^\circ(Se_2) = 75.7 \pm 2.5$, and $D_0^\circ(Te_2) = 61.3 \pm 1.1$ kcal mol^{-1}.

Recently[26], the same molecules were studied by photo-ionisation mass spectrometry. For $D_0^\circ(S_2)$ the value 101.0 ± 0.2 kcal mol^{-1} was found in good agreement with the other techniques; the result obtained, 62.3 ± 0.2 kcal mol^{-1}, for $D_0^\circ(Te_2)$ favours the thermochemical value given in Ref. 25, but the result found for $D_0^\circ(Se_2) = 78.6 - 83.2$ kcal mol^{-1} is rather uncertain, and in any case higher than established thermochemical values. However, Meschi and Searcy[27] recently measured the magnetic moments of S_2, Se_2, Te_2, Se_6 and Se_5 with a magnetic deflection method and from their result concluded that the value of 78.6 for $D_0^\circ(Se_2)$ is to be preferred. As previously reported values for the dissociation energy of F_2 were is disagreement[28–30], a new mass-spectrometric determination was undertaken[31], from which the value $D_0^\circ(F_2) = 37.5 \pm 2.3$ kcal mol^{-1} was obtained.

For Al_2 [32], the dissociation energy has been redetermined in the course of a mass spectrometric study of the gaseous aluminium chalcogenides.

8.3 HETERONUCLEAR DIATOMIC MOLECULES

Previous data on gaseous intermetallic and other hetero-nuclear diatomic molecules, obtained mainly from mass spectrometric studies of vapours in equilibrium with compounds, have been reviewed, and the general trends among their properties have been discussed for different groups of the periodic system [8, 10, 11, 33].

8.3.1 Intermetallic molecules

Several intermetallic molecules of high physical stability have been characterised in which gold, or silver, is a constituent atom. Dissociation energies for the molecules AuNi, AuCo and AuFe[34] have been obtained by studying the vapours in equilibrium over liquid solutions of gold and the transition metals above 1800 K. The dissociation energies of the molecules AuMn[35] and AgMn[36] were similarly determined. Bond energies of the gaseous rare-earth aurides[37] (in some cases exceeding 70 kcal mol^{-1}) have been obtained by studying gaseous equilibria of the type: MAu(g)+Au(g) = M(g)+Au(g), where M is either La, Ce, Pr, or Nd. Similarly, dissociation energies for TbAu[38], HoAu[38, 39], LuAu[38] and HoAg[39] have been determined. An upper limit of 50 kcal mol^{-1} has also been established for the dissociation energy of the NdAg molecule[17]. The dissociation energy for the first reported gaseous intermetallic compound with platinum, namely LuPt, with $D_0^\circ = 95 \pm 8$ kcal mol^{-1}, has been obtained by studying the equilibria LuPt(g)+Au(g) = LuAu(g)+Pt(g) and LuPt(g) = Lu(g)+Pt(g) over a Lu–Pt–Au alloy in the temperature range 2530–2782 K [40].

The dissociation energy of the molecule PdGe[41] has been determined and that of AlCu[42] redetermined.

The determination of the activities of the constituents of the gallium liquid intermetallic system with copper, silver, and gold, as well as of the dissociation energies of the CuGa, AgGa and AuGa[43] molecules have been made by Knudsen cell–mass spectrometry between 1300 and 1600 K.

8.3.2 Other molecules

The first reported transition metal monosilicide molecule, AuSi, was characterised by Gingerich[44] through the study of the equilibria AuSi(g)+Al(g) = AlAu(g)+Si(g), and Au_2(g)+Si(g) = AuSi(g)+Au(g) over the temperature range 1831–1989 K. Silicides of the first-row transition metals, namely FeSi, CoSi and NiSi, and some other transition metals, namely RuSi, RhSi, PdSi, and AuSi, were also studied and their dissociation energies determined[45, 46]. The vaporisation equilibria of a number of Group IIIB–VB compounds were investigated, and dissociation energies for AlP[47], GaAs[48], GaP[49, 50], and TlBi[51] were determined. The following paragraphs deal with selected classes of other heteronuclear molecules of interest in high temperature equilibrium studies.

8.4 HALIDES

The study of high-temperature equilibrium reactions involving halides has attracted much interest in recent years. These investigations are indeed relevant to the understanding of the behaviour at high temperature of fused salts. Reviews on the subject have appeared and correlations among the properties of the gaseous mono-halide species have been discussed[52–57].

8.4.1 Fluorides

Ehlert[58], using a Farlow-type reaction involving manganese trifluoride and graphite, investigated under equilibrium conditions the gaseous species $(CF_2)_n$ ($n = 1,2,3\ldots$), CF_3 and C_2F_2. A review of data on C_1 fluorides is reported in his paper, together with an evaluation of $D_0^\circ(CF_3{-}F)$, $D_0^\circ(CF_2{-}F)$, $D_0^\circ(CF{-}F)$ and $D_0^\circ(C{-}F)$ from new experimental data. In particular the C=C bond in difluoroacetylene was found to be weaker than that in acetylene by some 60 kcal mol^{-1}. Infrared spectroscopic studies by the matrix isolation technique on the CF_2 and CF_3 species were carried out by Snelson[59]. The equilibrium reaction $C_2F_4 \rightleftharpoons 2CF_2$ was studied[60] with the mass spectrometer and the heat of formation of CF_2 was calculated. Equilibrium reactions involving the gaseous fluoride species of the Group IVB elements are treated in the papers by Timms *et al.*[61], Thompson and Margrave[62], and Zmbov *et al.*[63]. The importance of the silicon difluoride species in providing new possibilities for syntheses of known silicon–fluorine compounds, and of novel types of organic and inorganic species has been emphasised. Bearing in mind the stability of the $(CF_2)_n$ polymers, the absence of $(SiF_2)_n$ polymers in the vapour phase at equilibrium, especially at higher temperatures, is difficult to understand. There is, however, a decreasing stability of the dimers in passing from $(SnF_2)_2$ to $(GeF_2)_2$.

The dissociation energy of CuF has been obtained[64] from the study of high-temperature gaseous equilibria in the Mg–Cu–F system. Vapour pressure measurements on MnF_2 by the Knudsen effusion method were carried out over a large temperature range by Hitchingham and Kana'an[65]. Mass spectrometric investigations of the equilibrium vapour over SbF_3 in a temperature range up to 529 K showed the only important species to be the monomer SbF_3 [66]. Bond-energy values in both gaseous and solid SbF_3 were discussed. The $(TlF)_n$ ($n = 1$–4) polymeric species were investigated by the mass spectrometric technique in the equilibrium vapour over thallous fluoride up to its melting point[67, 68]. Thermodynamic data for the sublimation process were obtained.

The enthalpies of formation and the dissociation energies of Sc, Y and La monofluorides were discussed by Krasnov[69] on the basis of a theoretical treatment and available experimental data. The existence of the gaseous La_2F_6 molecule was demonstrated by Skinner and Searcy[70], and the equilibrium reaction $La_2F_6 \rightarrow 2LaF_3$ was investigated in the course of a vaporisation study of lanthanum trifluoride. Infrared spectroscopic data for CoF_2, NiF_2 and ZnF_2 were obtained by Hastie *et al.*[71] and for CaF_2, SrF_2 and BaF_2 by Calder *et al.*[72], utilising the matrix isolation technique.

Wesley and De Kock[73] reported similar data for La, Ce, Pr, Nd, Sm, and Eu trifluorides.

8.4.2 Chlorides

The saturated vapour composition over $BeCl_2$ had been investigated by Ko and associates[74]. By comparing torsion and transpiration pressure measurements they obtained strong evidence for a predominantly dimeric vapour at molecular effusion pressures. Hildenbrand and Theard[75], using the mass spectrometric technique, investigated gaseous equilibria in the Be–Al–Cl system by passing HCl(g) over a Be–Al mixture in a Knudsen-cell molecular source. The enthalpies of reaction Be(g)+AlCl(g) → BeCl(g)+Al(g) and Be(g)+$BeCl_2$(g) → 2BeCl(g) were combined with the heat of sublimation derived from β-$BeCl_2$ ($\Delta H_{298\,K}$ = 32.5 ± 0.5) to yield the dissociation energies D_0°(BeCl) = 91.9 kcal and D_0°($BeCl_2$) = 219.4 kcal. The conclusions of these authors on the vapour composition over $BeCl_2$ differ significantly from those of Ko and associates, being consistent with a largely (70–90 mole %) monomeric vapour in the effusion pressure region. These results were confirmed unambiguously by accurate vapour molecular weight measurements and showed that a rapid increase in dimer concentration to *c.* 70 mole % occurs at the $BeCl_2$ boiling point (755 K)[76].

Equilibrium reactions involving gaseous sodium chloride trimers and tetramers were investigated by Feather and Searcy[77]. A comparison of the experimental data with theoretical values based on an ionic model is discussed, with particular emphasis on the new species, Na_4Cl_4. The disproportionation reaction Ge(s)+$GeCl_4$(g) → 2$GeCl_2$(g) was shown to be the only equilibrium involved in the vapour phase over the system Ge(s)+$GeCl_4$(g). No polymeric species could be observed[78].

Guido *et al.*[79], studied the equilibrium vaporisation of CuCl and showed that the vapour is made up of trimer, tetramer, and pentamer, the first two in comparable amounts. Dissociation energies for these polymers were also calculated.

The vaporisation of $MnCl_2$ was investigated by the mass-effusion and torsion-effusion techniques. Correlations between the thermodynamic properties of the $MnCl_2$ gaseous species were discussed[80]. Mass-effusion, torsion and transpiration techniques were used to investigate the sublimation of $FeCl_2$ [81], $CoCl_2$ [82] and $NiCl_2$ [83]. In particular the equilibrium reactions Co(c)+$CoCl_2$(g) → 2CoCl(g) and Ni(c)+$NiCl_2$(g) → 2NiCl(g) were considered in order to calculate the heat of formation and entropy of gaseous CoCl and NiCl.

Niobium pentachloride was the object of an interesting investigation of the vaporisation process up to the critical point[83], and the equilibrium reactions $\frac{1}{4}$Nb(s)+$NbCl_5$(g) → $\frac{5}{4}$$NbCl_4$(g) and $\frac{1}{3}$Nb(s)+$NbCl_4$(g) → $\frac{4}{3}$$NbCl_3$(g)[85] were studied by transpiration measurements of the total niobium and chlorine in the vapour in equilibrium with solid niobium at temperatures between 800 and 1400 K.

The infrared spectra and the geometries of 18 gaseous dichlorides are reported in papers by Hastie *et al.*[86, 87] and by Jacox and Milligan[88].

8.4.3 Bromides and iodides

Equilibrium reaction studies involving bromides and iodides have been but little investigated. The reaction of gaseous $GeBr_4$ in a Knudsen cell with Ge(s) gave rise by disproportionation to the species $GeBr_2$, but polymeric species were not observed[78]. The lowering of the molecular weight of unsaturated vapours of $SnBr_2$ at high temperature was attributed to the depolymerisation process, $Sn_2Br_2 \rightarrow 2SnBr$[89]. A mass spectrometric investigation of this system would be desirable. The same applies to the stannous bromide–caesium bromide system, in which the formation of a complex gaseous species, $CsSnBr_3$, has been suggested[90]. Additional investigation of the equilibrium reactions in the vapour phase over zirconium tribromide is called for according to the analysis of the vaporisation data made by Tsirelnikov[91].

The dissociation energies of monobromides and monoiodides of Sc, Y and La are discussed from both theoretical and experimental points of view by Krasnov[92].

8.5 OXIDES

Equilibrium studies involving oxides have been the object of many investigations. The oxides for the most part vaporise to elemental products and to gaseous MO and MO_2 species. In some cases vaporisation gives rise to polymeric species.

Schofield[93] discussed the dissociation energies of Group IIA diatomic oxides. An excellent review of the dissociation energies of gaseous monoxides has been published by Brewer and Rosenblatt[94]. They calculated also a complete set of free-energy functions based on the free divalent ion model. These tabulations enable calculation of the partial pressures of the MO species.

The dissociation energy of silicon monoxide has been redetermined[95] by studying the isomolecular exchange equilibrium: $Ge(g)+SiO(g) \rightarrow GeO(g)+Si(g)$, in the temperature range 1400–1560 K. The value found, $D_0^\circ(SiO) = 182.8 \pm 3$ kcal mol^{-1}, is in agreement with that obtained from spectroscopic studies[96], and is *c.* 10 kcal mol^{-1} lower than that previously accepted, which was based on thermochemical studies involving solid phases and obtained from weight-loss or transpiration techniques. If this value is proved correct, the dissociation energies of molecules such as LaO, YO, ScO, CeO, UO, UO_2 and BO calculated from exchange equilibrium reactions involving SiO, will need revision. However, taking into account the discussion by Uy and Drowart[97], one cannot consider this new value for $D_0^\circ(SiO)$ as completely established.

The vaporisation of $Na_2O(c)$ has been studied with a mass spectrometer[98], and from evaluation of equilibria involving $Na_2O(c)$, $Na_2O(g)$, Na(g), NaO(g) and $O_2(g)$ it was possible to derive the dissociation energy of NaO(g) and the heat of atomisation of $Na_2O(g)$. It is to be noted that the value $D_0^\circ(NaO) = 60.3 \pm 4$ kcal mol^{-1} is the first direct thermochemical value obtained, and it is in exact agreement with the value calculated using Rittner's electrostatic model. This suggests that the binding in NaO is ionic.

The thermodynamics of the evaporation of the system urania–uranium[99], and of Cm_2O_3 [100], were studied using an effusion method. Total effusion rates and, hence, total vapour pressures were measured. In the first system, additional mass spectrometric measurements showed that vaporisation occurs predominantly to UO(g), with smaller amounts of U(g) and UO(g), and enabled measurements to be made of the partial pressures of each species in equilibrium over the condensed system. In the second system, without mass spectrometric analysis of the effusing vapours, Cm_2O_3 was shown to vaporise congruently to CmO(g) and O(g) on the basis of estimates of the heat of formation of Cm_2O_3(s). This vaporisation behaviour is consistent with general trends in the 4f and 5f series of oxides. Recently, the equilibrium decomposition pressure of oxygen, and thermodynamic data for non-stoichiometric berkelium oxide have been determined[101].

The vaporisation of europium monoxide[102] has been studied by the target collection Knudsen-effusion technique. Additional x-ray powder-diffraction and mass-spectrometric analyses indicated vaporisation according to the equation: $4EuO(s) \rightarrow Eu_3O_4(s) + Eu(g)$. The presence of the Eu_3O_4 phase (tripositive state for europium) deviates from the vaporisation pattern expected for EuO, bearing in mind that divalent europium, along with samarium and ytterbium, behaves like the alkaline earths[103].

Measurements of the equilibrium properties of the monoxide–dioxide reactions, $MO_2(g) \rightarrow MO(g) + \frac{1}{2}O_2(g)$, were made for cerium, praseodymium and neodymium[104] at *c.* 2000 K, using a Knudsen cell with O_2 inlet, and a mass spectrometer. The comparison between the second M—O bond energies of these rare earth dioxides with the M—O bond energy in the monoxides, suggested that other dioxide molecules of the rare earth series might be stable molecules. The atomisation energy of NdO_2(g)[17] has recently been determined through the Knudsen-cell mass-spectrometric study of the reaction: $NdO_2(g) + Nd(g) \rightarrow 2NdO(g)$. The existence of the NdO_2 molecule, and an approximate evaluation of its physical stability had been predicted earlier on the basis of a parallelism with the NdC_4 molecule[105].

The phase behaviour, vaporisation characteristics and thermodynamic properties at high temperature for the lanthanum–oxygen and cerium–oxygen systems were studied by Ackermann and Rauh[106, 107].

Among the noble metal oxides, the system ruthenium–oxygen has been studied by the mass-spectrometric Knudsen-cell method[108]. Heats and entropies of formation were determined for the gaseous species RuO_3, RuO_2, and RuO.

The heat of formation as measured in transpiration and Knudsen-cell studies, were found to be in agreement, but some discrepancies exist in the associated entropies.

The volatility of thallous oxide was investigated[109] with an effusion method. The mass spectrum of the effusing vapours showed that decomposition is not important in the vaporisation process; the vapour pressure and the heat of sublimation of Tl_2O were then calculated.

Mass spectrometric and microbalance studies of the vaporisation of chromium trioxide, reported by McDonald and Margrave[110], showed that the vapour is constituted of polymeric species $(CrO_3)_n$ with $n = 3$–5. They also calculated the relative heats of polymerisation. In a similar and rather more complete mass-spectrometric study[111], consisting of double cell experiments, cell exhaustion experiments and heats of sublimation measurements, it was demonstrated that in addition to $(CrO_3)_n$, ($n = 3$–5) the monomer, dimer and hexamer are also present. The partial pressures of each component were measured.

The vapour pressures of ReO_2 and ReO_3 were measured by the Knudsen effusion mass spectrometric technique[112–114]. The vapour in equilibrium with the solid phase is primarily $Re_2O_7(g)$, with a small amount of $ReO_3(g)$. The heats of formation of the solid phases were also obtained. Saturated vapour pressures, the enthalpy of evaporation and other physicochemical properties of rhenium heptoxide were determined[115]. The enthalpies of formation of $ReO_3(g)$ and $Re_2O_6(g)$ were reported[116].

The thermodynamics of dissociation of copper and nickel oxides[117, 118] have been studied using the mass spectrometric–Knudsen cell technique. Mass spectrometric investigations of condensed phase–gas phase equilibria for PbO[119], WO_3 and MoO_3 [120] have shown that for these oxides, vaporisation gives rise to polymeric species. Partial and total pressures, and heats of sublimation were determined. The heat of vaporisation of MoO_3 has also been determined calorimetrically[121].

For FeO(g), a thermochemical value of $D_0^\circ(\text{FeO}) = 97 \pm 3$ kcal mol^{-1} was obtained[122] by studying the equilibrium reaction: $\text{FeO(g)} = \text{Fe(g)} + \frac{1}{2}\text{O}_2\text{(g)}$, over the temperature range 1747–2150 K. This experimental value is in good agreement with a previously estimated value[94].

A vast amount of experimental work has been done on the titanium–oxygen system[123–127]. Some discrepancies in respect of the dissociation energy of TiO(g), as derived from studies of the vaporisation of solid phases and also from studies of isomolecular exchange reactions, have been discussed.

The vapour pressure and molecular composition of selenium dioxide in the gas phase, and the heat of vaporisation have been determined[128, 129]. Both saturated and unsaturated vapours were found to consist only of monomeric SeO_2. A similar result was found from mass spectrometric studies of the sublimation of SeO_2 and SeO_3 [130]. SeO_2 has also been studied by infrared and matrix isolation techniques and the thermodynamic functions have been calculated[131].

The vaporisation process of phosphorus oxides were investigated by mass spectrometry[132] over a temperature range where various polymorphic modifications are stable. The equilibrium vapour was shown to consist of polymeric species, primarily $P_4O_{10}(g)$. From the heats of reaction of equilibrium processes involving the gaseous species, average values for the (P—O) and (P=O) bond energies were calculated. InGaO, $InGaO_2$ and InAlO [133] have been identified and characterised from vaporisation experiments of some Group IIIA metal–metal oxide systems.

8.6 CARBIDES

Carbides are an important group of refractory materials whose vaporisation behaviour at high temperatures is attracting rapidly increasing interest. More experimental work has been done in the last 2 years than ever before. High-temperature condensed-phase–gas-phase equilibria for these compounds have been studied by the Knudsen effusion-loss of weight or target collection methods, and in particular by the combination of mass spectrometric and Knudsen effusion or Langmuir techniques.

Over the temperature range so far investigated carbides vaporise to form the elements and, to lesser extent, to form the carbides MC, MC_2, and MC_4, MC_2 being the predominant gaseous metal-carbon species observed. Practically all the thermodynamic data for these systems have been obtained from mass spectrometric investigations of solid–vapour equilibria. In particular, the stabilities of the carbide molecules have been obtained by measuring the enthalpy of reactions of the type:

$$M(g)+C(s) \rightarrow MC(g)$$

$$M(g)+2C(s) \rightarrow MC_2(g)$$

$$M(g)+4C(s) \rightarrow MC_4(g)$$

$$MC_2(g)+2C(s) \rightarrow MC_4(g)$$

An almost complete set of data is now available for the lanthanun carbides. The thermodynamics of several rare earth–carbon systems have been reviewed by Balducci, De Maria and Guido[134]. They reported volatility and vapour composition, heats of formation of solid dicarbides, and bond and atomisation energies of $MC_2(g)$ and $MC_4(g)$ molecules as measured and predicted from bonding correlations. A general observation which could be made is that the rare earth dicarbides of those metals with lower vapour pressure vaporise to form the metal, gaseous MC_2 molecules, and (depending on the element and the temperature range explored) MC_4 molecules as well, whereas those elements with higher vapour pressure vaporise to form monatomic metal. Decomposition pressures, heats and entropies of vaporisation and heats of formation for the solid dicarbide phase were determined for $EuC_2(s)$ [135], $YbC_2(s)$ [136], $SmC_2(s)$ [137], and $TmC_2(s)$ [138]. The values of $\Delta H^\circ_{v,298}$ found for TmC_2 [138] extend the linear relationship observed between the enthalpy of vaporisation of lanthanide and alkaline earth dicarbides and the vapour pressure of the pure metal at 1500 K [136]. No MC_2 gaseous molecules were reported for these elements. Recently[139], however, the molecule $EuC_2(g)$ was identified over the system EuC_2–C, and its dissociation energy was determined. It is believed that $TmC_2(g)$ and $SmC_2(g)$ (predicted to be stable molecules[134]), should be detectable with the mass spectrometer. However, experiments so far on the TmC_2–C and SmC_2–C systems have done no more than set limits for carbide vapour species at $TmC_2/Tm = 0.005$ [138] and $SmC_2/Sm = 0.003$ [140].

The vaporisation of the carbon rich dicarbide phases of scandium, yttrium and lanthanum[141–143] at high temperatures up to 2600 K have been studied to Kohl and Stearns. They determined the stability of the dicarbide and tetracarbide molecules, and other thermodynamic properties, by studying the pressure independent reactions as described above, assuming unit activity for C(s). For the lanthanum–carbon system they claim to have observed the unexpected LaC_3 molecule, which is the only tricarbide metal molecule as yet reported in rare earth–carbon systems. The vaporisation behaviour and phase relationships of the Y–C system were also studied using a combination of mass spectrometric and thermal analysis techniques[144]. Gas phase–condensed phase equilibrium data of various Y–C phases were combined with results from thermal analysis, x-ray and high-temperature neutron-diffraction studies to obtain the phase relationships for this system.

This very complex and interesting study has been duplicated for the system Nb–C [145]. The aim of this type of study is to obtain an understanding of the chemical bonding in defect carbide systems.

De Maria and co-workers carried out a systematic study of the vaporisation behaviour of the rare earth–carbon systems. Both the dicarbide and tetracarbide molecules from cerium[146], neodymium[105], praseodymium[147], holmium[148], dysprosium[148] and lutetium[149], and the dicarbide molecules for gadolinium[134], erbium[150], terbium[134] and europium[139] were detected. For all these molecules, dissociation and atomisation energies were reported. The same authors have discussed and used the hypothesis of Chupka *et al.*[151] that the C_2-group has the character of pseudo-oxygen in order to correlate qualitatively the stabilities of the metal oxides and the corresponding carbides. The M—O bond strength is generally higher than that of M—C_2, the average difference being within 1 eV. A comparison between the bond energies in gaseous dicarbides and monoxides is reported in Table 8.2.

Table 8.2 Comparison of the bonding energies (in eV)* of gaseous monoxides† and dicarbides [From Table 5–8 of Ref. 8, revised]

Metal	D_0°(M—O)	D_0°(M—C_2)	D_0°(M—O) − D_0°(M—C_2)
Sc	6.7	5.8_6	0.87
Y	7.0	6.7	0.3
La	8.1	6.9	1.2
Ce	8.1	7.0	1.1
Pr	7.7	6.5	1.2
Nd	7.2	6.4	0.8
Gd	7.0	6.5_5	0.4_5
Tb	7.1	6.5	0.6
Eu	5.6	5.6	0.0
Dy	6.3	5.8	0.5
Ho	6.4	5.8	0.6
Er	6.3	5.9	0.4
Lu	6.8	6.3	0.5
Th	8.3	7.3	1.0
B	8.3	6.6	1.7
Ti	6.2_5‡	5.9	0.3_5
	6.8_5	5.9	0.9_5
V	6.6	5.9	0.7
Zr	7.8	5.9	1.9§
Hf	8.0	6.9	1.1§
Cr	4.7	4.6	0.1
Si	8.3	7.2	1.1
Ge	6.8	6.4	0.4

*1 eV = 23.05 kcal = 96.44 kJ
†From Ref. 94
‡From Ref. 127
§From Ref. 213

The analogy between MC_2 and MO species was used to derive an approximate expression[8] (which appears to be a useful guide in high temperature equilibrium studies of metal–carbon systems), relating the relative concentration of an MC_2 species with respect to the metal atom at a given temperature to the dissociation energy of the corresponding monoxide and other thermodynamic data.

Recently, Filby and Ames[152, 153] obtained dissociation energies of LaC_2, PrC_2, NdC_2, GdC_2 and TbC_2 from a Knudsen cell-mass spectrometric study of the isomolecular C_2-exchange reactions of the type: $M_aC_2(g) + M_b(g) \rightarrow M_bC_2(g) + M_a(g)$, where M_a and M_b are two rare earth metals. The enthalpy of the exchange reaction gives the difference in the dissociation energies of M_aC_2 and M_bC_2: $D_0^\circ(M_bC_2) = D_0^\circ(M_aC_2)\text{-}\Delta H_{o,r}^\circ$, provided that equilibrium is actually established between the gaseous species involved in the exchange reaction. The derived dissociation energy obviously requires an accurate knowledge of the dissociation energy of the reference molecule. It is to be noted that the dissociation energies obtained through the C_2-exchange reactions for PrC_2, NdC_2 and TbC_2 (151,151,150$\pm$3 kcal mol^{-1}) are substantially in agreement with values previously[134] determined by measuring the enthalpy of the reaction: $M(g) + 2C(s) = MC_2(g)$, combined with the heat of formation at 0 K of $C_2(g)$.

Of actinide elements, the vaporisation of various phases of the plutonium–carbon system have been investigated by Knudsen effusion-mass spectrometry[154]. The decomposition pressures of the monocarbide, sesquicarbide and dicarbide phases, and the standard heat and entropy of formation of the monocarbide and the sesquicarbide were determined. Combining the free energies for formation of the sesquicarbide and the dicarbide, the free energy for the transition sesquicarbide $\leftrightarrow$ dicarbide has been obtained, leading to a transition temperature of 1662 °C, which compares very well with the value of 1660$\pm$10 °C determined independently by DTA studies. PuC_2 molecules were not observed in the gaseous phase, the limit being Pu:PuC_2 = 1000:1 at 2100 K. Considering that high concentrations of dicarbide molecules have been found for uranium and thorium[155, 156] it is possible that $PuC_2(g)$ will be observed and characterised at higher tempera-tures in the vapour phase above the carbon-rich phase with the use of a more sensitive mass spectrometer.

Several transition metal–carbon systems have been submitted to high-temperature vaporisation studies. The vaporisation rates of Ti, Zr, Hf, Nb and Ta carbides were measured in the temperature range 2800–3400 K [157]. The evaporation coefficient was also determined at various temperatures. Rates of vaporisation of ZrC have also been determined using a neutron-activation analysis method[158]. Although the atomisation energy of the TiC_2 molecule has been previously reported[159], several investigations on the titanium–carbon system have not resolved the composition of the vapours over the condensed phase. Recently[160], the vaporisation of the titanium carbide was studied by Knudsen effusion–mass spectrometry at tempera-tures up to 2800 K. The molecules TiC_2 and TiC_4 were observed, and from study of the various equilibrium reactions it was possible to determine the dissociation energies of both the dicarbide and tetracarbide molecules.

The analogy between monoxides and dicarbides which has been noted for several Group IIIA, IVA and rare-earth elements seems to hold also for the elements of Group IVB. Although some discrepancies exist as regards the value of $D_0^\circ(TiO)$, the difference $D_0^\circ(Ti—O) - D_0^\circ(Ti—C_2)$ remains within the normal range. Other transition metal dicarbides for which dissociation energies have been determined are VC_2 and CrC_2 [161]. For vanadium, the molecule VC_4 was also observed in the vapour in equilibrium with the condensed phase. It is interesting to note that neither monocarbides nor higher carbides (other than dicarbides and tetracarbides) have as yet been observed for these transition metals. The molecular ion, CrC^+, was previously reported in a spark-source mass spectrometry experiment[162], but in view of certain features of this experiment, it could be attributed to fragmentation of a higher carbide molecule, e.g. the dicarbide. The difference $D_0^\circ(V—O) - D_0^\circ(V—C_2)$ follows the usual trend of other metal monoxides and dicarbides but the reported value of $D_0^\circ(Cr—C_2)$ seems to be an exception being slightly higher than that of Cr—O.

If this is the real situation with transition metal dicarbides, the dicarbides of Nb, Ta, and Zr should be stable molecules, as the corresponding monoxides have been observed. Actually, NbC_2 gaseous molecules were detected in the vapours over the Nb—C condensed system[145], but no quantitative value for $D_0^\circ(Nb—C_2)$ has been reported as yet.

The dissociation energies of a number of gaseous monocarbides of the noble metals were determined from equilibrium studies on the vapours above the condensed phases. It is interesting to note that, apart from metal atoms, only MC(g) (and no MC_2(g)) molecules were observed for these systems. The monocarbides of this group for which the dissociation energies have been determined are PtC[163], RhC[163], RuC[164], and IrC[164]. Upper limits only could be established for AuC and PdC[163]. The trends in the dissociation energies of both the monocarbides and monoxides are similar, but here the stabilities of the carbides are higher than those of the oxides.

An interesting high temperature equilibrium study has been made on the carbon-phosphorus system[165]. By studying the reactions:

$$C_nP(g) \rightarrow nC(s) + P(g) \qquad (n = 1,2) \text{ and}$$

$$C_nP_2(g) \rightarrow nC(s) + P_2(g) \qquad (n = 1,2)$$

by Knudsen effusion mass spectrometry at temperatures up to 2500 K, it was possible to determine the atomisation energies of the molecules CP, C_2P, CP_2 and C_2P_2.

8.7 SULPHIDES, SELENIDES AND TELLURIDES

Exchange equilibrium reactions of the type B(g) + YX(g) → Y(g) + BX(g), where X = O, S and Se, were studied by Uy and Drowart[97] from which the dissociation energies of the gaseous boron monochalcogenides were obtained. The equilibrium BO(g) + YS(g) → YO(s) + BS(g) was also investigated. Equilibrium studies based on the exchange reactions Pb(g) + BiX(g) → Bi(g) + PbX(g) (where X = O, S, Se or Te) were used by the same authors to derive the dissociation energies of gaseous BiO, BiS, BiSe and BiTe[166]. The equilibrium and free-surface sublimation pressures of red mercuric sulphide were determined by the torsion method over the temperature range 500–564 K [167]. Measurements of the sublimation pressure of zinc sulphide[168] and cadmium sulphide[169], both under equilibrium conditions, and from crystalline orientated planes, were made. Equilibrium and free surface sublimation pressures of synthetic and naturally occurring samples of lead sulphide were also determined by the torsion method[170].

Vapour pressures of As_2S_2, As_2S_3, As_2Se_3 and As_2Te_3 have been measured using a quartz membrane manometer[171]. From density measurements on overheated vapours, the composition of the vapour over As_2S_3 [169] was inferred to contain monomers and dimers at temperatures below the boiling point, whereas the vapours over As_2Se_3 are appreciably dissociated under these conditions. It may be helpful to make use of spectroscopic techniques to further study the composition of these vapours. Equilibrium sublimation pressures of cadmium selenide have been determined in the temperature range 942–1041 K [172]. The free-surface sublimation pressures over the basal-plane, measured by the same authors, showed an interesting difference in value between the non-metal and the metal side, the latter being less by a factor of 0.63. The equilibrium vaporisation process of manganese telluride was also studied[173], and it was shown that it vaporises congruently according to the relation: $MnTe(s) \rightarrow Mn(g) + \frac{1}{2}(1-\alpha)Te_2(g) + \alpha Te(g)$, with α equal to 0.66. The vaporisation of titanium monosulphide was investigated by the Knudsen cell–mass spectrometric technique[174]. The reaction $TiS(s) \rightarrow Ti(g) + S(g)$ was shown to be four times more important than the process $TiS(s) \rightarrow TiS(g)$, while $TiS(s) \rightarrow Ti(g) + \frac{1}{2}S_2(g)$ was even less important. The dissociation energy D°_{298} of TiS(g), was evaluated to be 101.8 ± 1.8 kcal mol^{-1}, and earlier vapour pressure data[175] were corrected for the partial dissociation of the vapour.

Isomolecular exchange equilibrium studies involving gaseous lanthanide monosulphide species have been studied mainly by Knudsen effusion and mass spectrometric techniques. From these investigations, the dissociation energies of gaseous ScS, YS, LaS, CeS, PrS, NdS, EuS, GdS, HoS and LuS were determined[176]. Vaporisation equilibrium data were also obtained for LaS[177], EuS[178] and it appears that along the rare-earth series, the behaviour is analogous to that observed for the vaporisation of monoxides and dicarbides[176, 179, 180]. The equilibrium vapour over Sc_2S_3 was investigated by the Knudsen cell-mass spectrometric technique[181]. A congruent vaporisation process occurs, according to two simultaneous reactions:

$$Sc_2Se_3(s) \rightarrow 2ScSe(g) + Se(g)$$

$$Sc_2Se_3(s) \rightarrow 2Sc(g) + 3Se(g)$$

The dissociation energies of several rare earth monoselenides and monotellurides, determined from studies of isomolecular exchange reactions, have been reported[182].

Heterogeneous equilibria in the boron phosphide–sulphur system were investigated by Grimberg *et al.*[183]. The temperature dependence of the equilibrium constants for the reactions $4BP(s) + 4B_2S_3(g) \rightarrow 12BS(g) + P_4(g)$ and $4BP(s) + 4BS_2(g) \rightarrow 8BS(g) + P_4(g)$ was evaluated.

8.8 NITRIDES AND BORIDES

Equilibrium vaporisation studies of nitrides and borides have been the object of some interest in the last few years. The vaporisation occurs mainly by a dissociative process and gaseous equilibria involving heteronuclear species of the type MeN were investigated. The gaseous reactions $N_2 + P_2 \rightarrow 2PN$, and $2P+N_2 \rightarrow 2PN$ were studied by Gingerich[184] to derive the dissociation energy of the heteronuclear species. The molecule TiN was identified in the vapour phase over titanium nitride at temperatures up to 2250 K [185]. The equilibrium reactions $TiN(g) \rightarrow Ti(s)+N(g)$, $TiN(s) \rightarrow TiN(g)$ and $Ti(s)+0.5\,N_2(g) \rightleftharpoons TiN(g)$ were also studied[186].

The equilibrium vapour over a ZrN–Nb mixture was investigated at temperatures above 2400 K by the Knudsen cell–mass spectrometric technique[187]. The ZrN gaseous molecule was identified, and its dissociation energy determined. The gaseous thorium mononitride molecule was identified in the equilibrium vapour over a nitrogen-containing thorium–boron–phosphorus alloy[188]. In the same paper the problem of estimating the dissociation energies of the diatomic nitrides of the Group III–IV transition metals is discussed. A mass spectrometric–Knudsen effusion investigation has shown that plutonium mononitride vaporises congruently in the temperature range 1658–1976 K [189], according to the reaction $PuN_{0.98}(s) \rightarrow Pu(g)+0.49\,N_2(g)$. Partial pressures and thermodynamic properties of the species are reported.

The vaporisation process of uranium mononitride was investigated in the temperature range 1919–2230 K [190]. The vaporisation occurs incongruently by preferential loss of nitrogen and the formation of two-phase system nitrogen-saturated liquid uranium–uranium-saturated non-stoichiometric uranium mononitride. No gaseous heteronuclear species were detected.

The equilibrium reactions $AuB(g)+Au(g) \rightarrow Au_2(g)+B(g)$, $AuB(g)+Ce(g) \rightarrow CeAu(s)+B(g)$, and $AuB(g) \rightarrow Au(g)+B(g)$ were studied by applying the Knudsen cell-mass spectrometric technique to the vapour phase over a Au–BN–C mixture and over a Au–Ce–CeS BN–C mixture[191]. The molecules CeB and UB have been observed similarly, and their dissociation energies measured[192]. The study of the gaseous equilibria over a thorium-boron alloy at *c.* 2800 K was used to determine the dissociation energy and the heat of formation of the gaseous ThB molecule[193]. Dissociation energies of the borides of a number of transition metals, namely PtB[194] and RuB, RhB, PdB, IrB and AuB[46] have also been determined.

8.9 MISCELLANEOUS

Equilibrium reactions involving a variety of inorganic materials have been studied at high temperature and a review on gaseous ternary compounds of the alkali metals appeared some years ago[195].

The thermal stability of barium hexafluorosilicate was investigated and the equilibrium decomposition pressures of $BaSiF_6$ in the reaction $BaSiF_6(g) \rightarrow BaF_2(s)+SiF_4(g)$ were measured by the transpiration, mass effusion and torsional recoil methods[196]. The use of this compound as a source of high-purity silicon tetrafluoride appears to be well substantiated.

Mixed gaseous halides AlClF(g), $AlCl_2F(g)$ and $AlF_2Cl(g)$ were identified and their thermodynamic properties measured in the simultaneous reactions of AlF_3 and $AlCl_3$ with Al [197]. Condensed phase–gaseous phase equilibria of sulphates of caesium, rubidium and thallium have been reported by Cubiciotti[198, 199]. The equilibrium vapour was found to consist predominantly of M_2SO_4 molecules, where M = Cs, Rb and Tl. In the case of thallous sulphate, small amounts of gaseous Tl_2O, SO_2 and O_2 were observed, together with a barely detectable fraction of higher polymers. Thermodynamic data for the gaseous sulphate molecules were reported. A negligible contribution from polymers was found in the equilibrium vapour over thallous nitrate[200], but the alkali nitrates gave rise to an appreciable fraction of dimers[201]. Equilibrium reactions involving oxyhalides of a number of metals have been much investigated in recent years[202–205]. Metal mono-, di- and tri-oxyhalides are known both in the condensed, and in the gas phases. They are technologically important both as reagents for the purification of the respective metals[206], and as limiting agents in the chloride process for the production of pure metals[207, 208]. An excellent review on this subject, giving the relevant physical properties and structural details, has been published by Ngai and Stafford[209].

High-temperature equilibria involving mineralogical compounds are yet in the embryonic stage. The equilibrium vapour over Apollo 12 and 11 lunar samples[210, 211] has been investigated by the Knudsen cell–mass spectrometric technique. A number of equilibria involving high temperature molecules were studied and the mode of vaporisation of the specimens was characterised.

8.10 CONCLUSIONS

The knowledge of the nature and physico-chemical properties of high temperature gaseous species is of primary importance in the investigation of homogeneous and heterogeneous high temperature equilibria. Although numerous and valuable results have been obtained, the direct dissociation of molecular species to form atoms is in many cases difficult to investigate. Isomolecular exchange reactions are more accessible and have been increasingly studied during recent years. This procedure depends on the existence of thermodynamic equilibrium among the gaseous species involved and, although convenient from the experimental point of view, the results ultimately are relative to the accuracy with which the standard dissociation energies of the reference molecules are known.

A number of correlations, which have been mentioned throughout this review, between different groups of molecular species, or among isosteric and isoelectronic molecules are known[12, 134, 176, 179, 180]. They may prove to be a useful guide in studying new high temperature reactions, particularly in predicting the experimental conditions for the observation of new species[8].

The coupling of the Knudsen effusion technique with the mass spectrometer is demonstrably one of the most powerful tools for studying high temperature equilibria.

Nevertheless, it has its limitations, and high-temperature optical spectroscopy is already proving to be a useful complementary technique especially for structural determinations.

The effort made during recent years to study various groups of binary compounds has been emphasised in this review. Much research work can be anticipated on high temperature reactions involving ternary species. Among these, the oxyhalides, either as mass or energy transport agents, or as troublesome contaminants in other materials, are of special interest not only from the theoretical standpoint, but also because of possible applications in modern technology.

The study of the mode of vaporisation at high temperatures of well characterised mineralogical compounds appears to be another promising line of research for high temperature equilibrium studies. Advancements in this field will greatly help to solve important aspects of cosmochemical problems, such as the treatment of condensation phenomena from primeval nebulae generally recognised to be of primary importance in the formation of the solid bodies we now observe in the solar system[212].

References

1. Brewer, L. and Searcy, A. W. (1956). *Ann. Rev. Phys. Chem.*, **7**, 259. (Palo Alto, California: Ann. Reviews Inc.)
2. Inghram, M. G. and Drowart, J. (1960). *High Temperature Technology*, 219 (New York: McGraw-Hill)
3. Gilles, P. W. (1961). *Ann. Rev. Phys. Chem.*, **12**, 355 (Palo Alto, California: Ann. Reviews Inc.)
4. Drowart, J. and Goldfinger, P. (1962). *Ann. Rev. Phys. Chem.*, **13**, 459 (Palo Alto, California: Ann. Reviews, Inc.)
5. Thorn, R. J. (1966). *Ann. Rev. Phys. Chem.*, **17**, 83 (Palo Alto, California: Ann. Reviews Inc.)
6. Margrave, J. L. and Mamantov, G. (1967). *High Temperature Materials and Technology*, 78 (London: Wiley)
7. Porter, R. F. (1967). *High Temperature Materials and Technology*, 56 (London: Wiley)
8. De Maria, G. (1970). *Chemical and Mechanical Behavior of Inorganic Materials*, 81 (New York: Wiley-Interscience)
9. Brewer, L. (1962). *Proceedings Robert A. Welch Foundation Conference on Chemical Research: VI. Topics in Modern Inorganic Chemistry*, 47 (Houston, Texas: Welch Foundation)
10. Drowart, J. (1967). *Phase Stability in Metals and Alloys* (P. S. Rudman, J. Stringer and R. I. Jafee, editors), 305. (New York: McGraw-Hill)
11. Drowart, J. and Goldfinger, P. (1967). *Angew. Chem.*, **79**, (13), 589
12. Verhaegen, G., Stafford, F. E., Goldfinger, P. and Ackerman, M. (1962). *Trans. Faraday Soc.*, **58**, 1926
13. Balducci, G., De Maria, G. and Guido, M. (1969). *J. Chem. Phys.*, **50**, 5424
14. Gingerich, K. A. and Finkbeiner, H. C. (1971). *J. Chem. Phys.*, **54**, 2621. See also Gingerich, K. A. (1968). *Chem. Commun.*
15. Kocke, D. L. and Gingerich, K. A. (1971). *J. Phys. Chem.*, **75**, 3264
16. Gingerich, K. A., Cocke, D. L., Finkbeiner, H. C. and Seyse, R. J. (1971). paper presented at *19th Ann. Conf. on Mass Spectroscopy and Allied Topics*, 186. (Atlanta, Georgia, May 2–7
17. Pupp, C. and Gingerich, K. A. (1971). *J. Chem. Phys.*, **54**, 3380
18. Kant, A., Lin, S. S. and Strauss, B. (1968). *J. Chem. Phys.*, **49**, 1983
19. Kant, A. and Lin, S. S. (1969). *J. Chem. Phys.*, **51**, 1644
20. Lin, S. S. and Kant, A. (1969). *J. Chem. Phys.*, **73**, 2450
21. Lin, S. S., Strauss, B. and Kant, A. (1969). *J. Chem. Phys.*, **51**, 2282
22. Kant, A. and Strauss, B. (1964). *J. Chem. Phys.*, **41**, 3806
23. Mellor, A. M. (1969). *J. Chem. Phys.*, **51**, 1678
24. Drowart, J. and Goldfinger, P. (1966). *Quart. Rev. Chem. Soc.*, **20**, 545
25. Budinikas, P., Edwards, R. K. and Wahlbeck, P. G. (1968). *J. Chem. Phys.*, **48**, 2859; **48**, 2867; **48**, 2870
26. Berkowitz, J. and Chupka, W. A. (1969). *J. Chem. Phys.*, **50**, 4245
27. Meschi, D. J. and Searcy, A. W. (1969). *J. Chem. Phys.*, **51**, 5134

28. Stricker, W. and Krauss, L. (1968). *Z. Naturforsch.*, **23a,** 486
29. Wagman, D. D., Evans, W. H., Parker, V. B., Halow, I., Bailey, A. M. and Schumm, R. H. (1968). *NBS Tech. Note,* **270-3**
30. Diebeler, V. H., Walker, J. A. and McCullough, K. E. (1969). *J. Chem. Phys.*, **50,** 4592
31. Decorpo, J. J., Steiger, R. P., Franklin, J. L. and Margrave, J. L. (1970). *J. Chem. Phys.*, **53,** 936
32. Uy, O. M. and Drowart, J. (1971). *Trans. Faraday Soc.*, **67,** 1293
33. Cheetham, C. J. and Barrow, R. F. (1967). *Advances in High Temperature Chemistry* (New York: Academic Press)
34. Kant, A. (1968). *J. Chem. Phys.*, **49,** 5144
35. Smoes, S. and Drowart, J. (1968). *Chem. Commun.*, 534
36. Kant, A. (1968). *J. Chem. Phys.*, **48,** 523
37. Gingerich, K. A. and Finkbeiner, H. C. (1969). *Chem. Commun.*, 901; Gingerich, K. A. and Finkbeiner, H. C. (1969). *J. Chem. Phys.*, **52,** 2956
38. Gingerich, K. A., Cocke, D. L., Finkbeiner, H. C. and Seyse, R. J. (1971). *Proceedings 19th Ann. Conf. on Mass Spectrometry and Allied Topics,* 186. (Atlanta, Georgia: ASTM-Committee E-14)
39. Cocke, D. L. and Gingerich, K. A. (1971). *J. Chem. Phys.*, **75,** 3264
40. Gingerich, K. A. (1971). *High Temp. Sci.*, **3,** 415
41. Peters, R., Vander Awera-Mahieu, A. M. and Drowart, J. (1971). *Z. Naturforsch.*, **26a,** 327
42. Uy, O. M. and Drowart, J. (1971). *Trans. Faraday Soc.*, **67,** 1293
43. Carbonel, M., Bergman, C. and Lafitte, M. (1971). *Colloques Int. du Centre National de La Recherche Scientifique, No. 201, Thermochemie.* (Marseille)
44. Gingerich, K. A. (1969). *J. Chem. Phys.*, **50,** 5426
45. Vander Awera-Mahieu, A., McIntyre, N. S. and Drowart, J. (1969). *Chem. Phys. Letters,* **4,** 198
46. Vander Awera-Mahieu, A., Peeters, R., McIntyre, N. S. and Drowart, J. (1970). *Trans. Faraday Soc.*, **66,** 809
47. De Maria, G., Gingerich, K. A. and Piacente, V. (1968). *J. Chem. Phys.*, **49,** 4705
48. De Maria, G., Malaspina, L. and Piacente, V. (1970). *J. Chem. Phys.*, **52,** 1019
49. Gingerich, K. A. and Piacente, V. (1971). *J. Chem. Phys.*, **54,** 2498
50. Piacente, V. and Gingerich, K. A. (1971). *High Temp. Sci.*, **3,** 219
51. De Maria, G., Malaspina, L. and Piacente, V. (1972). *J. Chem. Phys.*, **56,** 1978
52. Bauer, S. M. and Porter, R. F. (1964). *Molten Salts Chemistry* (M. Blander, editor). (New York: Interscience)
53. Hildenbrand, D. L. (1967). *Advances in High Temperature Chemistry* (L. Eyring, editor) Vol. 1. (New York: A. P.)
54. Zmbov, K. F. and Margrave, J. L. (1968). *Mass Spectrometry in Inorganic Chemistry, Advances Chemistry Series 72* (J. L. Margrave, editor) (Washington, D.C.: American Chemical Society)
55. Hastie, J. W. and Margrave, J. L. (1968). *Fluorine Chem. Rev.*, **2,** 77
56. Kiser, R. W., Dillard, J. G. and Dugger, D. L. (1968). *Advances Chemistry Series No. 72,* 153 (Washington, D.C.: American Chemical Society)
57. Berkowitz, J. (1971). *Advances in High Temperature Chemistry,* Vol. 3. (L. Eyring, editor) (New York: A. P.)
58. Ehlert, T. C. (1969). *J. Phys. Chem.*, **73,** 949
59. Snelson, A. (1970). *High Temp. Sci.*, **2,** 70
60. Zmbov, K. F., Uy, O. M. and Margrave, J. L. (1968). *J. Amer. Chem. Soc.*, **90,** 5090
61. Timms, P. L., Kent, R. A., Ehlert, T. C. and Margrave, J. L. (1965). *J. Amer. Chem. Soc.*, **87,** 2824
62. Thompson, J. C. and Margrave, J. L. (1967). *Science,* **155,** 669
63. Zmbov, K. F., Hastie, J. W., Hauge, R. H. and Margrave, J. L. (1968). *Inorg. Chem.*, **9,** 608
64. Hildenbrand, D. L. (1968). *J. Chem. Phys.*, **48,** 2457
65. Hitchingham, W. C. and Kana'An, A. S. (1969). *High Temp. Sci.*, **1,** 268
66. Cubiciotti, D. (1969). *High Temp. Sci.*, **1,** 268
67. Keneshea, F. J. and Cubiciotti, D. (1965). *J. Phys. Chem.*, **69,** 3910
68. Cubiciotti, D. (1970). *High Temp. Sci.*, **2,** 65
69. Krasnov, K. S. (1969). *Izv. Vyssh. Ucheb. Zaved. Khim. Tekhnol.*, **12,** 578
70. Skinner, H. B. and Searcy, A. W. (1971). *J. Phys. Chem.*, **75,** 108

71. Hastie, J. W., Hauge, R. H. and Margrave, J. L. (1969). *High Temp. Sci.*, **1**, 76
72. Calder, V., Mann, D. E., Seshadri, K. S., Allavena, M. and White, D. (1969). *J. Chem. Phys.*, **51**, 2093
73. Wesley, R. D. and De Kock, C. W. (1971). *J. Chem. Phys.*, **55**, 3866
74. Ko, H. C., Greenbaum, M. A., Farber, M. and Selph, C. C. (1967). *J. Phys. Chem.*, **71**, 254
75. Hildenbrand, D. L. and Theard, L. P. (1969). *J. Chem. Phys.*, **50**, 5350
76. Hildenbrand, D. L. and Knight, D. T. (1969). *J. Chem. Phys.*, **51**, 1260
77. Feather, D. H. and Searcy, A. W. (1971). *High Temp. Sci.*, **3**, 155
78. Uy, O. M., Muenow, D. W. and Margrave, J. L. (1969). *Trans. Faraday Soc.*, **65**, 1296
79. Guido, M., Balducci, G., Gigli, G. and Spoliti, M. (1971). *J. Chem. Phys.*, **55**, 4566
80. McCreary, J. R., Rassoul, S. A. and Thorn, R. J. (1969). *High Temp. Sci.*, **1**, 412
81. Kana'an, A. S., McCreary, J. R., Peterson, D. E. and Thorn, R. J. (1969). *High Temp. Sci.*, **1**, 222
82. Kulkarni, M. P. and Dadape, V. V. (1971). *High Temp. Sci.*, **3**, 277
83. Seshagiri Rao, B. and Dadape, V. V. (1971). *High Temp. Sci.*, **3**, 1
84. Johnson, J. W., Silva, W. J. and Cubiciotti, D. (1970). *High Temp. Sci.*, **2**, 20
85. Keneshea, F. J. and Cubicotti, D. (1969). *J. Phys. Chem.*, **73**, 3054
86. Hastie, J. W., Hauge, R. H. and Margrave, J. L. (1971). *High Temp. Sci.*, **3**, 56
87. Hastie, J. W., Hauge, R. H. and Margrave, J. L. (1971). *High Temp. Sci.*, **3**, 257
88. Jacox, M. E. and Milligan, D. E. (1969). *J. Chem. Phys.*, **51**, 4143
89. Karpenko, N. V. (1969). *Vestn. Leningrad. Univ., Fiz. Khim.*, **2**, 77
90. Karpenko, N. V. (1969). *Vestn. Leningrad. Univ., Fiz. Khim.*, **1**, 114
91. Tsirelnikov, V. I. (1969). *J. Less Common Metals*, **19**, 287
92. Krasnov. K. S. (1969). *Izv. Vyssh. Ucheb. Zaved. Khim. Khim. Teknol.*, **12**, 578
93. Schofield, K. (1967). *Chem. Rev.*, **67**, 707
94. Brewer, L. and Rosenblatt, G. M. (1969). *Advances in High Temperature Chemistry* (New York: A. P.), Vol. 2
95. Hildebrand, D. L. and Murad, E. (1969). *J. Chem. Phys.*, **51**, 807
96. Gaydon, A. G. (1968). *Dissociation Energies of Diatomic Molecules*, 3rd edn. (London: Chapman and Hall)
97. Uy, O. M. and Drowart, J. (1970). *High Temp. Sci.*, **2**, 293
98. Hildenbrand, D. L. and Murad, E. (1970). *J. Chem. Phys.*, **53**, 3403
99. Ackermann, R. J., Rauh, E. G. and Chandrasekharaiah, M. S. (1969). *J. Phys. Chem.*, **73**, 763
100. Smith, P. K. and Peterson, D. E. (1970). *J. Chem. Phys.*, **52**, 4963
101. Turkotte, R. P., Chikalla, T. D. and Eyring, L. (1971). *J. Inorg. Nucl. Chem.*, **33**, 3749
102. Haschke, J. M. and Eick, H. A. (1969). *J. Phys. Chem.*, **73**, 374
103. Haschke, J. M. and Eick, H. A. (1968). *J. Phys. Chem.*, **72**, 1697
104. Staley, H. G. and Norman, J. H. (1969). *J. Mass Spectrometry and Ion Phys.*, **2**, 35
105. Balducci, G., Capalbi, A., De Maria, G. and Guido, M. (1968). *J. Chem. Phys.*, **48**, 5275
106. Ackermann, R. J. and Rauh, E. G. (1971). *J. Chem. Thermodynamics*, **3**, 445
107. Ackermann, R. J. and Rauh, E. G. (1971). *J. Chem. Thermodynamics*, **3**, 609
108. Norman, J. H., Stanley, N. G. and Bell, W. E. (1968). *Mass Spectrometry in Inorganic Chemistry, Advances in Chemistry Series*, No. 72, p. 101 (Washington, D.C.: American Chemical Society)
109. Cubiciotti, D. (1970). *High Temp. Sci.*, **2**, 213
110. McDonald, J. D. and Margrave, J. L. (1968). *J. Inorg. Nucl. Chem.*, **30**, 665
111. Washburn, C. A. (1969). *Ph.D. Thesis*, UCRL-18685 (University of California: Berkeley, Calif.)
112. Battles, J. E., Gundersen, G. E. and Edwards, R. K. (1968). *J. Phys. Chem.*, **72**, 3963
113. Semenov, G. A. and Shalkova, E. K. (1969). *Vestn. Leningrad Univ., Fiz. Khim.*, **3**, 111
114. *Issled Protsessov Met. Tsvet. Redk. Metal.* (1969). *30*-4, Moscow: Izd. Nauk.)
115. Nielson, L. A., Nikolaev, R. K., Vasileuskaya, I. I. and Vasileva, A. G. (1969). *Zh. Neorg. Khim.*, **14**, 1136
116. Skinner, H. B. (1970). *Ph.D. Thesis*, UCRL-19645 (University of California: Berkeley, Calif.)
117. Kazenas, E. K., Chizhikov, D. M. and Tsvetkov, Y. V. (1969). *Izv. Akad. Nauk. SSSR. Metal.*, **2**, 60

118. Kazenas, E. K., Chishikov, D. M. and Tsvetkov, Y. V. (1969). *Issled. Protsessov. Met. Redk. Metal.*, 28 (Moscow: Izv. Nauka)
119. Chizhikov, D. M., Kazenas, E. K. and Tsvetkov, Y. V. (1969). *Izv. Akad. Nauk. SSSR. Metal.*, **5,** 57
120. Kazenas, E. K., Chishikov, D. M. and Tsvetkov, Y. V. (1969). *Issled. Protsessov. met. Tsvet. Redk. Metal.*, 19–27. (Moscow: Izd. Nauka.)
121. Charlu, T. V. and Kleppa, O. J. (1971). *J. Chem. Thermodynamics*, **3,** 697
122. Balducci, G., De Maria, G., Guido, M. and Piacente, V. (1971). *J. Chem. Phys.*, **55,** 2596
123. Waalbeck, P. G. and Gilles, P. W. (1967). *J. Chem. Phys.*, **46,** 2465
124. Drowart, J., Coppens, P. and Smoes, S. (1969). *J. Chem. Phys.*, **50,** 1046
125. Gilles, P. W., Hampson, P. J. and Wahlbeck, P. G. (1969). *J. Chem. Phys.*, **50,** 1048
126. Hampson, P. J. and Gilles, P. W. (1971). *J. Chem. Phys.*, **55,** 3712
127. Balducci, G., De Maria, G., Guido, M. and Piacente, V. (1972). *J. Chem. Phys.*, **56,** 3422
128. D'Yachkova, N. N., Vigdorovich, E. N., Ustyugov, G. P. and Kudryatsev, A. P. (1969). *Izv. Akad. Nauk. SSSR., Neorg. Mater.*, **5,** 2219
129. Sonin, V. I., Novikov, G. I. and Polyachenok, O. G. (1969). *Zh. Fiz. Khim.*, **43,** 2980
130. Ficalora, P. J., Thompson, I. C. and Margrave, J. L. (1969). *J. Inorg. Nucl. Chem.*, **31,** 3771
131. Cesaro, S. N., Spoliti, M., Hinchliffe, A. J. and Ogden, J. S. (1971). *J. Chem. Phys.*, **55,** 5834
132. Muenow, D. W., Uy, O. M. and Margrave, J. L. (1970). *J. Inorg. Nucl. Chem.*, **32,** 3459
133. Kohl., F. J. and Stearns, C. A. (1971). *NASA Rep.* Cesara, S.N. TN D–6318
134. Balducci, G., De Maria, G. and Guido, M. (1970). *"Proceedings 1st Int. Conference on Calorimetry and Thermodynamics"*, 415. (Warsaw: Polish Scientific Publishers)
135. Faircloth, R. L., Flowers, R. A. and Pummery, F. C. W. (1968). *J. Inorg. Nucl. Chem.*, **30,** 499
136. Haschke, J. M. and Eick, H. A. (1968). *J. Phys. Chem.*, **72,** 1697
137. Stout, N. D., Hoenig, G. L. and Nordine, P. C. (1969). *J. Amer. Ceram. Soc.*, **52,** 145
138. Seiver, R. L. and Eick, H. A. (1971). *High Temp. Sci.*, **3,** 292
139. Balducci, G., De Maria, G. and Guido, M. (1972). *J. Chem. Phys.*, **56,** 1431
140. Avery, D. F., Cuthbert, J. and Silk, C. (1967). *Brit. J. Appl. Phys.*, **18,** 1133
141. Kohl, F. J. and Stearns, C. A. (1971). *J. Chem. Phys.*, **54,** 1414
142. Kohl, F. J. and Stearns, C. A. (1971). *J. Chem. Phys.*, **52,** 6310
143. Stearns, C. A. and Kohl, F. J. (1971). *J. Chem. Phys.*, **54,** 5180
144. Storms, E. K. (1971). *High Temp. Sci.*, **3,** 99
145. Storms, E., Calkin, B. and Yencha, A. (1969). *High Temp. Sci.*, **1,** 430
146. Balducci, G., Capalbi, A., De Maria, G. and Guido, M. (1969). *J. Chem. Phys.*, **50,** 1969
147. Balducci, G., Capalbi, A., De Maria, G. and Guido, M. (1965). *J. Chem. Phys.*, **43,** 2136; see also Ref. (134)
148. Balducci, G., De Maria, G., Capalbi, A. and Guido, M. (1969). *J. Chem. Phys.*, **51,** 2871
149. Guido, M., Balducci, G., De Maria, G. (1972). *J. Chem. Phys.*, in press
150. Balducci, G., De Maria, G. and Guido, M. (1969). *J. Chem. Phys.*, **51,** 2876
151. Chupka, W., Berkowitz, J., Giese, C. F. and Inghram, M. G. (1958). *J. Phys. Chem.*, **62,** 611
152. Filby, E. E. and Ames, L. L. (1971). *High Temp. Sci.*, **3,** 41
153. Filby, E. E. and Ames, L. L. (1971). *J. Phys. Chem.*, **75,** 847
154. Kent, R. A. (1970). *Recent Developments in Mass Spectroscopy, Proc. Int. Conf. on Mass Spectroscopy*, 1124. (University Park Press: Tokio)
155. Norman, J. H. and Winchell, P. (1964). *J. Phys. Chem.*, **68,** 3802
156. Jackson, D. D., Barton, G. W., Jr., Krikorian, O. H. and Newbury, R. S. (1962). *Thermodynamics of Nuclear Materials*, 529. (IAEA: Vienna)
157. Fesenko, V. V. and Bolgar, A. S. (1969). *Teplofiz. Vys Temp.*, **7,** 244
158. Karnadkhova, N. M., Karazhanova, G. I., Luknitskii, V. A., Maslov, I. I., Terent'eva, Z. P. (1969). *Metody Primen Neitrowaktiv Anal.*, 167. (Riga: USSR, Izd. "Zinatne")
159. Drowart, J., Pattoret, A. and Smoes, S. (1967). *Proc. Brit. Ceram. Soc.*, **8,** 67
160. Stearns, C. A. and Kohl, F. J. (1970). *High Temp. Sci.*, **3,** 274
161. Kohl, F. J. and Stearns, C. A. (1970). *J. Phys. Chem.*, **74,** 2714
162. Doan, A. S. (1966). NASA Report TM X-52216

163. Vander Awera-Mahieu, A. and Drowart, J. (1967). *J. Phys. Letters,* **1,** 311
164. McIntyre, N. S., Vander Awera Mahieu, A. and Drowart, J. (1968). *Trans. Faraday Soc.,* **64,** 3006
165. Smoes, S., Myers, C. E. and Drowart, J. (1971). *Chem. Phys. Letters,* **8,** 10
166. Uy, O. M. and Drowart, J. (1969). *Trans. Faraday Soc.,* **65,** 3221
167. Mitchell, M. J. and Munir, Z. A. (1970). *High Temp. Sci.,* **2,** 265
168. Munir, Z. A. and Mitchell, M. J. (1969). *High Temp. Sci.,* **1,** 381
169. Munir, Z. A. (1970). *High Temp. Sci.,* **2,** 58
170. Hansen, E. E. and Munir, Z. A. (1970). *High Temp. Sci.,* **2,** 169
171. Kuadze, B. M., Ustyugov, G. P., Kudryavtsev, A. A. (1968). *Tr. Mosk. Khim. Tekhnol. Inst.,* **58,** 27, 29
172. Seacrist, L. and Munir, Z. A. (1971). *High Temp. Sci.,* **3,** 340
173. Wiedemeier, H. and Sadeek, H. (1970). *High Temp. Sci.,* **2,** 252
174. Edwards, J. G., Franzen, H. F. and Gilles, P. W. (1971). *J. Chem. Phys.,* **54,** 545
175. Franzen, H. F. and Gilles, P. W. (1965). *J. Chem. Phys.,* **42,** 1033
176. Smoes, S., Coppens, P., Bergman, C. and Drowart, J. (1969). *Trans. Faraday Soc.,* **65,** 682
177. Cater, E. D. and Steiger, R. P. (1968). *J. Chem. Phys.,* **72,** 2231
178. Hariharan, A. V. and Eick, H. A. (1971). *High Temp. Sci.,* **3,** 123
179. Cater, E. D., Holler, B. A. and Fries, J. A. (1967). Univ. of Iowa Report, No. 1182-15
180. Coppens, P., Smoes, S. and Drowart, J. (1967). *Trans. Faraday Soc.,* **63,** 2140
181. Leary, H. J., Jr. and Wahlbeck, P. G. (1969). *High Temp. Sci.,* **1,** 277
182. Bergman, C., Coppens, P., Drowart, J. and Smoes, S. (1970). *Trans. Faraday Soc.,* **66,** 800
183. Grinberg, Ya. Kh., Zhukov, E. G., Koryazhkin, V. A. and Medvedeva, Z. S. (1969). *Zh. Neorg. Khim.,* **14,** 2583
184. Gingerich, K. A. (1969). *J. Phys. Chem.,* **73,** 2734
185. Stearns, C. A. and Kohl, F. J. (1970). *High Temp. Sci.,* **2,** 146
186. Ryklis, E. A., Bolgar, A. S. and Fesenko, V. V. (1969). *Porosh. Met.,* **9,** 62
187. Gingerich, K. A. (1968). *J. Chem. Phys.,* **49,** 14
188. Gingerich, K. A. (1968). *J. Chem. Phys.,* **49,** 19
189. Kent, R. A. and Leary, J. A. (1969). *High Temp. Sci.,* **1,** 176
190. Gingerich, K. A. (1969). *J. Chem. Phys.,* **51,** 4433
191. Gingerich, K. A. (1971). *J. Chem. Phys.,* **54,** 2646
192. Gingerich, K. A. (1970). *J. Chem. Phys.,* **53,** 746
193. Gingerich, K. A. (1969). *High Temp. Sci.,* **1,** 258
194. McIntyre, N. S., Vander Awera-Mahieu, A. and Drowart, J. (1968). *Trans. Faraday Soc.,* **64,** 3006
195. Büchler, A. and Berkowitz-Mattuck, J. B. (1967). *Advances in High Temperature Chemistry* (New York: Academic Press), Vol. 1
196. Canka, Ya, K., Pretzer, W. R., Livingstron, W. A. and Kana'an, A. S. (1970). *High Temp. Sci.,* **2,** 322
197. Farber, M. and Harris, S. P. (1971). *High Temp. Sci.,* **3,** 231
198. Cubiciotti, D. (1971). *High Temp. Sci.,* **3,** 349
199. Cubiciotti, D. (1970). *High Temp. Sci.,* **2,** 389
200. Cubiciotti, D. (1970). *High Temp. Sci.,* **2,** 131
201. Büchler, A. and Stauffer, J. L. (1966). *J. Phys. Chem.,* **70,** 4092
202. Zmbov, K. F., Uy, O. M. and Margrave, J. L. (1969). *J. Phys. Chem.,* **73,** 3008
203. Gupta, S. K. (1969). *J. Phys. Chem.,* **73,** 4086
204. Srivastava, R. O. and Farber, M. (1971). *J. Phys. Chem.,* **75,** 1760
205. Gupta, S. K. (1971). *J. Phys. Chem.,* **75,** 112
206. Boesiger, D. D. and Stevenson, F. D. (1970). *Met. Trans.,* **1,** 1859
207. Nisel'son, L. A., Nikolaev, R. K. and Orshanskaya, Z. N. (1967). *Izv. Akad. Nauk SSSR, Metal,* **20,** 209
208. Sytnik, A. A., Furman, A. A. and Kulyasova, A. S. (1966). *Zh. Neorgan. Khim.,* **11,** 1004
209. Ngai, L. H. and Stafford, F. E. (1971). *Advances in High Temp. Chemistry,* Vol. 3, 213. (Academic Press: New York and London)
210. De Maria, G., Balducci, G., Guido, M. and Piacente, V. (1971). *Proceedings of the Second Lunar Science Conference,* **2,** 1367. (Cambridge, Mass.: M. I. T. Press)
211. De Maria, G., Balducci, G., Guido, M. and Piacente, V., to be published
212. De Maria, G. and Piacente, V. (1969). *Atti Accad. Naz. Lincei.,* **47,** 525
213. Stearns, C. A. and Kohl, F. J. (1971). *NASA Rep.* TM-X-67844

9
Thermodynamics of Crystals

E. F. WESTRUM, Jr.
University of Michigan

9.1 INTRODUCTION 232

9.2 MOLECULAR CRYSTALS 233
 9.2.1 *Molecular freedom: transitions* 234
  9.2.1.1 *Succinonitrile* 234
  9.2.1.2 *Methanol* 235
  9.2.1.3 *Bis(benzene)chromium* 236
 9.2.2 *Internal rotation* 237
 9.2.3 *Plastic crystals* 239
  9.2.3.1 *Bicyclo compounds* 239
  9.2.3.2 *Trimethylacetonitrile* 241

9.3 IONIC CRYSTALS 241
 9.3.1 *Ionic freedom* 241
 9.3.2 *Crystal field levels* 243
  9.3.2.1 *Lanthanide sesquioxides* 243
  9.3.3.2 *Lanthanide hexaborides* 245

9.4 COVALENT CRYSTALS 245

9.5 THERMOCHEMISTRY OF CRYSTALS 247
 9.5.1 *Enthalpy of formation* 247
 9.5.2 *Strain energies* 248

9.6 ELECTRONIC AND MAGNETIC CONTRIBUTIONS TO THERMAL QUANTITIES 248
 9.6.1 *Electronic contributions: Schottky anomalies* 248
 9.6.2 *Para- and ferro-electricity* 249
 9.6.3 *Semi-, super-, and metallic conductivity* 249
 9.6.4 *Magnons, spin waves, and magnetic transformations* 249

9.7 OTHER STABILITY—TRANSITIONAL PHENOMENA 250
 9.7.1 *Ordering transitions* 250
 9.7.2 *Phase transitions* 250
 9.7.3 *Energetics of phase behaviour* 251
 9.7.4 *Solid solutions* 251

9.8 CERTAIN THERMOPHYSICAL CONTRIBUTIONS 253
9.8.1 *Anharmonicity* 253
9.8.2 *Non-stoichiometry and defect structures* 253
9.8.3 *Evaluation of* θ_{Debye} *from spectral data* 253
9.8.4 *Constraints of pressure and volume* 254
9.8.5 *Miscellaneous* 255

9.1 INTRODUCTION

Even a cursory glance at the thermodynamics literature during the last few years will convince the reader of a remarkable change. Although the number of 'professional' thermodynamicists interested in the solid state is limited to a few hundred individuals the world over, the bulk of solid-state thermodynamic information now comes from the laboratories of a variety of solid-state scientists whose prime interest may be in some other field; but who, from time to time, generate thermodynamic data. At the same time, the concerns of the thermodynamicists have of necessity incorporated the recent advances and such adjuvant techniques as nuclear magnetic resonance, cold neutron scattering, optical spectroscopy, as well as magnetic, electrical, and other related techniques in supplement to appropriate calorimetric or equilibrium thermal and thermochemical measurements.

This review stresses particularly the significant developments in the biennium, giving somewhat less attention to the important advances in the preceding period of years which are often necessary to establish a point of departure for an advance.

To reduce the extent of the several thousand publications relevant to the title topic to more manageable proportions, this review emphasises the special interests of its author and by fairly arbitrary decisions minimises the coverage of recent theoretical developments, correlations, and calculations, and excludes any attempt at being truly encyclopaedic in scope. It stresses those aspects which represent especially significant or unique contributions or initiated new trends in the application of thermochemical and thermophysical techniques to the study of the crystalline state.

The pre-eminence of the heat capacity as a revealing parameter of energetics at the molecular and crystalline level is enhanced as the temperature drops below the thermal chaos. At higher temperatures the entropy – for example, of an order–disorder transition – becomes a significant and useful criterion, and at the modern higher temperature region it may even become the dominant factor in the Gibbs energy equation since its contribution is multiplied by the absolute temperature. In general, non-calorimetric approaches have been delineated only so far as necessary in order to put in context a particular development since these topics are generally treated in greater detail elsewhere in this series. As already indicated, the interactions, interfaces, and interests of the solid-state thermodynamicists are inextricably woven into the entire fabric of modern science. Brief mention of the thermodynamics of the amorphous or vitreous phase has been incorporated, as have the transformations between the crystal and other phases (e.g. melting and sublimation) and attention has been given to the important area of the thermochemistry of crystals (i.e. the determination of enthalpy of formation, etc.). Other equally important aspects, e.g. high-temperature galvanic cells, lattice dynamics, have been slighted.

The 150 references are far too few to constitute a full bibliography and those selected were chosen in such a way that at least one is included for each major topic which cites important contemporary and/or earlier papers. The reader wishing to gain access to the increment of published (or unpublished) data during the biennium on a particular substance, system, or area of thermodynamics is referred to the *Bulletin of Thermodynamics and Thermochemistry*[1] which by means of compound-property indexes, extensive bibliographies, lists of books and conference titles, as well as abstracts of unpublished results from laboratories around the world, provides an annual summary of thermodynamic progress.

Although full-time critical evaluation of data and the systematic compilation of thermodynamic tables is now approaching international stature, it is interesting that the two recent major compilations of thermophysical and thermochemical properties of organic substances are the result of individual enterprise, and are essentially the only extensive endeavour in this area since Parks and Huffman's pioneering monograph, *The Free Energies of Organic Compounds*, which appeared in 1932 and has been out of print since 1938. Stull, Westrum, and Sinke's monograph entitled *The Chemical Thermodynamics of Organic Compounds*[2] provides tabulated thermodynamic functions for 916 compounds from 298 to 1000 K together with enthalpies of formation for *c.* 4400 species on which adequate thermophysical data were not available for complete tables. In additional chapters it endeavours to educate the industrial chemist on *how* to apply thermodynamics in the laboratory and plant. It has been supplemented by Cox and Pilcher's *Thermochemistry of Organic and Organometallic Compounds*[3] which details the enthalpies of formation of some 3000 substances with estimates of error, details of the method of measurement and enthalpies of vaporisation (sublimation) whenever available. Both books provide examples of the determination, use, and application of chemical thermodynamic data, and the use of modern bond-energy schemes, as well as modern techniques for property measurement.

The heat capacity of solid nitrogen[4] provides considerable information on the rotational motion of the molecules; so-called orientation defects contribute substantially just below the transition temperature and nearly free rotation occurs in the higher temperature phase.

9.2 MOLECULAR CRYSTALS

Despite the undisputed importance of organic chemistry in the physical and biologico-medical sciences, until very recently few attempts have been made to summarise the thermodynamics of the organic solid state. The wide variety of investigators who have been attracted to the study of molecular crystals, and the diversity and complexity of the methods used have reduced the communication between investigators and have given rise to a claim that the area is 'weakly coupled' to other areas. An interesting attempt to survey present results, views, and speculations in this area of lattice dynamics, phonons, electronic transitions, and exciton coupling to radiation and phonon fields has been presented in the symposium proceedings entitled *Excitons, Magnons and Phonons in Molecular Crystals*[5] as well as in the textual treatment of *Excitons in Molecular Crystals*[6] by Craig and Walmsley. Bondi's recent endeavour to summarise thermal equilibrium properties elastic moduli, thermal conductivity, and molecular diffusion related properties in generalised coordinates is entitled *Molecular Crystals, Liquids, and Glasses*[7]. It involves the elimination of the non-specific effects of molecular structure on physical properties, emphasises the few instances of highly specific effects and of possible errors in experimental measurements, and, most importantly, areas of serious ignorance.

9.2.1 Molecular freedom: transitions

Abrupt changes of the molecular freedom of a molecule in a crystalline solid are among the most fascinating and still least understood of physical problems. Such transitions are often accompanied by dramatic changes in physical characteristics of the substance disapportionate to the subtle structural metamorphoses which occur at the molecular level. Although many experimental techniques provide insight into the nature and the mechanism of molecular freedom in crystalline substances, calorimetric studies of heat capacity and related thermal phenomena have contributed much to the understanding of molecular motion over the past two decades. Precise measurements of heat capacity and enthalpies of transition are now readily made by adiabatic calorimetry over the range 0.3–600 K and with only somewhat increased difficulty to temperatures as high as 1500 K.

9.2.1.1 Succinonitrile

An example of the use of one approach to interpretation of the entropy of transition is that of succinonitrile, $[NC(CH_2)_2CN]$[8]. x-Ray diffraction analysis indicates Crystal II to be completely ordered monoclinic (consisting only of *gauche* molecular forms), whereas infrared data shows Crystal I to be a highly disordered, body-centred cubic phase (composed of a temperature-dependent, equilibrium mixture of *gauche* and *trans* geometrical isomers). In solid–solid transitions in which order–disorder effects predominate, the entropy change may be approximated by $\Delta S_t = R \ln Q_h/Q_l$ in which Q_h/Q_l is the ratio of the number of orientational states statistically occupied in the higher and lower phases. The disordering at the 223.3 K transition arises from four terms: (a) alignment of the principal axis of the molecule with the cube diagonals ($R \ln 4 = 2.755$ e.u.), (b) rigid rotation of the molecule about this axis ($R \ln 2 = 1.377$), (c) entropy of mixing *gauche* and *trans* isomers ($-R[X_t \ln X_t + X_g \ln X_g - \ln 2)] = 2.08$), (d) volume change of the lattice on transition ($R \ln [V_h/V_l]$; 'free' volume $= 0.18$). The sum of these four calculated contributions, $\Delta S_t = 6.39 \pm 0.09$ e.u., is in excellent accord with the experimental value, 6.35 ± 0.03 e.u. The low entropy of melting, 2.68 e.u., at T_m, 331.3 K, indicates that the Crystal I phase of succinonitrile is plastically crystalline. The agreement obtained between the third-law entropy of the gas, 79.04 ± 0.10 e.u. at 298.15 K, and the corresponding value obtained from the spectroscopic data, 79.09 ± 0.10 e.u., is ample indication of the absence of residual disorder in the structure in Crystal II and justifies the assumptions implicit in calculating the contributions to the entropy.

9.2.1.2 Methanol

A particularly interesting hydrogen-bonded substance for which data are presented in Figure 9.1 in the transition-melting region is methanol. Here the striking drop in heat capacity at the 155 K first-order transition (ΔS_t = 0.97 e.u.) is followed by melting at 175.6 K (ΔS_m = 4.4 e.u.)[9]. Timmermans empirical definition of a plastic crystal as a substance which has an entropy of melting less than $5 \text{ cal mol}^{-1} \text{ K}^{-1}$ would occasion classification of

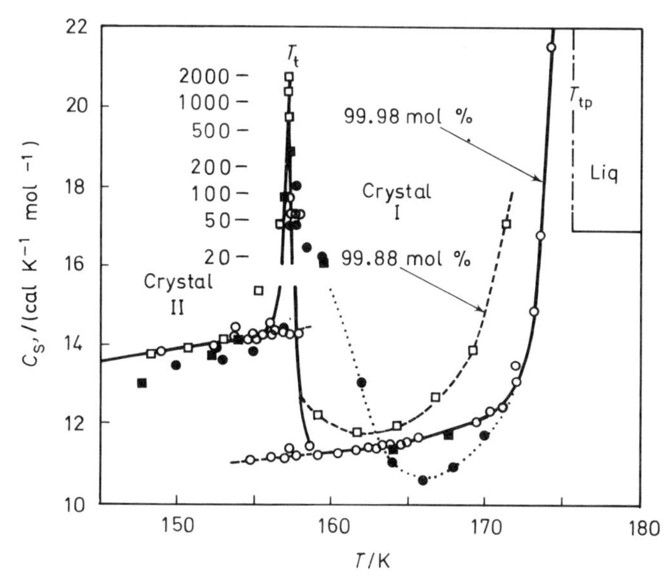

Figure 9.1 The heat capacity of methanol in the fusion region. Data on a high purity (o) and less pure (□) sample of Carlson and Westrum[8]. Those from earlier studies are indicated by ■ and ● (From Carlson and Westrum[9], by courtesy of the American Institute of Physics)

methanol as a plastic crystal. Other characteristic macroscopic properties of typical plastic crystals (plasticity, relatively low entropic transition into the plastically crystalline state, and more marked change in physical properties on transition than on melting) occur as a consequence of globularity of the molecule. However (as noted previously), persistence of hydrogen bonding in the liquid and gaseous phases causes liquid methanol to have unusual properties for a molecule of its small mass, e.g. a relatively low vapour pressure and an extended liquid range (163 K).

However, the abnormally small entropy of the liquid phase – a consequence of the ordering resulting from hydrogen bonding – is the prime cause for the small entropy of melting rather than the supposed plastically crystalline nature of Crystal I. This is confirmed by examination of the data in Table 9.1

Table 9.1 Thermodynamic properties of selected monosubstituted methanes

Compound	*Mole wt*	T_t/K	ΔS_t/ (cal mol^{-1} K^{-1})	T_m/K	ΔS_m/ (cal mol^{-1} K^{-1})	$S°$ (298.15 K)/ (cal mol^{-1} K^{-1}) *liq.*	$S°$ (298.15 K)/ (cal mol^{-1} K^{-1}) *ideal gas*	T_b*
Methylamine†	31.06	101.5	0.230	179.70	8.157	35.90	57.73	266.5
Methanol	32.04	157.3	0.966	175.59	4.377	30.40	57.24	337.65
Methane-thiol‡	48.10	137.6	0.350	150.16	9.399	—	60.86	280.6
Chloromethane§	50.49			175.44	8.760	36.74	55.94	248.78
Nitromethane‖	61.01			244.73	9.476	41.05	65.73	374
Bromomethane¶	94.95	173.8	0.653	179.44	7.964	—	58.61	276.56

*Ref. 10; †Ref. 11; ‡Ref. 12; §Ref. 13; ‖Ref. 14; ¶Ref. 15; **Ref. 16; ‡‡Ref. 17

comparing the behaviour of methanol with that of other small molecules. Hence, the small entropy of melting ($\Delta S_m = 4.377$ cal mol^{-1} K^{-1}) is a consequence of intermolecular structures (ordering in the liquid) rather than of independent molecular behaviour (reorientational–rotation) in the Crystal I phase.

The small entropy of transition ($\Delta S_t = 0.966$ cal mol^{-1} K^{-1}) and the decrease in heat capacity across the Crystal II $\longrightarrow$ Crystal I transition (similar to that in dimethylacetylene) is atypical for plastic crystals. Although the thermophysics of the methyl alcohol system have now been well defined under saturation pressure, the striking drop (3 cal mol^{-1} K^{-1}) in heat capacity at the transition cannot be accounted for in terms of molecular freedom. Hence, data on the isothermal compressibility and the thermal expansion for conversion of the heat capacity to constant volume as well as reconciliation of the thermal, x-ray, dilatometric, dielectric, and nuclear magnetic resonance data on the strongly hydrogen bonded Crystal I phase are clearly desiderata. It is to be further noted that Figure 9.1 reveals the crucial significance of high purity substances in thermal investigations!

9.2.1.3 Bis(benzene)chromium

The π-complexes of transition metals with aromatic molecules are sandwich compounds in which the aromatic ring may rotate about the ring–metal bond. A thermal investigation[18] was undertaken to ascertain the magnitude of the energy barrier to the rotation of the rings as well as evidence for distorted ring structure in this compound. The existence of ring distortion is of thermodynamic interest for it implies (a) that the distorted molecules might fit into the crystal lattice into distinct orientations related to one or another by the 60 degrees rotation about the threefold axis of the molecule, and (b) that the resulting crystal might be ordered or disordered with respect to these orientations. Such a disordered phase would have $R \ln 2$ more entropy than the ordered one. The spectroscopic entropy calculated for bis(benzene)-chromium on the assumption that benzene rings are not disordered (i.e. that the symmetry number is 12) agrees with the third-law value near 350 K to within the resultant standard deviation of the two entropy values[18]. Inclusion of a small barrier (i.e. < 1 kcal mol^{-1} which cannot be excluded by the magnitude of the precision index) would improve the fit of the spectroscopic and third-law entropies. If, however, the molecule were to be trigonally distorted as postulated by Jellinek[19] on the basis of x-ray diffractional analysis, it would require a spectroscopic entropy greater by $R \ln 2$. Since evidence for a phase transition was found, the phase existing at 5 K is that which is stable at 350 K. If the authors assume the molecule to be distorted and that the heat capacity measurements were in fact on the disordered phase, then the absence of a transition shows that the disorder was 'frozen' into their sample, and that $R \ln 2$ zero point entropy should be added to the third-law entropy. In this case the third-law and spectroscopic entropies (for the disordered molecule) would be in agreement. If, however, they assume that the molecule was distorted and if measurements were made on the ordered phase a discrepancy of $R \ln 2$ would now exist between the spectroscopic and third-law values. The thermodynamic evidence, therefore, favours a symmetric molecule in which the benzene rings are not disordered. This conclusion agrees with x-ray evidence on related molecules.

9.2.2 Internal rotation

An unusual transformation (depicted in Figure 9.2) takes place in 2-methyl-thiophene[20]. Although no evidence whatever for isothermal liberation of transformation energy was observed, a glass-type transition in which a continuously inflected curve with no evidence for maximum heat capacity in the transition region occurs. Thermal equilibration in the transformation region required more than 8 h; above and below this region less than an hour was required. At cooling rates as gradual as 10 K h^{-1} slow energy release took place over periods in excess of 20 h on heating into the region 150–156 K. Such energy release was independent of whether the sample was cooled previously from the melt or from temperatures between the anomaly and the melting point and was comparable in magnitude to that expected for the conversion of the undercooled form on the basis of the difference in the slopes of the heat capacity curves adjacent to the transition. The greater slope on the high-temperature side is consistent with the onset of an additional degree of freedom; indeed, the barrier to rotation of the methyl group attached to the thiophene nucleus may be reasonably assumed to be lowered with increasing temperature so that a less restricted rotation takes place. Transformations of similar appearance have been noted in 3-methyl-thiophene, in *sym*-dimethylhydrazine, and in thiazole with comparable rises in the heat capacities. The postulated change from librational to rotational motion of methyl groups obviously will not hold for thiazole, but increased librational motion of the aromatic rings (loosely packed by virtue of the protuberance) may be responsible in such cases.

Although the great success of thermodynamics in the determinations of molecular structure and internal rotation has been evident for nearly two decades, this work has been carried out largely on gaseous molecules. In the determination of structures from vapour heat capacities and third-law gaseous entropies, the statistical mechanical quantities are usually calculated on the measured value. However, utilisation of the third-law entropy of the gas requires that the low-temperature, solid-phase heat capacity be augmented by enthalpy of vaporisation or sublimation, vapour pressure measurements,

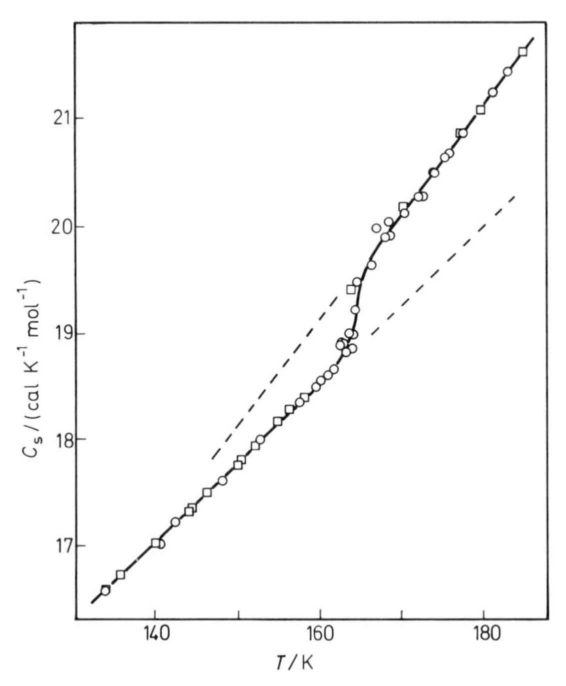

Figure 9.2 Heat capacity of 2-methylthiophene in the transformation region
(From Carlson and Westrum[20], by courtesy of The American Chemical Society)

and equation of state data for the vapour. The advantages of determining barrier heights exclusively from crystalline-phase heat capacities have recently been demonstrated by Wulff[21]. His method involves the evaluation of the heat capacity contributions of a crystal from (a) the intermolecular (lattice) vibrations, (b) the intramolecular vibrations (calculated within the harmonic-oscillator approximation by combination of Einstein functions based upon the observed spectra), (c) $C_p - C_v$ (resulting from the expansion of the lattice as occasioned by the anharmonicities of the vibrations), and (d) the internal rotational freedom. Although barriers in the crystalline state potentially involve intermolecular forces which are absent from the gaseous phase, excellent accord between barrier values for gases and crystals has been obtained.

9.2.3 Plastic crystals

Molecular crystals may be divided into two classes; plastic and non-plastic. Plastic crystals are the high-temperature form of solids composed of near-spherical molecules. They lack long-range order in the solid state and flow easily under stress. Non-plastic solids have the usual characteristics of molecular solids. Diffusion rates are *c.* 10^4 times higher in the plastic solids than in the non-plastic solids. This difference can be ascribed to the occurrence of relaxation around the vacancies in the plastic crystal to yield a small heterophase region in which the molecules can move in a co-operative process.

9.2.3.1 Bicyclo compounds

Towards the end of the decade, systematic plastic-crystal studies tended to centre upon the bicyclo-heptanes, -octanes, and -nonanes[8, 22]. The resulting thermodynamic data on such nearly symmetrical top molecules are summarised in Table 9.2[23, 24].

Interpretation of the thermal data on plastic crystal transitions was initiated by Guthrie and McCullough[25] who treated ΔS_t for simple tetrahedral molecules stochastically and concluded that a minimum of about ten distinguishable molecular orientations exist at each lattice site for substances normally considered to be plastic crystals. The assumption that the normal phase is ordered (i.e. that the standard entropy at 0 K, S_0°, is zero and $Q_l = 1$) gives rise to an entropy increment for the transition to the plastically crystalline phase of $R \ln Q_h$. By a suitable choice of symmetry elements, agreement between the model and the observed thermodynamics of transition has been obtained for a number of systems. Extensions of the above interpretation have been engendered by two groups[22, 26] and their co-workers. To a first approximation, the sum of the entropies of transition and melting appears approximately constant for the almost homologous series shown in Table 9.2. Since there is yet no completely unifying theory which accounts quantitatively for the transitions, systematic investigations of carefully selected systems, quantitative establishment of the role of molecular and crystal symmetry, of the nature of the motion of internal groups and of whole molecules at the lattice sites, and the effect of the volume increment are certainly desiderata.

Darmon and Brot[27] have presented evidence for the rapid reorientation of molecules in plastically crystalline phases. They have considered the implications of the Frenkel model of step reorientations between successive potential minima and conclude that rapid molecular reorientation between potential minima (hence, a disordered crystal) is a valid characterisation of the plastically crystalline phase. The melting theory of Pople and Karasz[28] has been extended to include a larger number of molecules by imposing an

Table 9.2 Transition and melting data on almost symmetrical top molecules in the crystalline state

Compound	T_t/K	ΔS_t (cal mol^{-1} K^{-1})	T_m/K	ΔS_m/ (cal mol^{-1} K^{-1})	$\Sigma\Delta S_i$/ (cal mol^{-1} K^{-1})
Bicyclo[2.2.2]octane	164	6.7	447	4.5	11.2
Bicyclo[2.2.2]oct-2-ene	111	0.8	389	2.4	10.9
	176	7.7			
Bicyclo[2.2.1]heptane	131	7.5	360	3.0	10.5
	306	0.06			
Bicyclo[2.2.1]hept-2-ene	129	9.0	320	2.4	11.4
Bicyclo[2.2.1]hepta-2,5-diene	202	10.6	254	1.6	12.2
Nortricyclene	173	9.0	330	2.5	11.5
1-Azabicyclo[2.2.2]octane	198	6.3	433	3.2	9.5
1,4-Diazabicyclo[2.2.2]octane	351	7.2	433	4.1	11.3
3-Azabicyclo[3.2.2]nonane	298	11.6	467	3.6	15.2
3-Oxabicyclo[3.2.2]nonane	208	8.2	448	3.6	11.8
Exo-2-cyanobicyclo[2.2.1]heptane*	238	7.8	300	2.3	10.1
Endo-2-cyanobicyclo[2.2.1]heptane*	177	2.8	332	2.1	4.9
Exo-2-methylbicyclo[2.2.1]heptane*	152	7.4	387	1.4	8.8
Endo-2-methylbicyclo[2.2.1]heptane*	160	?	164	?	12.2

*The last four compounds were studied in the U.S.S.R. by Kolesov *et al.*[24]; all of the others are reported in Ref. 23.

Table 9.3 Standard zero point entropies for globular molecules predicted by correlation scheme*

Compound	$S^o_{(c)}$	$+\Delta S^o_{sub}$†	$= S^o_{(g)}$ (*calorimetric*)	$S^o_{(g)}$	$= S^o_{trans}$	$+ S^o_{rot}$	$+ S_{vib}$	$+ \Delta S_*$† (*statistical*)	$S^o{}_0$
Bicyclo[2.2.2]octane	50.2	27.3	77.5	78.3	40.0	24.0§	14.3		0.8±1
Bicyclo[2.2.2]octene	50.3	35.7	76.0	80.2	40.0	(26.2)§	(14.0)		4.2±2
1-Azabicyclo[2.2.2]octane	49.5	27.4	76.9	79.5	40.0	(25.4)	14.1		2.6±1
3-Azabicyclo[3.2.2]nonane	56.1	30.6	86.7	86.1	40.4	27.6	(16.1)	2.0	−0.6±2
3-Oxabicyclo[3.2.2]nonane	56.5	28.0	84.5	86.1	40.4	27.6	(16.1)	2.0	1.6±2
Bicyclo[2.2.1]heptane	48.4	25.6	74.0	77.1	39.6	25.0	(12.5)		3.1±2

*Unless otherwise indicated the entropy values (cal mol^{-1} K^{-1}) are at 298.15 K; (c) and (g) represent crystal and gaseous phases (cf. Ref. 22).

†ΔS^o_{sub} is the standard entropy of sublimation. ‡Additional estimated entropy increment for cage expansion. §Quantities in parenthesis are estimated values.

additional parameter which introduces orientational disorder to positional disorder.

Thermodynamic arguments may be insufficient to account for the observed phenomena. The schemes mentioned above have the implicit assumption that the third-law is valid for these systems; but data are accumulating from which the inference may be drawn that this assumption is not generally valid. The necessity for adjuvant measurements, e.g. vapour pressures, normal coordinate analyses of spectral data, cryogenic diffraction measurements, is obvious.

Ascertainment of the validity of the third-law for these systems requires a comparison of the thermal entropy, derived from data on the solid phases and the thermodynamics of sublimation, with an entropy calculated from such spectral and molecular constants as are available. That zero-point entropy may be a significant problem in the analysis of data and in the construction of model systems for these globular molecules is implied by the results of Table 9.3. A correlation scheme for the entropies of fusion and transition that leads to a similar pattern of zero-point entropies has been developed[22]. The comparison of the two sets of values of S°_{gas} points up the desirability of adjuvant non-thermal measurements previously noted for these globular molecules.

9.2.3.2 Trimethylacetonitrile

Evidence that trimethylacetonitrile, Me_3CCN, has a symmetry number of unity and undergoes two transitions with $\Sigma\Delta S_t = 2.12$, which is nearly $R \ln 3 = 2.18$, has been presented[29]. Transition II may be of higher than first order, but the large apparent heat capacities observed (>6 kcal mol^{-1} K^{-1}) suggest that transition I is essentially first order. Although delineation of a mechanism for the two transitions separately is precluded by the absence of structural and n.m.r. data, it seems plausible to assume that the three methyl groups are potentially distinguishable by their orientations. Since the potential energy barrier to internal rotation is *c.* 4 kcal mol^{-1} and RT at 300 K is only *c.* 0.6 kcal mol^{-1}, the methyl groups may be interlocked to each other like meshed gears, lacking the thermal energy to surmount the barrier. If this is the case, the $R \ln 3$ total transitional entropy increment is the entropy increase due to the rotation of the entire molecule about the C—CN axis. This agrees well with the observations drawn from dielectric work by Clemett and Davies, who concluded that trimethylacetonitrile may rotate about the C—CN axis in the Crystal I phase but not about the C—Me axis. They also noted that this phase is birefringent and anisotropic just below the melting point. Hence, this substance may be considered to be a 'uniaxial plastic crystal' with ΔS_m of 7.60 cal mol^{-1} K^{-1}, i.e. significantly larger than for an ordinary plastic crystal.

9.3 IONIC CRYSTALS

9.3.1 Ionic freedom

Although it has been established that the ammonium ion does not rotate freely in simple halide salts, the prospect of such motion is enhanced by an increase in the uniformity of the crystalline field to be expected in salts with high lattice symmetry containing polyatomic anions of low charge as in NH_4ClO_4. In principle, the thermal method for deducing the extent of ionic freedom simply involves comparing the heat capacities of the ammonium salt and the corresponding isostructural alkali salt[30]. Since the contributions from both the ammonium and alkali salts to the increment $(C_p - C_v)$, occasioned by the anharmonicity of the vibrational modes, are nearly the same—as are the internal vibrational and torsional modes of the anions, the

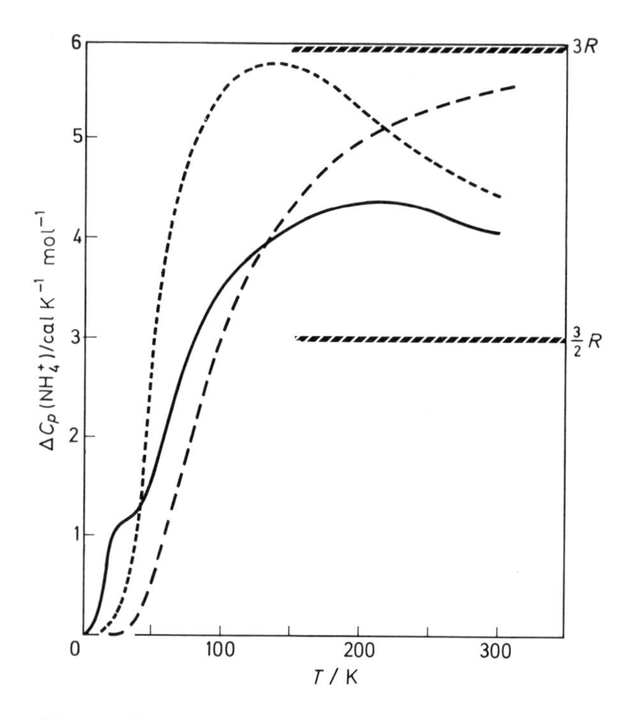

Figure 9.3 The —— curve is the ΔC_p as resolved from the heat capacity. The ---- curve represents an Einstein vibrator with $\Theta_E = 300$ k, and the ··· curve represents a librator with a 1.0 kcal mol^{-1} barrier
(From Westrum and Justice[30], by courtesy of the American Institute of Physics)

expression $[C_p(NH_4ClO_4) - C_p(KClO_4) - C_{int}(NH_4{}^+)]$, adjusted as it is for the almost negligible contribution from the internal vibrational modes of the ammonium ions, should represent the torsional contribution of the ammonium ions. $C_{int}(NH_4^+)$ can be evaluated from the frequency assignment. Comparison of the measured heat capacity of NH_4ClO_4 with that of $KClO_4$ led to the conclusion that rotation of the NH_4^+ ion in this salt is opposed by a potential energy barrier of less than 1.0 kcal mol^{-1} as depicted in Figure 9.3[30]. A combination of the thermophysical values with aqueous NH_4ClO_4

thermochemical data indicates the absence of zero-point energy and consequently of disorder.

9.3.2 Crystal field levels

Although resolution of the levels resulting from splitting of the ionic ground state by the crystal fields is usually achieved by spectroscopists, on occasion the cryogenic calorimeter can do yeoman service, especially in the absence of optical grade samples.

9.3.2.1 Lanthanide sesquioxides

Subsequent to earlier heat capacity measurements on La^{III} and on Nd^{III} oxides and deduced crystal field splitting of the ground electronic level of the former were reported, the spectrum of Nd^{III} ion in a La^{III} oxide host was observed by Henderson *et al.*[31]. They report five crystal field lines for the ground state ($^4I_{\frac{9}{2}}$) at 0, 23, 84, 253 and 496 cm^{-1}. The first two excited levels are in excellent agreement with corresponding values deduced from heat capacity data alone (0, 21, 81 and 400 cm^{-1}), and the centre of density of the higher levels also corresponds. However, the interpretation of the electronic heat capacity data of the cubic Gd^{III}, Dy^{III}, Ho^{III} and Er^{III} oxides were hampered by the lack of spectral data. These cubic lattices have two non-equivalent types of cation sites. Three-fourths of the cations have C_2 symmetry. The remaining cations possess C_{3i} symmetry. The spectra reported for Dy^{III}[32], Er^{III}[33, 34] and Tm^{III}[34] ions with C_2 symmetry in Yt^{III} oxide or in single crystal Ln^{III} oxides together with heat capacity data permitted derivation of the lattice heat capacity for the cubic oxides from Gd_2O_3 to Lu_2O_3 and energy levels for the C_{3i} ions in Dy_2O_3, Er_2O_3, and Tm_2O_3[35]. As an example of the success of the method consider Er_2O_3[35] as represented in Figure 9.4. Contributions to the electronic heat capacity, C_{el}, based on spectroscopic levels reported for the C_2-type ions in Er^{III} oxide[33, 34], consisting of doublets at 0, 39, 76, 89, 158, 258, 495 and 500 cm^{-1}, are represented by dashed curves in Figure 9.4. Figure 9.4 depicts C_{el} for the low-temperature region. These values were derived by subtracting the lattice contribution from the apparent heat capacities. The C_{el} remaining after removal of the contributions of the C_2 ions is attributed to the levels in the C_{3i} ions. The levels and their degeneracies are deducted in part from the position and magnitude of the Schottky peaks in the observed C_{el} (for separations of $<100\ cm^{-1}$) and in part from the splittings of the ground state term by fourth- and sixth-order terms of cubic crystal fields. The levels thus derived are doublets at 0, 12, 50, 55, 150 and 450 cm^{-1} and a quartet at 550 cm^{-1}. These correspond most closely to the levels in a cubic field with a ratio of sixth- to fourth-order terms (V_6/V_4) of -0.25. The Γ_8 level is split to 12 and 50 (or 55) cm^{-1}, and the second Γ_8 level is split to 150 and 450 cm^{-1}. The C_{el} data require levels at 12, 50, and 55 cm^{-1} ($\pm 10\%$ of ΔE).

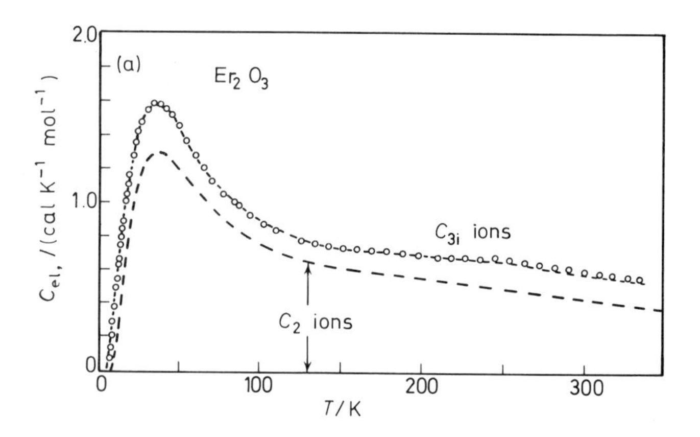

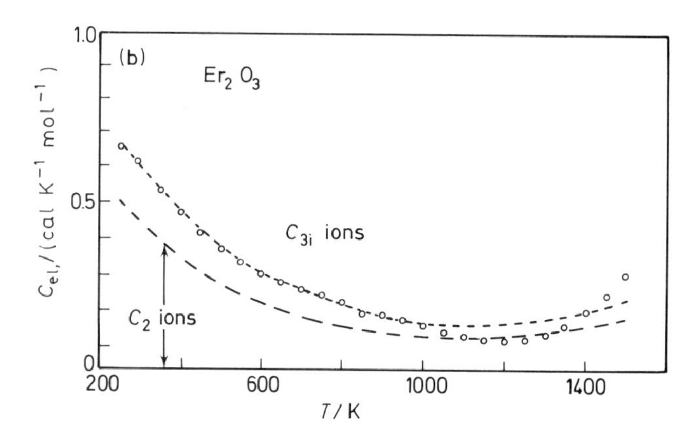

Figure 9.4 Schottky anomalies for Er^{III} ion in Er_2O_3. In (a) the circles represent data from Justice *et al.*[35]. The dashed curves represent the calculated electronic heat capacity of the C_2 ions based on the spectroscopic data of Kisliuk *et al.*, and Gruber *et al.*[34]; the dotted curve shows total electronic heat capacity by adding the contribution for the C_{3i} levels. In (b) the circles represent heat capacities from enthalpy increment determinations and the other curves have the same significance as in (a)
(From Justice *et al.*[35], by courtesy of the American Chemical Society)

Parallel studies on lanthanide chalcogenides (e.g. La_2Te_3[36], Nd_2Te_3[37] etc.) have been initiated at Akademgorodok.

9.3.3.2 *Lanthanide hexaborides*

These electrically conducting substances pose significant problems in the interpretation of their thermal anomalies, and consequently of their thermodynamics. The heat capacity of LaB_6 is of normal sigmate behaviour without thermal anomalies; the other hexaborides studied all show co-operative, anti- to para-magnetic λ-type transitions near 10 K followed by complex Schottky transitions with maxima between 20 and 80 K[38]. The resolution of the magnetic and electronic contributions for the lanthanide hexaborides was achieved by means of the heat capacity of diamagnetic LaB_6 as an estimate of the lattice vibrational heat capacity. For NdB_6, the Schottky anomalies were resolved in terms of energy levels (2, 4, 4 at 230 and 640 cal mol^{-1} K^{-1})[39] generated by the crystal field splitting of the 10-fold degenerate ground state of the assumed Nd^{III} ions[39]. The data on PrB_6 and CeB_6 (λ-transition at 7 K, Schottky at 35 K) appear to be interpretable on a basis similar to that for NdB_6[38].

However, the results for GdB_6 (λ-anomaly at 16 K, Schottky maximum at 30 K corresponding to a splitting of 61.3 cm^{-1}) are inconsistent with the much smaller expected crystal field splitting of an $^8S_{\frac{7}{2}}$ ground state. No cogent explanation of the anomaly has yet been postulated. The total transitional entropy in GdB_6 and other properties (lattice parameters, Mössbauer and adsorption spectra, magnetic susceptibility, etc.) cast considerable doubt that lanthanide metals are in fact in the Ln^{III} oxidation states in some of the hexaborides. Since EuB_6, YB_6, SmB_6 and the alkaline earth hexaborides all contain divalent metal ions, this state may obtain at low temperatures in other lanthanide hexaborides[40].

Related optical spectral studies of low 4f levels of the lanthanides have been summarised by (the late) G. H. Dieke[41]. The availability of a recent summary of the thermodynamic and magnetic properties of the lanthanide chalcogenides, the pnictides and binary semi-metallic compounds[42], as well as an excellent critical assessment of the enthalpies, entropies and Gibbs energies of formation of the lanthanide oxides compiled by Holley *et al.*[43], largely eliminates the need for comments at present except to note that thermal and thermochemical data on the lanthanide halides are still lagging and little progress was reported in the present biennium.

9.4 COVALENT CRYSTALS

As a consequence of silica's role as one of the principal rock-forming oxides, there has been a long-standing interest in its chemical thermodynamics on the part of chemists and geologists alike. Nevertheless, in spite of a series of independent thermochemical studies carried out over a period of more than 50 years, there is as yet no general agreement about the entropies of transformation between the various low-pressure polymorphs of silica or even about its enthalpy of fusion. In recent years the interest in the thermodynamics of the silica system has been heightened by the discovery of two new high-pressure polymorphs – coesite and an even denser modification, stishovite. Although produced by hydrothermal means, both forms have been discovered in nature in isolated sandstone in Meteor Crater, Arizona, formed under the transient high-temperature, high-pressure conditions generated upon meteoritic impact. Their temperature–pressure stability fields are of considerable interest. The quartz–coesite transformation may eventually offer a convenient basis for calibration of high-pressure–high-temperature apparatus. The interest in stishovite, on the other hand, is in part motivated by the prediction that this mineral plays an important role in the chemistry of the earth's mantle. Although a number of equilibrium investigations have

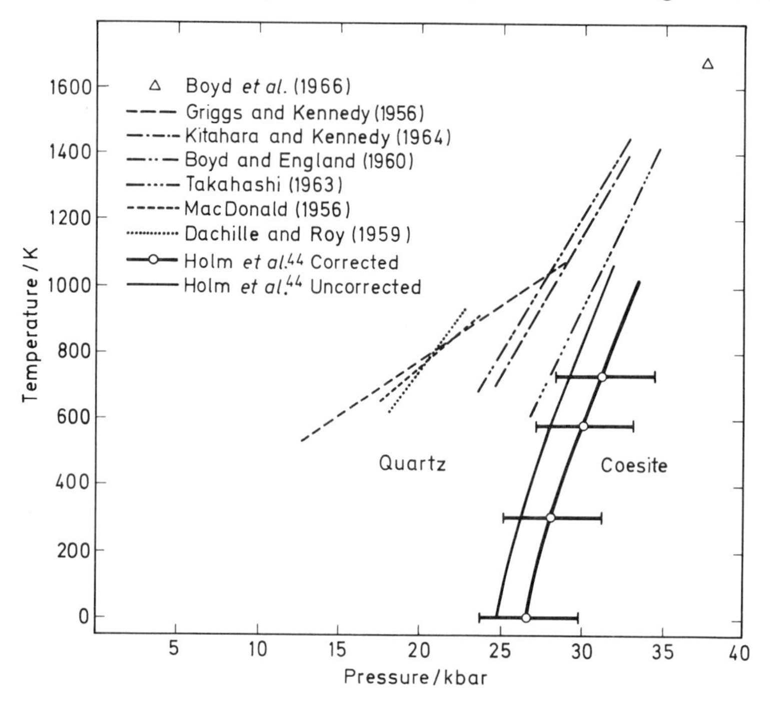

Figure 9.5 *P–T* stability fields for the quartz–coesite equilibrium calculated from thermodynamic data reported by Holm
(From Holm *et al.*[44], by courtesy of Pergamon Press)

been carried out, there is considerable disagreement among the various investigators. Hence, an independent approach to this problem, such as is provided by calorimetry, is of major significance. A recent communication[44] reports the results of a study of coesite and stishovite based on the combined use of adiabatic cryogenic calorimetry, high-temperature solution calorimetry, and drop calorimetry. Precise determination of the heat capacity at low temperatures permitted evaluation of the thermal properties to 350 K. The enthalpy of solution of coesite in a lead–cadmium–borate melt was measured at 700 °C and compared with that for pure, crystalline quartz.

Since stishovite tends to convert to a vitreous phase at high temperature, the enthalpy of transformation of this mineral to stishovite glass was determined by means of 'transposed-temperature' drop calorimetry. In this approach, a small sample of stishovite is first dropped from 300 K into the calorimeter at 1070 K and the total enthalpy increment determined. In the second experiment the enthalpy of the transformed stishovite glass is determined. The difference between the two results yields the enthalpy of transformation of stishovite to glass. Determination of the enthalpies of solution of quartz, cristobalite, silica glass, and stishovite glass in the oxide melt at 1070 K permits comparison of the new results with data obtained by earlier workers who used conventional calorimetry in aqueous hydrogen fluoride solution. The combined approach permitted the evaluation of the chemical thermodynamics of coesite and stishovite phases as well as the delineation of the $P-T$ stability fields. One of these is depicted in Figure 9.5.

9.5 THERMOCHEMISTRY OF CRYSTALS

9.5.1 Enthalpy of formation

Insofar as the present biennium in thermochemical measurements is concerned, one might observe that the present era is an optimistic one with moderately precise values available by oxygen or fluorine rotating-bomb, flame, or solution calorimetry. True, there do remain significant gaps in our knowledge of the enthalpies of formation of metal halides and even oxides. Even larger uncertainties are associated with the present data for many organometallic compounds and for organic compounds of silicon, for example, virtually no reliable data exist. One of the primary developments of the present era, however, has been the recognition of certain 'key compounds' and the endeavour to provide measured 'key' enthalpies of formation with a far lower precision index than would normally be acceptable. In view, however, of the number of interesting new compounds synthesised each year, the thermochemical 'gap' is widening precipitously with time. For these reasons the development of empirical procedures to enable accurate prediction of the enthalpies of formation of large varieties of organic substances has become an increasingly crucial problem. Another motivation for completion of such schemes comes from the desirability of being able to estimate reliable enthalpies of formation for strain evaluation. There is growing evidence of a forthcoming second generation of schemes which will enable the more reliable prediction of enthalpies of formation for both organic and organometallic compounds.

For those globular molecules *par excellence* – the carboranes – Gal'chenko *et al.*[45, 46] have reported enthalpies of formation of -40.6, -57.5, and -74.3 kcal mol^{-1} for the 1.2-*ortho*-, 1.7-*meta*-, and 1.12-*para*-carboranes, respectively. The unexpectedly large differences are difficult to accept unless occasioned by the explosive nature of the combustion and the difficulties of ascertaining the amount and nature of the impurities. A differential scanning calorimetric search[47] as low as 140 K revealed a transition at 277 K with a ΔS_t of 3.0 e.u. suggesting the probability of plastically crystalline behaviour.

A critical review of the enthalpies of formation and equilibrium data for over 1500 transition metal complexes involving inorganic and organic ligands has been provided by Ashcroft and Mortimer[48].

Scaling down both thermophysical and thermochemical calorimetry to accommodate smaller samples is a desideratum; a differential calorimeter for enthalpies of combustion on microgram quantities has been described[49].

9.5.2 Strain energies

An interesting by-product of precision enthalpy of combustion data are strain energies for molecules. A few recent values for bicyclic molecules are incorporated into Table 9.4[22]. Other strain energies for less-symmetrical bicyclo-alkanes have been obtained[50].

Table 9.4 Strain energies in bridged-ring molecules at 298.15 K*

Compound	*By Allen's Scheme*	*By* ΔH(hyd)	*Adopted*
Bicyclo[2.2.1]heptane	18	18.5	18.5
7-Oxabicyclo[2.2.1]heptane	16	6.5	6.5
Bicyclo[2.2.2]octane	9.7	9.2	9.2
Bicyclo[2.2.2]oct-2-ene	12	9.7	9.7
1-Azabicyclo[2.2.2]octane	9.5	7.9	9.5
1,4-Diazabicyclo[2.2.2]octane	9.5	7.9	9.5
3-Oxabicyclo[3.2.2]nonane	9.1	5	5
3-Azabicyclo[3.2.2]nonane	8.2	7	7

*Units: kcal mol^{-1} (cf. Ref. 22).

9.6 ELECTRONIC AND MAGNETIC CONTRIBUTIONS TO THERMAL QUANTITIES

9.6.1 Electronic contributions: Schottky anomalies

An excellent brief summary of the various cryogenic contributions which together with the lattice heat capacity combine to represent the total heat capacity curve has been summarised by Gopal[51], and their resolution discussed by Morrison *et al.*[52]. Deduction of electronic density-of-states parameters from heat capacity data has been analysed[53] in general and evaluated for clathrate salts[54]. Electronic heat capacity trends in the transition metals have been correlated with d electron configurations[55]. Moreover, recent developments on the Verwey transition in magnetite (Fe_3O_4)[56] and the Morin transition in hematite (Fe_2O_3)[57] have resulted in trends more complex than anticipated.

Enhanced appreciation of the significance of Schottky anomalies in transition metals and impurity modes is evident as had already been noted in Section 9.3.2. An especially interesting Schottky anomaly arises near 0.2 K as a consequence of the 10^{-9}% OH^- ion in NaF. The sixfold orientational degeneracy is split due to tunnelling into a singlet, triplet and doublet[58]. The data show how certain impurities, which at concentrations as low as 0.1 p.p.m.

would be very difficult to detect by other means (e.g. optical spectroscopy) can be observed and even quantitatively determined by heat capacity measurement. They also reflect the great care required for the preparation of heat capacity samples! Tunnelling modes for Li^+ ion in KCl : Li have also been reported[59].

9.6.2 Para- and ferro-electricity

The thermodynamics of low-temperature electrocaloric phenomena in alkali halides doped with polar impurities has been formulated[60] and found to accord with heat-capacity data on RbCl : CN[61], adiabatic depolarisation temperature changes and the electrocaloric coefficient for KCl : OH. Recent progress in the phase transitions of order–disorder ferroelectrics[62] and a general treatment of the thermodynamics of broadened ferroelectric phase transitions[63] have been reported.

9.6.3 Semi-, super-, and metallic conductivity

Thermodynamics of type II superconductors including the results of phenomenological theories and comparison with experimental values have been compiled[64] and predictions of the effect of impurities on the singularities in the heat capacity of superconductors[65] made by utilisation of Ginzberg–Landau parameters. The important semi-conductor ⇄ metal 'phase' transition has been explored by Hyland[66].

9.6.4 Magnons, spin waves, and magnetic transformations

Phonons and their Interactions[67] and *spin waves*[68] both represent timely surveys of this important area. The forefront of experimental progress includes spin-wave contributions in Mn, Fe^{II}, Co and Ni fluorides[69], in perovskite structures[70], and in EuTe[71]. Magnon heat-capacity contributions and their enhancement in ferromagnetic semiconductors have been treated[72]. Magnetic ordering[73], first-order phase transitions[74], magnetothermodynamics of $MnCl_2{\cdot}4H_2O$ [75], the Curie-point heat capacity of Gd metal[76], and the Néel-point heat capacity of CoO[77] are typical of recent endeavours.

An interesting trend in the temperature dependence of the heat capacity near 0 K has become evident in recent studies on $UO_{2.250}$ [78]. It is evident from Figure 9.6 that up to 10 K the heat capacity of $UO_{2.250}$ is much higher than that of UO_2 or ThO_2. The lattice contribution of $UO_{2.250}$ and of UO_2 in this temperature region can be well approximated by the heat capacity of ThO_2 since these compounds have similar structures and nearly the same lattice constants and cationic masses. The spin-wave contribution in UO_2 becomes very large as the temperature approaches 30.4 K, the Néel temperature. In $UO_{2.250}$ there is a contribution from the 0.250 moles of oxygen atoms, but this is probably negligible below 20 K. Another estimate of the lattice heat capacity of $UO_{2.250}$ may be obtained from values of the lattice constants at 300 K from which Θ_D is calculated as 220 K. Hence, the lattice heat capacity of $UO_{2.250}$ is 1.52 times the heat capacity of ThO_2, up to 10 K. Regardless of which estimate of the lattice heat capacity of $UO_{2.250}$ is taken, it is evident that there is a large heat-capacity contribution from other sources. presumably magnetic.

After deduction of the lattice contribution the remaining heat capacity,

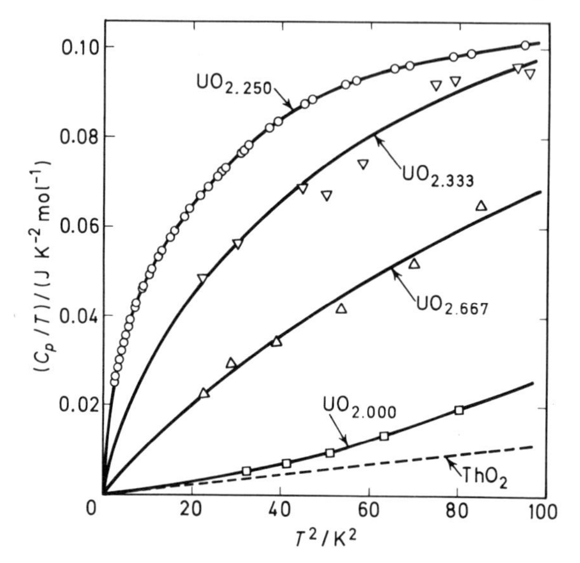

Figure 9.6 C_p/T *v.* T^2 for several actinide oxides[78]. The dashed curve for ThO_2 is drawn for $\Theta_0 = 253$ K (From Flotow, Osborne and Westrum[78], by courtesy of the American Institute of Physics)

C_{mag} for $UO_{2.250}$ is well represented by the relationship: $C_{mag} = (0.0155)T^2$J K^{-1} mol^{-1}. No theoretical explanation of this interesting temperature dependence is known.

9.7 OTHER STABILITY–TRANSITIONAL PHENOMENA

9.7.1 Ordering transitions

Order-disorder transitional models have been examined for molecular crystals[79], for polymers[80], for alkali graphites[81] and for alloys[82], as well as for the paraelectric phase of $(NH_4)_2SO_4$ [83]. In addition, earlier sections of this review (e.g. 9.2.1.1, 9.2.2.1, 9.3.1, 9.5.1, 9.6.4) have dealt in detail with selected examples of such transitions.

9.7.2 Phase transitions

The energetics of crystals are related to those of other phases through phase transitions, e.g. sublimation, melting, solid-state, crystal-glass, etc. Discussion of several of these has already occurred in this review; in addition such studies as sublimation pressures on alkaline earth bromides[84], platinum[85] and pentadienes[86] provide useful data leading to enthalpies of vaporisation. Direct determination of the enthalpy of vaporisation may be made in a calorimeter suitable for use over a vapour pressure range of 10^{-3}–200 Torr[87]. The thermodynamic theory and description of phase transitions has been extensively treated[88–93] and special attention provided for solid–solid transitions[94] and for melting[95]. The tremendous growth in accurate knowledge about the structure and properties of crystals has not been paralleled by developments on the mechanism of melting. In fact, Ubbelohde's *Melting and Crystal Structure*[96], in which liquids are described in terms of 'melts' of quasi-crystalline models and anti-crystalline models related to them, is still in many respects a reasonable delineator of the state of the art. Drop-calorimetric enthalpies of melting for III–V, IV–V, and V–VI compounds have been reported[97]; a dearth of adiabatic calorimetric data on the details of the melting transition still persists.

Comparison between experiment and theory on *Phase Transformations*[98] in metals and alloys in which extensive agreement has been attained does not, however, imply that the essential features of the field can be considered to be understood. Related progress on many types of crystalline materials has been detailed in other symposium proceedings[99–101].

9.7.3 Energetics of phase behaviour

Brewer's proposed methods for predicting the crystal structures, mutual solubilities, and multi-component phase diagrams for the 30 elements of the first three transition series[102] involves correlation of electronic and crystal structures, spectroscopically derived energies of various electronic states of gaseous atoms, as well as certain aspects of regular solution theory. His treatment provides elegant examples of the interrelation of energetics and bonding as well as considerable thermodynamic insight into the importance of various factors involved in metallic systems. Phase diagrams and thermodynamic properties have been computer evaluated by Kaufman[103] for systems of refractory metals on the basis of phase competition, valence, size effects, and lattice stability.

9.7.4 Solid solutions

A novel approach to the study of the thermodynamics of mixing in the crystalline state has been reported[103, 104]. Studies on the system tetramethylmethane–tetrachloromethane were devised to test the several theories of *liquid* mixtures in the crystalline state. Since the crystalline phase has many properties in common with the liquid phase (i.e. a large number of first neighbours, local order, and a determinable lattice structure), measurement of the excess functions of crystalline solutions may permit a better test of these theories of mixtures than is possible with liquids. The approach required measurement of the heat capacity of a system composed of molecules of sufficient similarity in size and shape to permit the existence of a continuous range of solid solutions and yet have sufficient difference to occasion deviation from ideal behaviour and for which impeccable mixing enthalpies were available for the liquid phase. Plastically crystalline substances were selected to provide sufficiently rapid thermal equilibration to make measurements practicable and provide 'spherical' molecules so that the inter-molecular forces may be considered to be central.

To ascertain the temperature dependence of the enthalpy of mixing including quantitative phase equilibrium data and enthalpies of transition, the heat capacities from 5 to 300 K for five compositions of the system were measured using low-temperature adiabatic calorimetry. The changes in temperature, magnitude, and shape of the anomalies in the heat capacity *v.* temperature curves occasioned by composition variation indicated that changes in molecular environment exert considerable effects on the re-orientational–rotation transitions of the pure components. The melting point measurements are consistent with the existence of a continuous series of solid solutions. The expected lowering of the 'rotational' transition of tetrachloromethane upon addition of tetramethylmethane is observed. The constancy of the tetramethylmethane transition temperature, 142.5 K, indicated phase separation at this temperature; eutectoid formation appears likely judging from the occurrence of an intermediate transition at 191.6 K.

The results demonstrate that the binary phase diagram for simple globular molecules can be complex, that equilibrium measurements can be achieved even at low temperatures on crystals of globular molecules if sufficient

Table 9.5 Comparison of derived experimental excess enthalpies with calculated values at 230 K*

			Mole fraction		
	0.200	0.334	0.501	0.666	0.826
			Excess enthalpy		
Harmonic oscillator cell model	43	78	102	101	78
Hildebrand's empirical approach					
Internal pressure considerations	63	88	101	92	61
Distortion energy	34	49	58	53	36
Total	97	137	159	145	97
Average potential model					
CMe_4 as reference	105	146	164	146	94
CCl_4 as reference	111	154	172	154	99
Average	108	150	168	150	97
Derived experimental	177	241	279	121	71

*Units: cal mol^{-1} of solution.

patience is available, and that the adiabatic calorimeter is a powerful tool for the study of phase diagrams at low temperatures. A comparison of the experimental excess enthalpies is revealed in Table 9.5.

Heat capacities of four crystalline argon–nitrogen mixtures[105] have been investigated also and the melting values compared with the theory of Pople and Karasz[28]. Quadrupole ordering of N_2 molecules was shown to be incomplete at low N_2 concentrations.

9.8 CERTAIN THERMOPHYSICAL CONTRIBUTIONS

9.8.1 Anharmonicity

Contributions from anharmonic vibrations to thermal properties are important for space-age problems in high-temperature thermal studies on refractory metals[106, 107], on Pb and Al [108, 109], and in discussion of heat-capacity deviations from Dulong–Petit values[110, 111]. Adiabatic calorimeters for service as high as 1300 K have been described by several authors[112–114].

9.8.2 Non-stoichiometry and defect structures

The metallurgical thermochemists' preoccupation with transition-metal alloy systems, oxides, nitrides, carbides, silicides and borides has been well served by an extensive review[115]. Computations of thermodynamic activities across entire non-stoichiometric existence ranges[116] as well as heat-capacity contributions from vacancies[109, 117] have been discussed. Although deviations from stoichiometry in AgS produced no unusual effects[118], anomalously high heat capacities for non-stoichiometric, annealed AgI, retained energy for quenched stoichiometric material, and elevation of the critical temperature for disordering have been claimed by Perrott and Fletcher[119]. However, Noelting and Rein[120] confirmed their own earlier measurements but were unable to confirm the anomalously high heat-capacity values of Perrott and Fletcher.

An advanced treatment of non-stoichiometric and other extended-defect systems which directs attention to numerous areas in which uncertainty, controversy, and need for clearer recognition of the real problems exist, has been compiled[121]. Particular attention is directed towards the thermodynamic factors which determine the relative stability of phases.

9.8.3 Evaluation of Θ_{Debye} from spectral data

For many years thermodynamicists and others have freely used Debye and Einstein functions to represent the heat capacities of crystalline solids. Such representations have the great advantage of not behaving wildly when extrapolated beyond the region of fit. Except in the simplest cases, however, expressions of this type were formulated merely as curve-fitting devices, with little attention being paid to actual frequencies of vibration in the crystal because the large number of vibrational branches defies the capability of accurate curve-fitting. Recently, however, advances on several fronts in solid-state spectroscopy have made available, in ever-increasing abundance, a wealth of very detailed information on vibrational frequencies in quite complicated crystals. Two recent papers have attempted to incorporate such information in the analysis of the heat capacities of $CaWO_4$ and $CaCO_3$ (calcite), respectively.

The earlier of these papers by Lyon and Westrum[122] was concerned with the problem of evaluating meaningful Θ_{Debye} for $CaWO_4$ and with the question which, if any, of these values were comparable to Θ_{Debye} values as determined from spin–lattice relaxation measurements in rare-earth doped $CaWO_4$. After subtraction from the measured heat capacity of all contributions due to both internal and lattice optical branches, it was found that effective Θ_{Debye} calculated from the acoustical branch-heat capacity resulted in a limiting value, Θ_0, quite comparable to the value from spin relaxation experiments.

The paper of Staveley and Linford[123] attempted the rather ambitious project of using only spectroscopic and elastic information to generate a Debye–Einstein representation of the heat capacity. This calculation, although only moderately successful, points out well some of the difficulties remaining with treatments of this sort, among these being the inability at present to incorporate anharmonic contributions without extensive calculational complications and the probable oversimplification of using Debye functions for the acoustical branches. Both reports concur in the view that calculation of a meaningful Θ_{Debye} for the acoustical branches of a crystal requires a careful accounting of acoustical degrees of freedom for the primitive cell contents.

9.8.4 Constraints of pressure and volume

High pressure has been a largely neglected constraint in the measurement of heat capacity and transition thermophysics. Certainly formidable problems exist, especially for transitions where significant volume increments occur. Some initiatory endeavours are, however, worthy of note. For example, the heat capacity anomaly in ammonium chloride near 243 K, discovered in 1922 by Simon, has been a prototype for the study of order–disorder transitions in crystals for many years. Although a wealth of experimental and theoretical studies have been devoted to the transition and the basic mechanism is well understood, none of the attempts to classify and totally explain the characteristic features are completely successful. In an exciting pioneering investigation by Trappeniers[124], the influence of pressure on the λ-point was studied by means of the differential thermoanalysis technique under conditions of hydrostatic pressures up to 3 kbar. They conclude that the λ-transition collapses under pressure and as a result the enthalpy change and the associated entropy change are strongly reduced. The latter drops to less than 25 % of its value at one atmosphere within the 2 kbar range; at still higher pressures the effect of the collapse is seen to level out. Unfortunately, similar studies are not available for a variety of other transitions in molecular and ionic crystals, nor are the measurements reported to date truly equilibrium values. They are suggestive, however, and are indicative for the urgency of further measurements of thermoproperties at elevated pressures. Techniques for the measurement of the heat capacity of metals from 77 to 300 K [125] at high pressure together with measurements on α-cerium and uranium metal from 1 to 6 K at pressures of 10 kbar[126] have been published.

Heat capacities at constant volume for water[127] and for carbon tetrachloride[128] have been described and the discrepancies occasioned by different methods considered.

9.8.5 Miscellaneous

A nuclear Schottky anomaly in antiferromagnetic V_2O_3[129] has been reported below 0.5 K. More exotic results include Van Hove singularities in the heat capacities of crystals[130] and systems with negative heat capacities[131].

Although melting to a smectic or a nematic liquid crystal phase[132] is one of several ways a crystal may depart from its phase, all too little equilibrium calorimetry has been performed to ascertain the energetics, order and thermodynamic behaviour through such transformations. The relatively new method of measuring heat capacities by the laser technique[133, 134] has been applied to the measurement of heat capacities at high temperatures as well as to the determination of thermodiffusivity and conductivity of solids. Although some problems of the method involve the accurate measurement of the energy density of the laser beam and the adjustment of differences in the reflectivity of the sample surface as well as the accurate measurement of the temperature rise of sample at high temperatures, the study of the heat capacity of synthetic sapphire has demonstrated the adaptability of the laser calorimeter to small samples at high temperatures. A subsequent development by Professor Takahashi involves application of the method at cryogenic temperatures as well.

A measurement of the trend in the heat capacity of nickel in the region of its Curie point in increments of less than 0.01 K has been achieved by combining heating by chopped infrared light, sensing of temperature variations by Peltier effect, and phase-locked amplification.

Ten of the fourteen chapters of the thermophysicists 'Bible' – *Experimental Thermodynamics, Volume I*[135] – pertain to a survey of the techniques of the calorimetry of solids at all temperature ranges including that of high-speed high-temperature thermodynamic measurements.

Interest in the vitreous state, and its relevance to inclusion with crystal thermodynamics, centres primarily around the low-temperature heat-capacity difference between glasses and crystal, and in the glass-crystal transformation and residual entropy. Recent heat capacities at low temperatures have been measured on vitreous glycerol[136] (used as a thermal contact medium), vitreous dimethylbutane[137], cyclohexanol[138], crystalline and 'glassy-crystalline hexanol[137], AsF_3[139], Ge[104], and BeF_2[141], as well as on high polymers[142]. The thermal properties[143] and the thermodynamics[144] of polymers have also been extensively reviewed elsewhere. Finally, some comment on the enhanced role of differential thermal analysis and differential scanning calorimetry is needed. These techniques of tremendous significance in survey measurements are, however, occasionally interpreted as being more definitive than circumstances warrant. Recent advances have been summarised[145, 146] and evaluated[147].

References

1. Westrum, E. F. Jr. (ed.) *Bulletin of Thermodynamics and Thermochemistry,* (Published annually, cf. especially (1969) **12** and (1970) **13**)
2. Stull, D. R., Westrum, E. F. Jr. and Sinke, G. (1969). *The Chemical Thermodynamics of Organic Compounds,* (New York: John Wiley)
3. Cox, J. D. and Pilcher, G. (1970). *Thermochemistry of Organic and Organometallic Compounds,* (New York: Academic Press)
4. Bogatskii, M. I., Kucheryavyi, V. A., Manzhelii, V. G. and Popov, V. A. (1968). *Phys. Status Solidi,* **26,** 453
5. Zahlan, A. B. (1968). *Exitons, Magnons and Phonons in Molecular Crystals,* (Cambridge: University Press)
6. Craig, D. P. and Walmsley, S. H. (1968). *Excitons in Molecular Crystals,* (New York: W. A. Benjamin)
7. Bondi, A. (1968). *Physical Properties of Molecular Crystals, Liquids and Glasses,* (New York: John Wiley)
8. Westrum, E. F. Jr. (1966). *J. Chim. Phys.,* **63,** 46
9. Carlson, H. G. and Westrum, E. F. Jr. (1971). *J. Chem. Phys.,* **54,** 1464
10. Timmermans, J. (1950). *Physico-Chemical Constants of Pure Organic Compounds,* (New York: Elsevier)
11. Aston, J. G., Siller, C. W. and Messerly, G. H. (1937). *J. Amer. Chem. Soc.,* **59,** 1743
12. Russell, H. Jr., Osborne, D. W. and Yost, D. M. (1942). *J. Amer. Chem. Soc.,* **64,** 165
13. Messerly, G. H. and Aston, J. G. (1940). *J. Amer. Chem. Soc.,* **62,** 886
14. Jones, W. M. and Giauque, W. F. (1947). *J. Amer. Chem. Soc.,* **69,** 983
15. Egan, C. J. and Kemp, J. D. (1938). *J. Amer. Chem. Soc.,* **60,** 2097
16. Pitzer, K. S. and Weltner, W. Jr. (1949). *J. Amer. Chem. Soc.,* **71,** 2842
17. Halford, J. O. (1950). *J. Chem. Phys.,* **18,** 361; Halford, J. O. and Miller, G. A. (1957). *J. Chem. Phys.,* **61,** 1536
18. Andrews, J. T. S., Westrum, E. F. Jr. and Bjerrum, N. (1969). *J. Organometallic Chem.,* **17,** 293
19. Jellinek, F. (1960). *Nature (London),* 871
20. Carlson, H. G. and Westrum, E. F. Jr. (1968). *J. Chem. Eng. Data,* **13,** 273
21. Wulff, C. A. (1963). *J. Chem. Phys.,* **39,** 1227
22. Westrum, E. F. Jr. and Wulff, C. A. (1970). Chapter in *Proceedings of the First International (1969) Conference on Calorimetry and Thermodynamics,* 161. (Warsaw: Polish Scientific Publishers)
23. Westrum, E. F. Jr., Wong, W. K. and Morawetz, E. (1970). *J. Phys. Chem.,* **74,** 2542
24. Kolesov, V. P., Seregin, E. A. and Skuratov, S. M. Personal Communication (1969)
25. Guthrie, G. B. and McCullough, J. P. (1961). *J. Phys. Chem. Solids,* **18,** 53
26. Amzel, L. M. and Becka, L. N. (1969). *J. Phys. Chem. Solids,* **30,** 521
27. Darmon, I. and Brot, C. (1967). *Molec. Cryst.,* **2,** 301
28. Pople, J. A. and Karasz, F. E. (1961). *J. Phys. Chem. Solids,* **18,** 53
29. Westrum, E. F. Jr. and Ribner, A. (1967). *J. Phys. Chem.,* **71,** 1216
30. Westrum, E. F. Jr. and Justice, B. H. (1969). *J. Chem. Phys.,* **50,** 5083
31. Henderson, J. R., Muramoto, M. and Gruber, J. B. (1967). *J. Chem. Phys.,* **46,** 2515
32. Henderson, J. R., Muramoto, M., Henderson, T. M. and Gruber, J. B. (1967). *J. Chem. Phys.,* **47,** 5097
33. Kisliuk, P., Krupke, W. F. and Gruber, J. B. (1964). *J. Chem. Phys.,* **40,** 3606
34. Gruber, J. B., Krupke, W. F. and Poindexter, J. M. (1964). *J. Chem. Phys.,* **41,** 3363
35. Justice, B. H., Westrum, E. F. Jr., Chang, E. and Radebaugh, R. (1969). *J. Phys. Chem.,* **73,** 333
36. Paukov, I. E., Nogteva, V. V. and Yarembash, E. I. (1968). *Zhur. Fiz. Khim.,* **42,** 998
37. Nogteva, V. V., Paukov, I. E. and Yarembash, E. I. (1969). *Zhur. Fiz. Khim.,* **43,** 2118
38. Westrum, E. F. Jr. (1970). *Les Elements Des Terres Rares,* 443. (Paris: Editions du Centre National de la Recherche Scientifique)
39. Westrum, E. F. Jr., Clever, H. L., Andrews, J. T. S. and Feick, G. (1966). *Rare Earth Research III,* (L. Eyring, editor), 597. (New York: Gordon and Breach)
40. Westrum, E. F. Jr. and Lyon, W. G. (1968). *Thermodynamics of Nuclear Materials, 1967,* 239. (Vienna: International Atomic Energy Agency)

41. Dieke, G. H. (1968). *Spectra and Energy Levels of Rare Earth Ions in Crystals,* (New York: Interscience)
42. Westrum, E. F. Jr. (1968). *Progress in the Science and Technology of the Rare Earths, Vol. 3,* 459 (L. Eyring, editors) (Oxford: Pergamon Press)
43. Holley, C. E. Jr., Huber, E. J. Jr. and Baker, F. B. (1968). *Progress in the Science and Technology of the Rare Earths, Vol. 3,* 343, (L. Eyring, editor) (Oxford: Pergamon Press)
44. Holm, J. L., Kleppa, O. J. and Westrum, E. F. Jr. (1967). *Geochim. Cosmochim. Acta,* **37,** 2289
45. Gal'chenko, G. L. and Stanko, V. I. (1969). *Vest. Mosk. Univ. Khim.,* **24,** 3
46. Gal'chenko, G. L., Martynovskaya, L. N., and Stanko, V. I. (1969). *Dokl. Akad. Nauk SSSR,* **186,** 1328; (1970). **188,** 587; (1970). *Zhur. Obshch. Khim.,* **40,** 2415
47. Baughman, R. H. (1970). *J. Chem. Phys.,* **53,** 3781
48. Ashcroft, S. F. and Mortimer, C. T. (1970). *Thermochemistry of Transition Metal Complexes,* (New York: Academic Press)
49. Mueller, W. and Schuller, D. (1971). *Ber. Bunsenges. Phys. Chem.,* **75,** 79
50. Chang, S-J., McNally, D., Shary-Tehrany, S., Hickey, M. J., and Boyd, R. H. (1970). *J. Amer. Chem. Soc.,* **92,** 3109
51. Gopal, E. S. R. (1966). *Specific Heats at Low Temperatures,* (New York: Plenum Press)
52. Morrison, J. A. and Newsham, D. M. T. (1968). *Proc. Phys. Soc., London, (Solid State Phys.)* [2], 1, 370
53. Beck, P. A. and Claus, H. (1970). *J. Res. Nat. Bur. Stand.,* **74A,** 449
54. Conway, M. M., Phillips, N. E., Geballe, T. H. and Kuebler, N. A. (1970). *J. Phys. Chem. Solids,* **31,** 2673
55. Neshpor, V. S. and Samsonov, G. V. (1969). *Izv. Vyssh. Ucheb. Zaved., Fiz.,* **12,** 23; (1968). *Fiz. Metal. Metalloved,* **25,** 1132
56. Westrum, E. F. Jr. and Grønvold, F. (1969). *J. Chem. Thermodyn.,* **1,** 543
57. Herbert, D. C. (1970). *J. Phys. C,* **3,** 891
58. Harrison, J. P., Lombardo, G. and Peressini, P. P. (1968). *J. Phys. Chem. Solids,* **29,** 557
59. Harrison, J. P., Peressini, P. P. and Pohl, R. O. (1968). *Localized Excitations Solids, Proc. Int. Conf., 1st 1967,* 474, (New York: Plenum Press)
60. Lawless, W. N. (1969). *J. Phys. Chem. Solids,* **30,** 1161
61. Harrison, J. P., Peressini, P. P. and Pohl, R. P. (1968). *Phys. Rev.,* **167,** 856; Field, G. R. and Sherman, W. F. (1967). *J. Chem. Phys.,* **47,** 2378; Dreyfus, R. W. (1968). *J. Phys. Chem. Solids,* **29,** 1941
62. Reese, W. (1969). *Phys. Rev.,* **181,** 905
63. Rolovs, B. and Romanovskis, T. (1970). *Izv. Akad. Nauk SSSR, Ser. Fiz.,* **34,** 2492
64. Milleron, P. F. and Fournet, G. (1970). *Elektrotech. Cas.,* **21,** 257
65. Grossmann, S. and Richter, P. H. (1970). *Phys. Letters,* **33A,** 39
66. Hyland, G. J. (1970). *J. Solid State Chem.,* **2,** 318
67. Enns, R. H. and Haering, R. R., eds. (1969). *Modern Solid State Physics, Vol. 2; Phonons and Their Interactions,* (New York: Gordon and Breach)
68. Akhiezer, A. I., Bar'Yakhtar, V. G. and Peletminskii, S. V. (1968). *Spin Waves,* (New York: John Wiley)
69. Begum, N. A., Cracknell, A. P., Joshua, S. J. and Reissland, J. A. (1969). *Proc. Phys. Soc., London (Solid State Phys.),* **2,** 2329
70. Rosenberg, H. M. (1969). *Spin Waves and Spin-phonon Interactions, U.S. Government Res. Develop. Rep.* AD-684401, (Springfield, Virginia: U.S. Clearinghouse Fed. Sci. Tech. Inform. CFSTI); **69,** 74
71. Masset, F. and Callaway, J. (1970). *Phys. Rev. B,* **2,** 3657
72. Shah, S. S. (1969). *Proc. Nucl. Phys. Solid State Phys. Symp. 13th, 1968,* **3,** 330
73. Cooke, A. H. (1969). *Proc. Int. Conf. Low Temp. Phys., 11th, 1968,* 1, 57, (St. Andrews, Scot.: Conf. Organizing Comm.)
74. Grazhdankina, N. P. (1969). *Sov. Phys. Usp.* **11,** 727; (1968). *Usp. Fiz. Nauk,* **96,** 291
75. Reichert, T. A. and Giauque, W. F. (1969). *J. Chem. Phys.,* **50,** 4205
76. Lewis, E. A. S. (1970). *Phys. Rev. B,* **1,** 4368
77. Salamon, M. B. (1970). *Phys. Rev. B,* **2,** 214
78. Flotow, H. E., Osborne, D. W. and Westrum, E. F. Jr. (1968). *J. Chem. Phys.,* **49,** 2438
79. Badalyan, D. A., Khachaturyan, A. G. and Kitaigorodskii, A. I. (1969). *Kristallografiya,* **14,** 404; Badalyan, D. A. (1969). *Kristallografiya,* **14,** 14
80. Stuart, H. A. (1970). *Ber. Bunsenges. Phys. Chem.,* **74,** 739

81. Parry, G. S., Nixon, D. E., Lester, K. M. and Levene, B. C. (1969). *J. Phys. C. (Solid State Physics)*, **2,** 2156
82. Cohen, J. B. (1970). *Phase Transform., Pap. Semin. Amer. Soc. Metals 1968,* 561, (Metals Park, Ohio: ASM)
83. Hamilton, W. C. (1969). *J. Chem. Phys.,* **50,** 2275
84. Peterson, D. T. and Hutchison, J. F. (1970). *J. Chem. Eng. Data,* **15,** 320
85. Plante, E. R., Sessoms, A. B. and Fitch, K. R. (1970). *J. Res. Nat. Bur. Stand., sec A,* **74,** 647
86. Osborn, A. G. and Douslin, D. R. (1969). *J. Chem. Eng. Data,* **14,** 208
87. Morawetz, E. (1968). *Acta Chim. Scand.,* **22,** 1509
88. Rice, S. A. (1971). *Phase Transitions, Proc. 14th, Conf. Chem., 1969,* 23, (London: Interscience)
89. Coopersmith, M. H. (1967). *Phys. Letters A,* **24,** 700
90. Coopersmith, M. H. (1970). *Advan. Chem. Phys.,* **17,** 43
91. Coopersmith, M. H. (1970). *Pure Appl. Chem.,* **22,** 311
92. Rao, C. N. R. and Rao, K. J. (1967). *Progr. Solid State Chem.,* **4,** 131
93. Garland, C. W. (1970). *Phys. Acoust.,* **7,** 51
94. Shmidt, N. E. (1970). *Zhur. Fiz. Khim.,* **44,** 1339
95. Borelius, G. (1967). *Rev. Int. Hautes Temp. Refract.,* **4,** 87
96. Ubbelohde, A. R. (1965). *Melting and Crystal Structure,* (Oxford: Clarendon Press)
97. Blachnik, R. and Schneider, A. (1970). *Z. Anorg. Allgem. Chem.,* **372,** 314
98. *Phase Transformations* (1970). Paper Sem. Amer. Soc. Metals 1968. (Metals Park, Ohio: ASM)
99. Carter, R. S. and Rush, J. J. (1969). *Molecular Dynamics and Structure of Solids,* (Washington, D. C.: U.S. Government Printing Office)
100. *Phase Transitions* (1971). *Proceedings of the Fourteenth Conference on Chemistry at the University of Brussels, May 1969.* (London: Wiley)
101. (1970). *Proceedings of the First International Conference on Calorimetry and Thermodynamics at Warsaw, 1969,* (Warsaw: Polish Scientific Publishers)
102. Brewer, L. (1965). Chapter in *High-Strength Materials,* (Zachay, V. F., editor) (New York: John Wiley)
103. Kaufman, L. (1970). Chapter in *Phase Diagrams Vol. 1,* 45 (Alper, A. M., editor) (New York: Academic Press)
104. Chang, E. T. and Westrum, E. F. Jr. (1970). *J. Phys. Chem.,* **74,** 2528; (1965), **69,** 2176
105. Pace, E. L., Smith, J. H. and Jepson, B. E. (1969). *J. Chem. Phys.,* **50,** 312
106. Leadbetter, A. J. and Settatree, G. R. (1969). *Proc. Phys. Soc., London, (Solid State Phys.),* **2,** 1105
107. Hoch, M. (1969). *High Temp.-High Pressures,* **1,** 531
108. Leadbetter, A. J. (1968). *Proc. Phys. Soc., London, (Solid State Phys.)* **1,** 1489
109. Brooks, C. R. and Bingham, R. E. (1968). *J. Phys. Chem. Solids,* **29,** 1553
110. Ida, Y. (1970). *Phys. Rev. B,* **1,** 2488
111. Mazur, P. (1969). *J. Appl. Phys.,* 40, 482
112. Westrum, E. F. Jr. (1967). Chapter in *Advances in High Temperature Chemistry,* **I,** (L. Eyring, editor), 239, (New York: Academic Press)
113. Stansbury, E. E. and Brooks, C. R. (1970). *High-Temp.-High Pressures,* **1,** 289
114. West, E. D. and Westrum, E. F. Jr. (1968). Chapter in *Experimental Thermodynamics* Vol. 1, 333 (McCullough, J. P. and Scott, D. W., editors) (London: Butterworths)
115. Kubaschewski, O. (1969). *Progr. Mater. Sci.,* **14,** 1
116. Libowitz, G. G. (1969). *J. Solid State Chem.,* **1,** 50
117. Nechaev, U. S. (1970). *Izv. Vyssh. Ucheb. Zaved., Chern. Met.,* **13,** 138
118. Perrott, C. M. and Fletcher, N. H. (1969). *J. Chem. Phys.,* **50,** 2344
119. Perrott, C. M. and Fletcher, N. H. (1968). *J. Chem. Phys.,* **48,** 2143, 2681; (1969) **50,** 2770; (1970) **52,** 3368; 3373
120. Noelting, J. and Rein, D. (1969). *Z. Phys. Chem. (Frankfurt am Main),* **66,** 150
121. Eyring, L. and O'Keeffe, M. (1970). *The Chemistry of Extended Defects in Non-Metallic Solids,* (New York: American Elsevier)
122. Lyon, W. G. and Westrum, E. F. Jr. (1968). *J. Chem. Phys.,* **49,** 3374
123. Staveley, L. A. K. and Linford, R. G. (1969). *J. Chem. Thermodyn.,* **1,** 1
124. Trappeniers, N. J. (1966). *Ber. Bunsenges Phys. Chem.,* **70,** 1080
125. Jura, G. and Stark, W. A. Jr. (1969). *Rev. Sci. Instrum.,* **40,** 656

126. Phillips, N. E., Ho, J. C. and Smith, T. F. (1968). *Phys. Lett.*, A. **27,** 49. Smith, T. F. and Phillips, N. E. (1970). *Les Propriétés Physiques des Solids Sous Pression* (Paris: Centre National de la Recherche Scientifique)
127. Dass, N. and Varshenya, N. C. (1968). *J. Phys. Soc. Jap.*, **25,** 1452
128. Dass, N. and Mitra, S. K. (1969). *J. Phys. Soc. Jap.*, **27,** 254
129. Andres, K. (1970). *Phys. Rev., B,* **2,** 3768
130. Gilat, G. (1969). *Phys. Rev. Letters,* **23,** 78
131. Thirring, W. (1970). *Z. Phys.,* **235,** 339
132. Brown, G. H. (1970). *Liquid Crystals* (Cleveland: Chemical Rubber Co.)
133. Murabayashi, M., Takahashi, Y. and Mukaibo, T. (1970). *J. Nucl. Sci. Tech.*, **7,** 316
134. Namba, S., Kim, P. H. and Kinoshita, N. (1967). *Sci. Pap. Inst. Phys. Chem. Res.*, **62,** 8
135. McCullough, J. P. and Scott, D. W., eds. (1968). *Experimental Thermodynamics, Vol.* **I,** (New York: Plenum Press)
136. Leadbetter, A. J. and Wycherley, K. E. (1970). *J. Chem. Thermodyn.*, **2,** 855
137. Adachi, K., Suga, H. and Seki, S. (1971). *Bull. Chem. Soc. Jap.*, **44,** 78
138. Adachi, K., Suga, H. and Seki, S. (1968). *Bull. Chem. Soc. Jap.*, **41,** 1073
139. Tarasov, V. V. and Zhdanov, V. M. (1970). *Zhur. Fiz. Khim.*, **44,** 2384
140. Leadbetter, A. J. and Wycherley, K. E. (1971). *Phys. Chem. Glasses,* **12,** 41
141. Chen, H. S. and Turnbull, E. (1969). *J. Appl. Phys.*, **40,** 4214
142. Rehage, G. (1970). *Ber. Bunsenges. Phys. Chem.*, **74,** 796
143. Anderson, D. R. and Action, R. U. (1970). in *Encyclopedia of Polymer Science, Vol. 13,* 7 (New York: John Wiley)
144. Joshi, R. M. (1970). in *Encyclopedia of Polymer Science and Technology, Vol. 13,* 789 (New York: John Wiley)
145. Schwenker, R. F. and Garn, P. D. (1969). *Thermal Analysis, Vols. 1 & 2* (New York: Academic Press)
146. Porter, R. S. and Johnson, J. F. (1968). *Analytical Calorimetry,* (New York: Plenum Press)
147. (1970). *Status of Thermal Analysis.* Proceedings of a Symposium on the Current Status of Thermal Analysis held at Gaithersburg, Md., April 21–22, 1970. (O. Menis, editor). National Bureau of Standards Special Publication 338. (Washington, D.C.: U.S. Government Printing Office)